The Profession and Practice of Horticultural Therapy

The Profession and Practice of Horticultural Therapy

Edited by
Rebecca L. Haller
Karen L. Kennedy
Christine L. Capra

CRC Press
Taylor & Francis Group
Boca Raton London New York

CRC Press is an imprint of the
Taylor & Francis Group, an **informa** business

CRC Press
Taylor & Francis Group
6000 Broken Sound Parkway NW, Suite 300
Boca Raton, FL 33487-2742

© 2019 by Taylor & Francis Group, LLC
CRC Press is an imprint of Taylor & Francis Group, an Informa business

No claim to original U.S. Government works

Printed on acid-free paper

International Standard Book Number-13: 978-1-138-30869-5 (Paperback)
International Standard Book Number-13: 978-1-138-30874-9 (Hardback)

Library of Congress Cataloging-in-Publication Data

Names: Haller, Rebecca L., author, editor. | Kennedy, Karen L., 1963- author, editor. | Capra, Christine L., author, editor.
Title: The profession and practice of horticultural therapy / authors: Rebecca L. Haller, Karen L. Kennedy, Christine L. Capra.
Description: Boca Raton, FL : CRC Press, Taylor & Francis Group, 2019. | Includes index.
Identifiers: LCCN 2018038384| ISBN 9781138308695 (pbk. : alk. paper) | ISBN 9781138308749 (hardback : alk. paper)
Subjects: LCSH: Gardening--Therapeutic use.
Classification: LCC RM735.7.G37 H35 2019 | DDC 615.8/515--dc23
LC record available at https://lccn.loc.gov/2018038384

Visit the Taylor & Francis Web site at
http://www.taylorandfrancis.com

and the CRC Press Web site at
http://www.crcpress.com

Dedication

To the late Richard H. Mattson, who had the foresight as a floriculture professor to dive into a new field of study and practice, the character to understand the significance of plants to human well-being, the heart to connect with people of all abilities, and the passion for a new undertaking called horticultural therapy before most others saw its potential. He was a pioneer who encouraged and supported budding practitioners over the years, and developed and taught a respected academic and experiential program of study at Kansas State University. The curriculum he created became a model upon which other initiatives were based, including professional competencies, a standardized core curriculum, and training around the world. The "products" of his work include many of today's leaders in horticultural therapy as well as the improved lives of countless clients. He was a connector and tireless advocate, and is missed.

To Paula Diane Relf, a virtuoso force in advancing the knowledge and profile of horticultural therapy, who has been at the forefront of defining and distinguishing the profession globally. A tireless organizer, she was a founder and president of the first association in the field (National Council for Therapy and Rehabilitation Through Horticulture) and started the People Plant Council (later to become the International People Plant Council) to "document and communicate the effect that plants have on human well-being and improved life-quality." As a professor at Virginia Tech, she mentored graduate students to do research and publish those results widely. In "retirement," her reach is ongoing and broad. She continues to collaborate, coax, and convince innumerable people to step up, be heard, take on the work to advance the profession, and become leaders. Thank you, Diane!

Contents

Section III: Practice within program models

Section IV: Tools for the therapist

Foreword

As a member of what might be called the early transition team moving the practice of gardening as a way for volunteers to provide a higher quality of life to people in health care facilities to the profession of horticultural therapy, I find it critical to express how important this book is to take the next big step in advancing the profession. To give context to my praise, I read *Therapy through Horticulture* by Alice Burlingame while in high school when it was first published, and I was convinced that horticultural therapy would be my career area, only to discover that it was a gift by volunteers to those in need. Widely known as garden therapy, it was supported by many farm and garden organizations and made major contributions to the quality of life of thousands of individuals. It was viewed by hospitals, rehabilitation centers, and other medical facilities as a leisure/social pursuit that brought members of the community to patients in positive interactions. In the 1960s and 1970s, the need for nearby nature began to be recognized by both the public and a few health-related facilities. Throughout this time, I studied horticulture and worked at nurseries and garden centers, but I continued my dream of working in horticultural therapy.

During the early part of this period, health facilities of all types that had taken their garden volunteers for granted began to recognize that their contributions had larger significance than previously understood. Professionals in different fields added a garden, greenhouse, or indoor plants to their therapeutic activities. More and more often, facilities found funds to hire part-time staff members transitioning to full-time but who were still untrained. I was given permission at the University of Maryland to explore the application of horticulture as therapy and eventually received both my MS and PhD with research and studies relevant to this area.

The great significance of the individuals who have written this book is that the centers where they work and the institutes where they teach, such as the Horticultural Therapy Institute, have been the leaders and the power behind professional education and the development of horticultural therapy as a genuine career over the last twenty years. I have had the opportunity to interact with many of the authors of this book over the years and have great respect for them and for their skill and knowledge in the use of the care and nurturing of plants to assist people addressing health issues. Their leadership in the American Horticultural Therapy Association has been particularly significant as that organization has changed and grown. This book is so important as it provides a challenge, incentive, and base upon which research and understanding of the value of horticultural therapy can advance. The professional quality and knowledge brought together to explore and develop theoretical models based on the most current understanding of the attributes of therapy will facilitate participation in research and sharing of critical information among a wide group of individuals at health care facilities, institutes, and universities.

I sincerely thank and congratulate Christine L. Capra, Rebecca L. Haller, and Karen L. Kennedy for putting together this important work, and I look forward to future publications.

Diane Relf, PhD, HTM
Professor Emeritus, Virginia Tech University;
Fellow, American Society for Horticultural Science

Preface

Horticultural therapy is a practice that benefits the health and well-being or rehabilitation of people through garden-based participatory interventions. It has been used for many decades and is applied in various forms in many parts of the world. Outcomes vary, from those related to recovery from illness or injury to employment and inclusion, to overall wellness. The needs and strengths of the people served as well as the intended outcomes influence the choice of treatment methods and levels of care. Horticultural therapy uses a person-centered, holistic approach within therapeutic, vocational, and wellness models to serve people who are ill, injured, disadvantaged, or socially excluded. As a developing profession, new resources are also emerging in research, theoretical papers, program descriptions, and discussion of applications in practice. This book is designed to bring many of these resources together into one comprehensive text to provide a foundation for study and practice in horticultural therapy and related fields.

The Profession and Practice of Horticultural Therapy describes the theories on which horticultural therapy is based and details models of practice and program types. It is intended to enhance understanding of the profession, provide insight into methods of practice, and provide a basis for education of new and experienced practitioners. Students in the field as well as health care and human service professionals can use this reference in the development and management of effective therapy and rehabilitation of inmates, residents, patients, trainees, students, and others who benefit from horticultural therapy programming.

Four sections provide the foundations: overview of horticultural therapy practice, theories to support practice, program models, and tools for the therapist. Chapters are authored by many different people in order to convey the broad range of approaches currently seen in the profession. Additional exhibits, tables, and images add tangible information to the discussions through real-life practice examples, perspectives, and photos of horticultural therapy in action.

While not intended to be a publication about therapeutic garden design, some information is included that may inform design choices, and examination of the importance of the benefits of the nature connection experienced in a garden space is found throughout the book. The text is not intended to be a manual of practice, focusing on practice protocols and methods. Many of these topics are covered in *Horticultural Therapy Methods: Connecting People and Plants in Health Care, Human Services, and Therapeutic Programs* (Haller and Capra 2017). Yet the reader will find guidance and tutelage for successful program applications and adaptations in many of the chapters.

For several reasons, the book also does not provide details on the full range of populations who may be aided in horticultural therapy programs. First, the populations who participate in this form of treatment are constantly expanding and diversifying globally. Second, facts on the characteristics, terminology, treatment, and attributes of individuals change, with the most current information widely and reliably available through online and print media. In addition, the editors recognize that there is considerable overlap among populations in horticultural therapy within each of the models of practice. For example, vocational programming applies principles and techniques that are similar in working with at-risk youth or those with developmental disabilities. Working with these populations in therapeutic or wellness programs would necessitate different approaches. Case and program examples illustrate distinctions and specifics of a spectrum of program participants throughout the book.

Throughout the text exhibits, terms for the individuals who are served vary with the examples given, including *client, participant, patient,* and others. These terms are conceptually interchangeable because the concepts and examples can be applied in many different settings that use varied vocabulary.

The editors of this book together have over 70 years of practice. This lifelong devotion to the field of horticultural therapy stems from the belief that it touches people deeply and can have profound positive impacts on program participants. As gardeners, the anticipation of growth, observation of change, and the reward of harvest are motivating and inspiring. As horticultural therapists, we are passionate about the importance of human-nature connections. Witnessing the human-nature growth first-hand is a source of sincere joy and reward.

This book is designed to provide relevant, current information on the breadth and depth of the field of horticultural therapy. It is our intent to inspire best practices and creative programs, to extend the reach and rewards of this work.

Acknowledgments

We are grateful to many individuals who gave their time and talents to bring this important body of work to fruition, including the authors and photographers who readily responded when our call went out. Those who lent their invaluable expertise by reading chapters for content include Beverly Brown, Pamela Catlin, Nancy Chambers, Elizabeth Diehl, and Reece Nelson. All of their enthusiasm and dedication to the field of horticultural therapy is admirable. Copy editors include Cait Anastis and Shawn Cremer, whose attention to detail was invaluable. We also thank the students of the Horticultural Therapy Institute, both past and present, who make our daily work in this field memorable and gratifying. They take this information and go out and do great deeds, changing the lives of so many who long for a connection to the garden and the earth. They, along with the clients we have served over the years, challenge and inspire us.

I personally thank my father and mother, Gerald and Josephine Capra, as well as my two loving daughters, Hannah and Kateri Kramer. Your support has meant everything to me.

Christine L. Capra

Many thanks to my husband, Dave, and daughter, Lydia Kennedy, for their support, encouragement, and patience throughout the many hours devoted to the development of this book. And to my parents, Jack and Carol Smith, for letting me dive into this profession as it was just budding and urging me to follow my passion.

Karen L. Kennedy

Thanks to Moss and Shawn Cremer, who have provided advice and perspective and served as patient sounding boards and steadfast champions throughout the formation of this book.

Rebecca L. Haller

Editors

Rebecca L. Haller, HTM, has practiced and taught horticultural therapy since receiving a master's degree in horticultural therapy from Kansas State University, Manhattan, Kansas, in 1978. She is the director of the Horticultural Therapy Institute in Denver, Colorado, and a faculty at Colorado State University, Fort Collins, Colorado. Her professional interests include training and education for excellence in practice, curriculum development, people–plant connections, the design and use of gardens as clinical spaces for horticultural therapy, and those endeavors that support professional development in this emerging field. Horticultural therapy practice experience includes serving a wide array of people and program types at Denver Botanic Gardens, where she also designed and taught a series of professional courses in horticultural therapy, managed the sensory garden, and created programs and access for people with disabilities. She established a vocational horticultural therapy program for adults with developmental disabilities in Glenwood Springs, Colorado, which is still thriving after more than 25 years of operation. She has served as a president, secretary, and board member of the American Horticultural Therapy Association (AHTA) and has worked on teams related to education and professional standards. In 2005, she received the Horticultural Therapy Award from the American Horticultural Society. Awards from the AHTA include: The Publication Award in 2009 and the Rhea McCandliss Professional Service Award in 2015. She has authored articles and chapters in the following publications: *Horticultural Therapy Methods: Connecting People and Plants in Health Care, Human Services, and Therapeutic Programs; HortTechnology; Horticulture as Therapy: Principles and Practice; Horticultural Therapy and the Older Adult Population; Towards a New Millennium in People–Plant Relationships; AHTA News Magazine;* and *Public Garden.*

Karen L. Kennedy, HTR, https://orcid.org/0000-0002-4430-9144, develops programs and provides services to individuals with a wide variety of disabilities, illnesses, and life situations. Since 1986 she has been a Registered Horticultural Therapist (HTR) and an active proponent of creatively using the interaction between people and plants to improve human health and well-being. After managing the Horticultural Therapy and Wellness Program at The Holden Arboretum for 23 years, Karen now works as a private contractor providing horticultural therapy and consulting services, developing educational materials, and teaching. As a faculty member of the Horticultural Therapy Institute, Denver, Colorado, she teaches the programming course and works on other educational projects. In addition, she nurtures her love of plants through writing, teaching, and facilitating webinars as the Education Coordinator for The Herb Society of America. She is a frequent presenter at regional and national professional conferences. She has served on the board of directors and committees of the American Horticultural Therapy Association (AHTA). She has authored chapters and articles in the following publications: *Horticulture as Therapy: Principles and Practice* (1998), *Horticultural Therapy Methods: Making Connections in Health*

Care, Human Service and Community Programs (2006, 2017), *Public Garden Management* (2011), *AHTA Magazine, Public Garden, The Herbarist,* as well as local periodicals. Karen received the 1994 Rhea McCandliss Professional Service Award from AHTA and the 2009 Horticultural Therapy Award from the American Horticulture Society. She holds a bachelor of science degree in horticultural therapy from Kansas State University.

Christine L. Capra, BA, https://orcid.org/0000-0002-7770-1548, is the program manager and co-founder of the Horticultural Therapy Institute in Denver, Colorado, where accredited horticultural therapy education is offered to students from across the United States and abroad. She is an active member of the American Horticultural Therapy Association, serving on a variety of work teams and contributing to its publications. Previously she helped manage the horticultural therapy educational program at the Denver Botanic Gardens. She has a bachelor of arts in journalism from Metropolitan State University in Denver. She is co-editor of the book *Horticultural Therapy Methods: Making Connections in Health Care, Human Service and Community Programs* (Taylor & Francis, 2006) and editor of the online HTI newsletter, "Making Connections." She has won numerous writing awards and has been published in *OT Weekly, Mountain Plain and Garden, Green Thumb News, People–Plant Connection, AHTA News Magazine, GrowthPoint, The Community Gardener, Health and Gardens, Colorado Gardener, Denver Catholic Register,* and *Our Sunday Visitor.* She was a reporter for the *Denver Catholic Register* for many years and received several writing awards from the Society of Professional Journalists, the Catholic Press Association, and the Colorado Press Association. In 2009, she was awarded the Publication Award from the American Horticultural Therapy Association.

Chapter Authors

Beverly J. Brown, PhD, HTR, https://orcid.org/0000-0001-9361-587X, has always had an affinity for the environment, especially plants. When she was young, her father would take the children in the family on rambles through fields, forests, and meadows near their home. The ease with which her father navigated in nature and the small, delightful bits of plant knowledge he shared sparked her lifelong interest in the green life surrounding us. In 1999, this interest culminated in a PhD in Ecology from Kent State University. Her research focused on the impact of showy invasive plant species on pollination of native species. Her interest in the people–plant interface grew as she saw the positive impact plants had on family members during long illnesses. In 2011, she became a Registered Horticultural Therapist and began teaching horticultural therapy courses online at Nazareth College in Rochester, New York. She directs the Horticultural Therapy Clinic at the college and also co-treats with clinicians from other disciplines, notably working with clients regaining brain function. Brown has served on numerous committees of the American Horticultural Therapy Association and was a board member for several years. In addition to teaching courses, she is also responsible for maintaining the greenhouse and associated gardens on campus. The gardens are designed to teach about all aspects of plants, their role in horticultural therapy, and their importance in our daily lives. She looks forward to continuing research on the positive impact that horticultural therapy can have on wellness and healing.

Gwenn Fried, BS, https://orcid.org/0000-0002-7771-1842, is the manager of the Horticultural Therapy Services Department at New York University Langone Medical Center, Rusk Rehabilitation. She has been a horticultural therapist for over 20 years. Fried specializes in program development, rehabilitation, and vocational training for children, adults, and seniors. A member of the faculty of Chicago Botanic Garden's Health care Garden Design Program, Fried teaches, consults, and lectures internationally on therapeutic garden design and horticultural therapy. Fried has been a member of and serves on the Board of Directors of Metro Hort Group, a professional horticultural organization in the New York metropolitan area.

Jonathan Irish, HTR, MA, LPC, https://orcid.org/0000-0002-1613-9037, is a Licensed Professional Counselor and a Registered Horticultural Therapist. He earned his master of arts at Denver Seminary and attended the Horticultural Therapy Institute in 2014. He is employed at Rogers Memorial Hospital in Oconomowoc, Wisconsin, as the Horticultural Therapy Coordinator. Irish facilitates horticultural therapy groups for inpatient and residential patients with eating disorders, trains other therapists, creates and modifies curriculum for the various patients served by Rogers, and helps design therapeutic landscapes.

He also has a small private practice working with individuals, couples, and families and incorporates horticultural therapy whenever possible. When not at work, he is active outdoors and enjoys gardening, camping, and hiking with his wife and children.

Barbara Kreski, MHS, OTR/L, HTR, https://orcid.org/0000-0002-6336-9054, supervises the horticultural therapy services department at the Chicago Botanic Garden, including oversight of delivery of services to off-site health and human services agencies, horticultural therapy programming at the Buehler Enabling Garden, and therapeutic garden design consultation. Kreski is the lead instructor of the horticultural therapy certificate program, and she is on the faculty of the health care garden design certificate. Kreski's background spans a wide range of clinical settings, including psychiatric hospitals, veteran's administration hospitals, skilled nursing facilities, public schools, and pediatric rehabilitation. She holds a master's degree in health sciences with a concentration in neuroscience. Kreski is an occupational therapist as well as a horticultural therapist.

K. René Malone, MS, CTRS, HTR, https://orcid.org/0000-0002-5721-2253, has a master's degree in therapeutic recreation with a minor in gerontology from the University of Tennessee at Knoxville. She currently lives in Bloomington, Indiana, and, after working for four years in long-term care in which she led horticultural therapy programs at three sister facilities, she started her contractual business, The Therapeutic Vine, in 2008. Populations served include children and adults with intellectual and physical disabilities, seniors, adults with substance use and mental health disorders, and children with behavioral disorders. She has been actively involved in the American Horticultural Therapy Association for over eight years, serving on the executive committee and on the board of directors. She currently focuses her efforts on the credentialing and magazine work teams.

Susan Conlon Morgan, MS, https://orcid.org/0000-0001-7358-5941, is a therapeutic horticulture practitioner and owner of The Horticultural Link, LLC, outside Dallas, Texas. She has delivered therapeutic horticulture services and workshops for a variety of groups, including older adults living with dementia, young men incarcerated at a boot camp jail, and the general public. Morgan holds a master of science degree in public horticulture from the University of Tennessee, and her thesis research evaluated the long-term effects of a horticultural program for youth at risk at the Brooklyn Botanic Garden in New York. After years working at public gardens, she earned her horticultural therapy certificate from the Horticultural Therapy Institute, with Colorado State University, and an international diploma in botanic garden education from the Royal Botanic Gardens, Kew, in England. She honed her practice at the Chicago Botanic Garden and Dallas Arboretum; at the latter she developed and facilitated a vocational horticulture program for resettled refugees.

Jay Stone Rice, PhD, https://orcid.org/0000-0002-5832-1574, interweaves psychology, earth-based wisdom traditions, and nature in his counseling, consulting, mentoring, and writing. He was the principal investigator for an exploratory study of the effectiveness of the San Francisco Sheriff's Department's innovative horticultural therapy program. Rice co-edited *The Healing Dimensions of People–Plant Relations* and has written about the social ecology of inner-city family trauma, trauma's relationship to substance abuse and crime, horticultural therapy's influence on brain/mind/body integration, and gardening as a treatment intervention. Rice has been on the faculty of the Horticultural Therapy Institute since its inception.

Jane Saiers, PhD, HTR, AAS, https://orcid.org/0000-0002-0417-6439, is a Registered Horticultural Therapist through the American Horticultural Therapy Association. Saiers has a doctorate in psychology and neuroscience from Princeton University and an associate of applied science degree in sustainable agriculture from Central Carolina Community College. She has provided horticultural therapy services to people with traumatic brain injury, substance use disorder, mental illness, dementia, and to people who are homeless. With her husband, Jane grows fruits, vegetables, and mushrooms and raises chickens for eggs at RambleRill Farm, their certified organic farm in Hillsborough, North Carolina. She offers farm-based horticultural therapy and wellness programs to individuals and groups.

Emilee Vanderneut, BSW, https://orcid.org/0000-0001-6326-5995, had a connection to gardening germinate while observing and interacting with her grandmother's deep love and respect for the natural world. She began her career in the horticulture industry while working toward her social work degree and horticultural therapy certificate. In both the private and public sectors, Vanderneut has developed horticultural therapy programs for incarcerated youth, adolescents who experience mental illness and substance abuse issues, military veterans, individuals who are developmentally and/or cognitively impaired, and elders. She has enjoyed a lifelong fascination and passion for program and curriculum development and program analysis, providing client-centered services and teaching. Using her 19 years of horticulture experience and 13 years of horticultural therapy experience, Vanderneut is the Program Manager for a horticultural therapy program at a psychiatric hospital managed by the state of North Carolina.

Matthew J. Wichrowski, MSW, HTR, https://orcid.org/0000-0002-7934-1064, has been practicing horticultural therapy at Rusk Rehabilitation New York University–Langone Medical Center for over 20 years and is currently Clinical Assistant Professor. He teaches in the Horticultural Therapy Certificate Program at New York Botanical Garden, presents regularly at national and international conferences, and has won many awards for his work. Wichrowski is also Editor-in-Chief of the American Horticultural Therapy Association's *Journal of Therapeutic Horticulture*. His research interests center around the effects of nature on human health and wellness, and his work has been published in a variety of media, including textbooks, journals, and magazines.

Pamela Young, HTR, https://orcid.org/0000-0001-9896-2192, coordinates the Sydney Thayer III Horticultural Therapy Center at Bryn Mawr Rehabilitation Hospital in Malvern, Pennsylvania. Since 1995, she has provided horticultural therapy services to those recovering from stroke, brain injury, spinal cord injury, orthopedic injuries, and other neurological disorders. In support of the profession of horticultural therapy she is active on a national level as reviewer for Professional Registration for the American Horticultural Therapy Association and locally serves the Mid Atlantic Horticultural Therapy Network as newsletter editor. She worked as a consultant for horticultural therapy start-up programs in the Philadelphia area, has presented and exhibited at the Philadelphia Flower Show, and takes an active role in connecting her community to gardening through volunteer work with the Chester County Food Bank.

Contributors

Thank you to the following contributors, who provided expertise, photos, and real practice examples to enhance the text.

Authors of exhibits or tables: Nicole Accordino, Angie Andrade, Erin Backus, Wendy Battaglia, Melissa Bierman, Isabelle Boucq, Patricia Cassidy, Nancy Chambers, Sally Cobb, Patricia Czarnecki, Elizabeth Diehl, Debra Edwards, John Fields, Cathy Flinton, Maria Gabaldo, Joni Gabriel, Kristina Gehrer, Norman Goodyear, Kristen Greenwald, Colleen Griffin, Kathryn Grimes, Susie Hall, Kirk Hines, Roberta Hursthouse, Abby Jaroslow, Eugene Jones, Hilda Krus, Stephanie Lanel, Carol LaRocque, Jean Larson, Mike Maddox, Amanda Mammas, Nancy Minich, Hailey Moses Striebich, John Murphy, Marion Myhre, Damien Newman, Andreas Niepel, Gina Owens, Sin-Ae Park, Kathryn Perry, Heather Quinlan, Diane Relf, Libba Shortridge, Daniela Silva-Rodriquez Bonazzi, Tamara Singh, Travis Slagle, Derrick Stowell, Philip Tedeschi, Jonathan Truath, Matthew Wichrowski, Joanna Wise, Chin-Yung Wung, and JoAnn Yates.

Photographs: Nicole Accordino, Isabelle Boucq, Beth Bruno, Patty Cassidy, Pam Catlin, Moss Cremer, Elizabeth Diehl, Debra Edwards, Joni Gabriel, Kristen Greenwald, Katie Grimes, Kirk Hines, Abby Jaroslow, Eugene Jones, Nancy Minich, Hailey Moses, John Murphy, Anna Maria Palsdottir, Janet Schoniger, Libba Shortridge, Travis Slagle, and JoAnn Yates.

section one

Overview of horticultural therapy practice

This section provides foundational information and perspectives on the profession of horticultural therapy, including background, history, and evolution, as well as assets and challenges of the field to inform the reader of key events, definitions, and perspectives that influence the current status of the profession globally. Comparisons to related fields are intended to provide a framework for distinguishing the unique aspects of horticultural therapy as well as the connections to other initiatives and practices. A "Therapeutic Use of Horticulture Spectrum" is introduced to further explore the range of interventions using plants. Likewise, the approach and methods described in later chapters may be based on the principles involved in therapeutic relationships discussed in Chapter three. Section one is intended to be the basis for interpreting subsequent sections of the book.

chapter one

Introduction to the profession of horticultural therapy

Christine L. Capra, Rebecca L. Haller, and Karen L. Kennedy

Contents

Horticultural therapy can literally change the world. In its wide-ranging past, present, and future applications, it has the potential to be one of the most effective catalysts for positive human growth and well-being. At its core, horticultural therapy works with the needs and assets of individuals in order to enhance overall wellness. That may take many different forms and be accomplished in ways that are as diverse as are people and types of horticulture. The scope of practice is broadened by the imagination and resourcefulness of practitioners and by those who support it. Many human issues of an individual, regional, or global nature may be addressed in part by this developing approach to therapy. For example, connection to nature through goal-directed horticultural therapy may help to alleviate stress and stress-related illness brought on by increasing global urbanization. Food production that is incorporated into programming in any model of horticultural therapy—vocational, therapeutic, or wellness—may have an impact on nutrition, food security, and skills, which transfer to daily living outside a health care facility. A person or family who is displaced or homeless may find a place of "rootedness" in a garden, which brings hope and solace. A person who struggles with addiction may find a healthier "high" in the act of cultivation. Gardening with the guidance of a skilled professional therapist offers real work opportunities and can address coping skills, offer acceptable outlets for anger management, and teach nurturance and self-care. These and other outcomes are not only possible but to be expected through the use of horticultural therapy.

"As horticultural therapists we are 'agents of change.' We mediate the process through which people recover, adjust, or cope …" Richard H. Mattson (1992). Horticultural therapists bring unique and multidisciplinary skills to the process of change, emphasize the strengths in those served, and offer an intimate connection with nature through engagement in gardening. The therapy may result in small, profound, and/or even life-altering results for program participants (clients). Whatever the outcomes, the process offers many gifts to the therapist as well as the client. In addition to the satisfaction of seeing incremental improvements in a client's quality of life, therapists also reap the benefits of working in a plant-rich environment and enjoy self-nurturing through the cultivation of plants. They regularly witness the wonders of nature as well as human capacity and resilience (Figure 1.1).

Throughout history the connection between plants and people has permeated life. This chapter will discuss the profession of horticultural therapy—a unique modality— that bridges the people–plant connection. In addition to an overview of the profession, a brief history of the development of horticultural therapy will be discussed, as well as characteristics, program models, general benefits, and educational options in the field.

Assertions of the beneficial worth of horticulture trace back to ancient Egypt and continue through to modern times, and understanding that gardens, in and of themselves, have the capacity to bring about solace is not hard to appreciate. Yet incorporating intimate and active engagement in plant cultivation adds another dimension to the beneficial relationship. Bringing this engagement with plants to a level of therapy is one that requires knowledge and skill in plant cultivation as well as human service. Horticultural therapy is the modality that seeks to bring these together—utilizing the innate pull of our connection to the active participation in gardening in a formal and therapeutic setting where goals and objectives are utilized to meet the needs of the individual. It is this unique connection that sets horticultural therapy apart.

Figure 1.1 Experiencing the garden. (Courtesy of Kirk Hines.)

What is horticultural therapy?

"Goals of the programs differ, but the basic premise behind horticultural therapy is that working with and around plants brings about positive psychological and physical changes that improve the quality of life for the individual" (Relf 1992). Since the formation of the horticultural therapy profession, a full gamut of definitions and models have been used, ranging from strict descriptions of its use in health care fields to those that are broadly expressed, and they may even include any beneficial horticultural experience (Dorn and Relf 1995). As a young profession, definitions have evolved and continue to do so. Some historical and contemporary descriptions are chronicled here in order to provide perspective on the evolution of the field and an understanding of the genesis of current descriptions.

In *Therapy Through Horticulture*, published in 1960, Donald Watson and Alice Burlingame explained the purpose of horticultural therapy as "the improvement of physical and mental health," adding that it may also prevent and aid in the recovery from illness as well as provide vocational rehabilitation. Unlike more contemporary definitions, the horticultural therapy they portrayed was sponsored by an organization such as a garden club or auxiliary, staffed by volunteers who loved gardening, and was not to be considered a "distinct therapy." Rather it was considered an activity to support other forms of treatment (Watson and Burlingame 1960). This book illustrates *therapeutic activity* prior to future presentations of the field as an *emerging profession*. In Watson and Burlingame's descriptions, those who conducted the gardening activities did not require professional training or skills in therapy; rather, they acted as volunteers who brought gardening activity to patients and others. Understanding this lack of training and the *activity* roots of the profession sheds light on the ongoing challenge to maintain a professional image of horticultural therapy practice and employment. (Chapter 4 details this and other issues faced in the continued development of horticultural therapy as a profession.)

Coinciding with the emergence of the field in the 1970s was the first textbook in horticultural therapy written by Damon Olszowy. It explained the principles that made horticultural therapy a "distinct discipline" and provided guidelines and methods for the establishment of programs (Olszowy 1978). Olszowy described modern therapy as related to "total personality needs," "concerned first with the individual and second with the disability itself," with the aim to improve both mental and physical well-being. Rehabilitation similarly had the objective to help a person improve mentally, physically, socially, and vocationally, in order to lead a healthy and full life. Horticultural therapy was described as addressing therapy and rehabilitation in a manner that emphasized planning to meet therapeutic outcomes, along with specific activities to address patient needs. The role of the horticultural therapist was to help the client form effective relationships to support positive individual change. This description by Olszowy outlined a new way of viewing horticultural therapy, which has persisted in its basic form to present day.

In the decades following Olszowy's book, numerous authors around the world have explored definitions and models for horticultural therapy, which generally include at least three elements: a trained professional, a client in some form of treatment with defined goals, and the cultivation of plants. While terminology may vary somewhat, these elements have been emphasized by numerous authors (Hefley 1973; Mattson 1982; Haller 2006; Haller and Capra 2018). Additionally, Diane Relf presented a model that illustrated the therapeutic benefits of horticultural therapy and has provided a foundation for programs, research, and documentation of the value of horticultural therapy to the present day (Relf 1981). It categorized the benefits to program participants as emotional or psychological, social, intellectual or cognitive, and physical. Richard Mattson described a series of models to depict the interactions between the client, plant and horticultural therapist, and the various types of bonding that occur between each of these elements in actual practice. The bonding is influenced by the status of the client at any particular time, the setting, and plants, as well as the actions of the therapist. He also modeled the client at various levels of engagement in the horticulture activity and addressed group situations in which the therapist adopts directive and participatory approaches (Mattson 1982).

Another definition that combines the elements involved in the process along with the outcomes to be achieved was offered by Rebecca Haller in 2006: "Horticultural therapy is a professionally conducted client-centered treatment modality that utilizes horticulture activities to meet specific therapeutic or rehabilitative goals of its participants. The focus is to maximize social, cognitive, physical and/or psychological functioning, and/or to enhance general health and wellness" (Haller and Kramer 2006; Haller and Capra 2018).

A model presented to illustrate this definition depicts the client, goals, therapist, and plant as the elements involved in the process of horticultural therapy (Figure 1.2). Note that the client is central—all the other elements exist to bring goal-directed positive change to the individual participant (client), with the guidance of a skilled professional (therapist) who utilizes hands-on horticulture (plants) with the client.

Matthew Wichrowski created a model of horticultural therapy practice, which is presented in Chapter 7, depicting "the interaction of beneficial features of the environment, therapeutic use of self, and utilization of gardening and horticultural activities as prescribed exercise, which meet the needs of the participant" (see Figure 7.1).

Definitions and descriptions of horticultural therapy vary around the world, yet they generally include the elements just described. Some emphasize the clinical nature of the work, while others are more inclusive. Exhibit 1.1 describes a somewhat broader definition in German-speaking countries of "garden therapy." It shows the perspective of the *Internationale Gesellschaft GartenTherapie* (IGGT), which reflects their actions in defining the scope of their work and the rationale behind their philosophies. Also described is a practice example from the author reinforcing the holistic nature and range of benefits achievable in nature-based practices. While similar in concept to the definition used above, this one also takes into account nature's passive influence on a person.

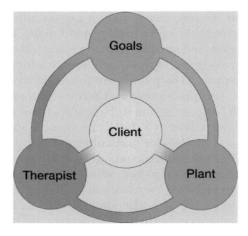

Figure 1.2 Horticultural therapy elements and process. Model by Rebecca L. Haller, Illustration by Moss Cremer. (With kind permission from Taylor & Francis Group: *Horticultural Therapy Methods: Connecting People and Plants in Health Care, Human Services, and Therapeutic Programs*, 2nd ed., 2017, 7, Haller, R.L. and Capra, C.L.)

EXHIBIT 1.1

Perspectives: The Idea of Garden Therapy in German-Speaking Countries

For some time, the *Internationale Gesellschaft GartenTherapie* (IGGT) has worked to create a formal definition of garden therapy. That sounds so easy, but it isn't. The situation in German-speaking countries, with different health systems utilizing many therapeutic disciplines and approaches to health, has made it challenging to create this definition.

Garden therapy is a technical measure where plant and garden-oriented experiences and activities are used to initiate specifically interactions between a person and nature, with the aim of the support of quality of life and the preservation and restoration of functional health, which includes: Healing or relief of disturbances with illness value; preservation and support of self-determined social participation, as well as supporting the overall wellness of this person.

The IGGT definition utilizes a basic model of human-nature interaction. This model looks at the interaction between a person and his or her natural environment, and it proclaims that this interaction has an effect on this person. This interaction takes part in different dimensions. One is that nature has a passive influence on the person, and the other is that this person will feel and see an effect through his or her actions on nature. A garden is an iconic model for this. The availability of these two possibilities of interaction is seen as a component of human health. It's about "doing in the garden" and "being in the garden."

In this way, the well-known biopsychosocial model of health is extended by the IGGT into a biopsychosocial-ecological model. We assume that all these dimensions, including the ecological, have an influence on the health of the individual. Many studies and experiences support this approach. We include now this fourth dimension in our therapeutic model. We utilize ecological therapies to complement the biological, psychological and social therapies—a field which is known as "green care." Garden therapy is part of green care and is evolving its own approach as a new ecological therapy. When put into practice this model has the potential to impact many dimensions of human function and provide a wide array of benefits.

The following two examples are intended to illustrate what this can mean in practice and how it extends the usual inventory of garden therapy.

Within the framework of interdisciplinary cooperation, a physical therapist and I co-treated an older female patient with the aim of early mobilization after having a stroke. Initially she was able to sit at the edge of her bed, in order to make small simple flower arrangements. So the horticultural activity was an aid to achieve another defined therapeutic goal (International Classification of Functioning, Disability and Health, ICF: D410—to change one's body position). One day, when the weather was particularly nice, we decided to offer her the opportunity to transfer into a wheelchair to go outside into the garden. She broke into tears and reported that she had not been out there for almost a year. Previously, she had been staying in a retirement home when she had the stroke.

For a garden therapist, this patient undoubtedly experienced something that could be described as "nature deficit," and the result of the walk was a continuous "blossoming." In later appointments, bringing plants to her became central to treatment and became increasingly important. One can say that the contact with nature

(Continued)

EXHIBIT 1.1 (Continued)

Perspectives: The Idea of Garden Therapy in German-Speaking Countries

was no longer a means of attaining another goal but was its own goal. This is not yet defined as such in the International Classification of Functioning, but in a four-dimensional health model, it would certainly be so.

The natural contact itself was often on the passive side. She did not have to plant something to restore this contact. This would often motivate her, even during this early phase of treatment. If we had only looked at active gardening and decided that it wasn't appropriate for her to participate in garden therapy, this would have been a great pity. The fact that we are now able to use garden therapy targeted for seriously ill patients is due to this expansion of our own professional concept.

But, of course, there is also the importance of the active dimension, which is to exert influence on the environment. Ultimately this is indeed the primal meaning of a garden, namely, to give people as natural beings the chance to shape the garden, to control things, and to experience directly the effects of their action.

As an example, let's look at the garden therapeutic work we do on the locked neuropsychiatric department of our hospital. The stay on such a closed ward means a significant loss of control. In addition, patients often experience negative emotions, the loss of orientation, loss of identity, and social disintegration as one goes into the "patient" role. The experience also presents less opportunity for physical exercise.

Therefore, the goal of health promotion is to create positive emotions (P), to create situations where one can orient oneself (O), social integration (S), identity-preserving measures (I), and the supply of movement (T = Tonus). Accordingly, there is garden therapy within the context of our "POSIT-ive basic therapies." In these therapies, we can address all dimensions with the tools of garden therapy. We create absolutely positive experiences with fragrances and colored flowers. We keep seasonal and regionally suitable plants and experiences, especially on the holidays to aid with orientation. We encourage the patients to socialize with each other. Personal identity and personality are reinforced when working with familiar plants from one's personal biographical garden and plant experiences, while working in the garden naturally promotes movement and exercise. Above all, patients create a sense of control over their own life situation by actively cultivating their own therapeutic garden.

Garden therapy within a four-dimensional health model, defined by two different modes of interaction, clearly expands the possibilities of therapy, thereby expanding the potential target groups and therapeutic goals. This may be the most important thing. Within this model, garden therapy is no longer only a method with the focus on achieving goals, as in other therapies such as occupational therapy, physical therapy, etc., but in addition, it provides opportunities for contact with nature, which in itself is a specific unique goal, having its own range of benefits. So garden therapy has evolved the potential to meet a wide range of human needs in many different settings.

Andreas Niepel
Garden Therapist, Helios Klinik Hattingen, President IGGT

At the Twelfth International People Plant Symposium in Australia in 2014, researchers from Konkuk University in South Korea presented a model of horticultural therapy that was intended to exhibit an international standard for the profession (Son 2016). This model encompasses concepts from three perspectives that have been described in various ways since the profession began. The three perspectives are *therapeutic benefits*, *therapeutic dynamics*, and *therapeutic roles*. To understand the basis for this model, it is helpful to summarize the three perspectives upon which it is constructed. *Therapeutic benefits* are based primarily on Diane Relf's early work that categorized the therapy as having an impact on participants (clients) in emotional, social, physical, and intellectual areas (Hefley 1973). *Therapeutic dynamics* were also first modeled by Relf (1981). She described the importance of caring for living plants and the various action, interactions, and reactions among the elements of client, plants, and therapist. In the early 1980s, Kansas State University conducted extensive research to revise the academic curriculum, profile practitioners, and process descriptions used in horticultural therapy (Mattson and Shoemaker 1982). As part of that research, Richard Mattson (1982) developed a hierarchy of graphic models based on three elements or *therapeutic roles*—a *client* with a defined need, *plants* or horticulture, and a professionally trained *therapist* (one who "is capable of facilitating therapeutic and/or rehabilitative processes with the client using a plant media"). The models were the first to graphically depict both therapeutic dynamics and therapeutic roles, and to show the actions and interactions that take place in various types and stages of programming (Figure 1.3).

Figure 1.3 Being close to familiar plants offers elder visitors an opportunity to recall memories of gardens past. (Courtesy of Patty Cassidy.)

Distinguishing horticultural therapy from similar practices

The term *horticultural therapy* is increasingly reserved for those therapeutic efforts that are clinical in nature, with alternative terminology used to represent activities that involve the use of gardening for human development and health promotion that does not involve goal directed treatment interventions. Nonclinical programs have been referred to as *community horticulture* (Haller 2006) and *therapeutic horticulture* (Point 1999; American Horticultural Therapy Association 2007). *Community horticulture* describes programs without treatment objectives and that seek human or community development, such as those found at school or community allotments, inner-city greening sites, or in some healing gardens. *Therapeutic horticulture* programs generally are distinguished from horticultural therapy by a lack of clinical goals or documentation of client behaviors and progress. They may also focus on passive interaction with plants. Additionally, there has long been a recognition that gardening has positive effects on all members of the community. In Japan, for example, the term *Engei Fukushi* (EF), or "horticulture well-being" is used to characterize the use of horticulture to enhance the quality of life for the benefit of all people, while "horticultural therapy" is a type of EF that requires a trained professional to address the needs of people in treatment (Matsuo 1999, 2012). The expression *horticultural well-being* is also used in South Korea to portray the myriad benefits gained through horticulture by anyone who experiences it, including those who are not in medical or social service programs. Note that each of these alternate terminologies is used to describe actions or activities that only address some of the key elements regarded as essential to horticultural therapy.

Another approach to terminology has been adopted in the United Kingdom by Thrive, a national charity that uses "gardening to bring about positive changes in the lives of people living with disabilities or ill health, or who are isolated, disadvantaged or vulnerable." *Social and therapeutic horticulture* is an umbrella label used by Thrive for gardening for therapy, including horticultural therapy as well as less clinical applications. Social and therapeutic horticulture is described as "… the process of using plants and gardens to improve physical and mental health, as well as communication and thinking skills. It also uses the garden as a safe and secure place to develop someone's ability to mix socially, make friends and learn practical skills that will help them to be more independent" (Thrive 2018b).

Previously, Thrive defined *therapeutic horticulture* as "the process by which individuals may develop well-being using plants and horticulture" and distinguished it from horticultural therapy by the type of goals and involvement (Point 1999). In 2007, the American Horticultural Therapy Association (AHTA) defined *therapeutic horticulture* as:

> a process that uses plants and plant-related activities through which participants strive to improve their well-being through active or passive involvement. In a therapeutic horticulture program, goals are not clinically defined and documented but the leader will have training in the use of horticulture as a medium for human well-being. This type of program may be found in a wide variety of health care, rehabilitative, and residential settings (AHTA 2007).

See Chapter 2 for more discussion about therapeutic horticulture and its relationship to horticultural therapy practice.

Training and education in horticultural therapy

History

With these varied definitions, the profession itself began to be recognized as early as the 1940s, around the time the term *horticultural therapy* itself was first used (Olszowy 1978). In 1952, the first training workshop in horticultural therapy took place at Michigan State University, with the first master's degree program in horticultural therapy for occupational therapists offered in 1955. Concurrently, in England the field developed with an emphasis on providing gardening assistance to those with disabilities, including the design of gardens and tools for those with physical challenges. Several books in the field were published, including: *Therapy through Horticulture* by Watson and Burlingame (1960), *The Easy Path to Gardening* by A.S. White et al. (1972) in the United Kingdom (with gardening techniques to accommodate gardeners with disabilities), *Hoe for Health* by Alice Burlingame (1974), and *Horticulture for the Disabled and Disadvantaged* by Damon Olszowy (1978). Many of these books described processes and characteristics of the people served. The 1970s saw accelerated progress in the advancement of the profession. In 1972, Kansas State University, in cooperation with the Menninger Foundation in the United States, began an undergraduate program in horticultural therapy (Odom 1973; Shoemaker 2002). This innovative collaboration began in support of the mental health field and helped not only those clients at the Menninger Clinic but also provided university students with formal training in psychology and horticulture (Lewis 1976). Soon after, other universities, such as the University of Maryland and Clemson University, offered both undergraduate and graduate degree programs in the field.

University degrees

By 1981, eight US universities offered bachelor's and/or master's degrees (NCTRH 1981). According to AHTA, in 2018 six universities in the United States offered either a concentration, a minor, a graduate certificate, or an "option" in horticultural therapy. Three additional universities offered at least one course in horticultural therapy, and seven university-affiliated certificate programs throughout the United States offered horticultural therapy education (AHTA 2018). University programs have changed over the years, with a current focus on certificate programs as the primary source of training in horticultural therapy. While there has been no growth in the number of degree programs in the United States, and the universities that offer degrees have shifted, education is often provided through certificate programs in collaboration with universities.

Many other institutions throughout the world have developed educational programs in the field of horticultural therapy, pointing to a global interest in learning how this modality can improve the lives of individuals. Organizations that offer training include Thrive in the United Kingdom, Trellis in Scotland, Awaji Landscape Planning & Horticulture Academy (ALPHA) in Japan, the Korean Horticultural Therapy and Well-Being Association in South Korea, Learning Cloud in Australia, and a variety of botanic gardens in Canada (Kent 2012; Park 2012; Koura 2012; Fung 2012). These and other educational programs vary in the type of training provided and may offer a range in levels of training, such as introductory seminars, workshops that train professionals to use horticulture with specific populations, and university degrees.

Certificates

As previously stated, those in the United States who desire education in horticultural therapy frequently find coursework through university-affiliated certificate programs. This type of educational program offers focused study in the field and allows a person who does not seek a degree to attend. A significant portion of those who enroll in courses in horticultural therapy have already earned one or more college degrees, so for them, the focused training provided by certificate programs is advantageous and desirable (Haller 2016). Certificates are used by employers and clients as validation of specialized training and knowledge. Earning a certificate demonstrates both initiative and willingness to make the time commitment necessary for professional and career development. In most cases horticultural therapy certificate programs are accredited by AHTA and work in conjunction with a university offering academic credit. These partnerships add credibility to this method of accomplishing horticultural therapy education.

Internships

In addition to the opportunities that university degree and certificate programs provide for education, internships play an important role in giving students real-life experience with the guidance of an experienced horticultural therapist. Internships are required in the United States in order to achieve professional registration (Horticultural Therapist Registered) and prepare the intern for practice through the direct use of horticulture as a treatment modality. For example, interns may hone skills in horticulture, interdisciplinary communications, professional conduct, session planning, and the many techniques used in sessions with clients.

Organizations

Co-occurring with the development of educational programs was the formation of associations around the world that serve to develop, inform, and/or promote the advancement of horticulture as therapy.

American Horticultural Therapy Association

In 1973, the National Council for Therapy and Rehabilitation through Horticulture (NCTRH) formed in the United States with the goal of promoting and enriching the profession of horticultural therapy as a therapeutic intervention and rehabilitative medium (Lewis 1976). Later the organization was renamed the American Horticultural Therapy Association (AHTA). The mission of the nonprofit national group is "to promote and advance the profession of horticultural therapy as a therapeutic intervention and rehabilitative modality" through disseminating information, encouraging professional development, providing standards, and maintaining a professional registration system. The organization also promotes research and educational opportunities in the field (AHTA 2018). See Exhibit 1.2 for a description by Diane Relf of the early years of NCTRH and AHTA through the lens of a pioneer and leader instrumental in the formation of the first professional organization in the United States.

EXHIBIT 1.2

Perspective: Purpose and History of a New Profession

Although horticulture had been used as an important part of the therapeutic setting since before psychiatry became a science, it only began to be recognized for its significant therapeutic value in the 1960s. In 1968, a survey of hospitals across the United States was conducted by Rhea McCandless under the auspices of the Menninger Foundation in conjunction with Kansas State University. Of the 216 hospitals responding, 64% had some type of "garden therapy" program. Many of the 36% without a program indicated they would start one if they had the professional staff to do so. Based on this survey and the identified need for "horticulturists with training to understand and respect patients and therapeutic processes," a degree program was established at Kansas State University. A survey of prisons conducted in 1970 through California State Polytechnic College indicated that at least one correctional institute in each state had a horticultural therapy program. Michigan State University, through the work of Alice Burlingame, offered coursework to occupational therapists in the field of horticulture to expand their skill in this area.

While working on my PhD in horticultural therapy at the University of Maryland in 1971, I conducted extensive library searches, including a new and very innovative computer-based one at the National Institutes of Health. Although there was little published beyond general program descriptions and anecdotes on the value and impact of volunteers, it was possible to identify the names of several dozen programs around the United States. With an exchange of letters, I found that most of them felt they were a unique example of a new treatment modality. Over the next year I began to work with Melwood Horticultural Training Center for Mentally Handicapped (now named simply Melwood), one of the most widely recognized programs using horticulture for rehabilitation. Their contacts in rehabilitation identified several more organizations in the mid-Atlantic region that were expanding programs. With these contacts and identified need, we decided to gather people together to exchange information and help in the development of horticulture as a therapeutic tool.

On April 9, 1973, a planning session was held at Melwood near Washington, DC, with participants from horticultural training and treatment facilities for people with intellectual disabilities, blindness, and those who were in psychiatric treatment. Participants also included representatives from the US Department of Health, Education and Welfare and from universities. As a result of this meeting, a decision was made to form the National Council for Therapy and Rehabilitation through Horticulture (NCTRH). The council was headquartered at Melwood and staffed by their employees and volunteers. In the following twenty months, under the guidance of individuals such as Earl Copus, executive director of Melwood and president of NCTRH, and myself, who prepared the bylaws and articles of incorporation and conducted the legal action to incorporate NCTRH, the council developed as a fully functioning organization. The name chosen was based on the significant differences between therapy and rehabilitation, with the aim to be as inclusive as possible for all individuals who supported the concept that horticulture was a valid treatment tool in diverse settings.

(Continued)

EXHIBIT 1.2 (Continued)

Perspective: Purpose and History of a New Profession

The first conference of the council was held in Washington, DC, on November 5 and 6, 1973. More than 200 participants attended from the United States, Asia, and Europe to learn about programs and make contacts to advance their knowledge and skill. The second annual conference was held October 8 and 9, 1974, again in Washington, DC, and again under the auspices of Melwood. By this time the council had 335 members from forty US states, Canada and England. With many talented professionals ready to assume responsibilities for the functioning of the council work, groups were established to undertake these tasks: (1) development of program establishment guidelines (2) development and affiliation of state chapters, (3) establishment of a job placement bank, and (4) professional registration. The development of a voluntary professional registration procedure for horticultural therapists was seen as critical by those who worked in hospitals and similar facilities, which required their staff to have professional credentials, in order to gain respect for their area of expertise. Although it was acknowledged that there was not yet a published body of knowledge based on research and program experience, it was anticipated that this would grow and become the basis for the advancement of the field.

At the same time that the NCTRH was establishing its operational base, the field of horticulture was experiencing phenomenal growth. Horticulture departments expanded in enrollment fivefold at many universities. The idea of using plants to improve health and quality of life caught on with the public: restaurants and public buildings were filled with indoor plants, urban parks were established, and community gardens became increasingly important. This rapid growth in interest in plants was accompanied by an equally rapid growth in the use of horticulture for therapy and rehabilitation. Six universities established degree programs in horticultural therapy. Noted centers of horticultural therapy, such as Melwood (Maryland), Menninger Clinic (Kansas), Clemson University (South Carolina), and Clinton Valley Center (Michigan), received increasing numbers of requests for information on developing programs, and visitors from all parts of the country came to observe their programs.

In 1975, NCTRH joined the American Horticultural Society and the American Society for Horticultural Science at the Center for American Horticulture in Virginia. Melwood agreed to loan me, a Melwood staff member and secretary for NCTRH, to the council as executive secretary for six months. At that time, the council assumed full responsibility for all administrative and operational financial needs.

Recognizing that the financial strength of an organization lies in membership, significant focus was placed on a membership drive. The purpose of the drive was twofold: (1) to increase membership to fund the employment of council staff and (2) to bring awareness of the importance of horticulture as a therapeutic medium to individuals throughout the therapeutic, rehabilitative, and horticultural fields.

During the early years of NCTRH the membership was as diverse as the people who believed that horticultural was therapeutic, including:

(Continued)

EXHIBIT 1.2 (Continued)

Perspective: Purpose and History of a New Profession

- Volunteers from groups such as the Men's Garden Club of America, Garden Club of America, the National Garden Club, Women's Farm and Garden Club, and Cooperative Extension Master Gardeners.
- Professionals from human service fields such as occupational therapy, physical therapy, recreation therapy, psychiatry, vocational rehabilitation, corrections, and education.
- Professionals in horticulture including landscape design, community gardening, greenhouse management, and association management.
- Individuals conducting horticultural programs in different treatment settings.

In 1988, a decision was made to change the name of the organization to the American Horticultural Therapy Association and focus primarily on the needs of professionals working in the field as horticultural therapists.

Diane Relf, PhD, HTM

Professor Emeritus, Virginia Polytechnic Institute & State University

Horticultural therapy around the world

As previously described, interest and practice in horticultural therapy as a profession are found in countries around the world. The history, status, and organizations of many were represented at the Tenth International People–Plant Symposium (IPPS) held in 2010 in Canada (Goodyear and Shoemaker 2012). For example, in Canada, horticultural therapy has been used since the 1950s. Gardening as therapy has long been used in psychiatric settings and physical rehabilitation in Canada. As in the United States, programming expanded to serve veterans returning from World War II in rehabilitation programs (Kent 2012). The Canadian Horticultural Therapy Association was established in 1987; it promotes the use of horticultural therapy and offers voluntary professional registration and educational accreditation (Kent 2012). The Hong Kong Horticultural Therapy Center began offering programs and training in 2005 (Fung 2012). In Japan, the Japanese Society of People–Plant Relationships focuses on research, while the Japanese Horticultural Therapy Association manages a national qualifications system for horticultural therapists (Koura 2012). South Korea introduced the practice in the 1980s and recognized a distinction between using horticulture for well-being versus treatment in the 1990s: Both are encompassed in the Korean Horticultural Therapy and Well-Being Association (Park 2012). The International Association for HortTherapy (IGGT) provides a common platform for therapists in Austria, Germany, and Switzerland (Neuberger 2012). (See Exhibit 1.1 for a description of the *Internationale Gesellschaft GartenTherapie* (IGGT) in German-speaking countries.) In addition to those represented at the International People–Plant Symposium, other parts of the world have created horticultural therapy programs and/or established associations and training for practitioners, including Israel, China, Australia (Cultivate NSW 2018), Sweden (Soderback et al. 2009), Denmark (Corazon et al. 2012), Norway, Italy (EUGO 2018),

Figure 1.4 Le Jardin D'epi Cure serves those with brain injuries outside Paris. (Courtesy of Stephane Lanel.)

and the United Kingdom (Figure 1.4). The *Fédération Française Jardins Nature et Santé* was formed in France in 2018 (Isabelle Boucq, correspondence with the author, February 22, 2018). In the United Kingdom, horticultural therapy has a history that spans the same decades as that of the United States, with the Society for Horticultural Therapy and Rural Training founded in 1978, later renamed simply Horticultural Therapy, and rebranded as Thrive in 1998. The organization provides services to a variety of people with disabilities or disadvantages It also provides training and education, consultation, and advice on how to garden for health and well-being (Thrive 2018). The prevalence of programs and organizations is expanding and includes places beyond those listed here.

Benefits

Beneficial aspects of using horticulture as a therapeutic modality include its universal appeal, and the fact that it encourages human growth, advances restoration, addresses innate psychological needs, offers versatility, has meaning and purpose, and impacts others (Haller and Capra 2018). While these are just a few of the important attributes that characterize the use of horticultural therapy, it's important to understand that the practice of horticultural therapy enhances the client, the facility, and even the therapist. It can be compared with the first green shoot in the garden. While one can see the potential striking out from the barren earth, there is much more below ground that make this modality a compelling instrument for change in the lives of others. Depending on the setting and clients of any given program the tangible benefits vary. Yet horticulture as a vehicle can be used to benefit an individual physically, cognitively, emotionally, and socially (Hefley 1973; Simson and Straus 1998) (Figure 1.5).

For a person with a developmental disability, working in a vocational horticulture program brings a sense of purpose and belonging, for the client with an eating disorder,

Figure 1.5 Student in BOOST program at Bullington Gardens. (Courtesy of John Murphy.)

using plant activities in a therapeutic setting offers a process to work through anxiety and unhealthy relationship dynamics, and for the elder with dementia a wellness horticultural therapy program offers an opportunity to participate with others in a nonthreatening activity with plants. As these examples illustrate, the program model in which horticultural therapy is practiced results in the specific characteristics and benefits achieved. The three main models of practice, which are detailed in Section Three: Practice within Program Models, are therapeutic, vocational, and wellness. These models may look different depending on the type of program, such as day treatment versus inpatient or residential, hospital versus community organization (nonclinical), one session per week versus full immersion in daylong sessions. Each model has a primary focus for outcomes, yet often there is overlap between the models. For example, the desired outcome of a trainee in a vocational program is skills needed to sustain supported or independent employment. Within vocational horticultural therapy sessions, wellness or therapeutic goals are invariably also addressed as needed. While the focus may be employment, horticultural therapy takes an individualized and whole-person approach to treatment. Skills for coping with anxiety and stress, improving physical stamina, or even cultivating a leisure interest exemplify some of the areas that may be addressed within that vocational program. A program of this nature may take place in a rehabilitation or habilitation organization employing many different types of therapies, be it in a facility where horticulture is the primary modality (i.e. in a greenhouse or on a farm) or within a horticulture business in the community.

With this evolving history and educational support for the field of horticultural therapy, the variety and types of programs have spread worldwide, as have the therapeutic methods employed to benefit those served. A growing body of research looks at how horticultural therapy activities, conducted with a trained horticultural therapist, enhance well-being. Reports on the research are found in conference papers, books, and journals. For example, AHTA produces the *Journal of Therapeutic Horticulture* publishing research and practice cases in the field. In England, the association Thrive has partnered with Loughborough University to gather and disseminate research and offer a practice guide for practitioners of social and therapeutic horticulture (Sempik et al. 2003, 2005a, 2005b). The International People–Plant Council's biannual meetings promote and disseminate research, and discussion on horticultural therapy as well as the broader topic of people and plant interactions. Bringing research into practice and programming is the role of the well-trained therapist. (See Chapter 14 for more about evidence-based practice.)

The skilled horticultural therapist

Within these varied program models, the trained horticultural therapist works to support each client or participant to achieve measurable goals and objectives. The effective horticultural therapist practices in a variety of settings and often as a member of clinical and rehabilitation teams. In addition, horticultural therapists work as educators, community providers of horticultural therapy services, and as independent contractors or consultants.

Skills and knowledge

The measure of a successful horticultural therapy program is often based on the training and skills of the therapist leading the program itself. In order for horticultural therapists to be effective, they must have multidisciplinary skills and knowledge, encompassing horticulture, human services, and horticultural therapy.

In 1982, Kansas State University, with funding from the National Institute of Mental Health, published the first job analysis, which categorized the essential skills for horticultural therapists as those related to therapy and those related to horticulture (Kuhnert et al. 1982). They include *therapeutic* skills such as:

- Motivation management.
- Communication: observation and recording and assessment of client behavior in both written and oral modes.
- Goal setting: long- and short-term treatment goals.
- Activity programming.
- Theoretical knowledge of client medication, types of disabilities, intervention theories, theories of child and personality development, and knowledge of the ethics of client-therapist relations.
- Group counselling
- General *horticulture* skills, such as:
  - Management of plant environments, both indoor and out.
  - Theoretical knowledge of pesticides, fertilizers, soils, organic gardening techniques, alternative energy techniques for plant culture, and horticultural containers.
  - Business management for indoor and outdoor crop production.
  - Landscape design, also floral arrangement and artistic interpretation of plants.

According to another job analysis conducted in the United States and reported in 2014, a variety of more comprehensive and detailed skills and knowledge were identified as important for competent practice (Starling et al. 2014). These include knowledge, skills and abilities in horticulture, therapeutic, and other job responsibilities. Some of the skills receiving the highest importance in the analysis included general horticulture skills, such as propagating plants, transplanting, planting seeds, raised bed gardening, growing plants for sale, production or installation, pest identification, understanding the toxicity of plants, and selecting appropriate plant material. Therapy skills included monitoring and addressing safety concerns; evaluating program effectiveness for individuals served; observing, analyzing, and evaluating client participation for progress; implementing individual and group sessions; and adapting, grading, and/or modifying activities and/or instructions.

While the terminology varies between the two job analyses, they generally report similar skills and knowledge requirements, and indicate that horticultural therapy is interdisciplinary in nature. Practitioners need to be proficient in horticulture, therapy, and other areas. These job analyses have both informed and validated the training and coursework recommended for horticultural therapists by AHTA.

Professional credentials

In the United States, a credentialing system has been in place since September 1975 (Shoemaker 2012). AHTA is the only organization in the United States that recognizes and registers horticultural therapists through a voluntary professional registration program. The designation Horticultural Therapist Registered (HTR) ensures that professional standards have been achieved based on academic requirements and professional training. The AHTA has identified key courses and topics of study and has adopted a core curriculum for guidance. The organization hopes that the identification of core knowledge and course topics will eventually lead to the development of an exam to assess the competency of practitioners and create a credentialing system that mirrors other therapies. Because of the multidisciplinary nature of the field, coursework is required in horticulture, human services, and horticultural therapy to achieve professional registration (HTR) status. This specific training is vital to becoming an effective horticultural therapy practitioner. Those who complete such training have an enhanced understanding of how to provide horticultural therapy services most effectively. They are in the best position to:

- Identify specific outcomes.
- Record and communicate those outcomes.
- Provide focused and successful horticultural therapy sessions.
- Utilize horticultural resources effectively.
- Obtain financial and administrative support.
- Articulate the value of horticultural therapy.
- Be involved in treatment or care planning.
- Manage sustainable programs (Haller 2012).

Personal traits

In addition to education, skills, and knowledge, horticultural therapy practitioners tend to possess the personal traits that match the demands of the vocation. A vocational interest inventory profile reported in 1982 (based on the work of J. L. Holland), showed the following occupational and personality traits as prevalent in registered horticultural therapists in the United States: social, investigative, and enterprising (Czerkies 1982). The data presented was intended to be used for career counseling and to inform others of horticultural therapy as a career path. In addition to this study, others describe common traits of horticultural therapists in order to inform prospective students and practitioners. These traits include: flexible, compassionate, organized, engaging, creative, analytical, constructive, empowering, and communicative. It is recommended that horticultural therapists be "motivated, versatile and able to communicate needs and benefits of this emerging type of human service" (Horticultural Therapy Institute [HTI] 2018). It is also critical to success that the therapist be motivational—that is, be able to effectively motivate participants—a task made easier by the intrinsically engaging nature of gardening processes and a plant-based modality.

Of course, characteristics or traits of individual therapists are naturally varied. And because the types, models, locations, and program participants involved in the work vary widely, professionals are commonly able to find a good match in some niche of this field of practice. Many times, a program is not only based on standard treatment processes and best practices but also reflects the personality of the person who develops and implements it. The following chapters illustrate many of the approaches and models of horticultural therapy, yet some are still to be created, formed, and launched through the passion, personalities, and determination of future practitioners.

Summary

To be effective at helping people improve their lives through horticultural therapy, it is important to recognize the innate connection humans share with nature and also understand the complexities of this treatment modality. The focus of horticultural therapy remains on maximizing the social, cognitive, physical, and/or psychological functioning and/or to enhance general health and wellness of the people served (Haller and Capra 2018). As the profession evolves, documents its efficacy through research, and gains acceptance worldwide, more trained professionals will enter the field, and program participants can reap the positive benefits of human interaction with plants and gardens.

References

American Horticultural Therapy Association (AHTA). 2007. "Definitions and Positions."

American Horticultural Therapy Association (AHTA). 2018. "Definitions and Positions," Accessed March 1, 2018. http://www.ahta.org/ahta-definitions-and-positions.

American Horticultural Therapy Association. "Missions and Vision of the AHTA," Accessed March 1, 2018. http://www.ahta.org/mission-vision.

Burlingame, A. 1974. *Hoe for Health: Guidelines for Successful Horticultural Therapy Programs.* Birmingham, MI: Alice Wessels Burlingame.

Corazon, S. S., U. K. Stigdotter, M. S. Moeller, and S. M. Rasmussen. 2012. "Nature as Therapist: Integrating Permaculture with Mindfulness and Acceptance-based Therapy in the Danish Healing Forest Garden Nacadia." *European Journal of Psychotherapy & Counselling* 14(4): 335–347.

Cultivate NSW, Accessed April 6, 2018. http://www.cultivatensw.org.au

Czerkies, V. L. 1982. "The Horticultural Therapy Vocational Interest Inventory Profile," in *Defining Horticulture as a Therapeutic Modality Part 1: Profiles in Horticultural Therapy*, edited by J. Shoemaker, and R. S. Mattson. Manhattan, KS: Department of Horticulture, Kansas State University.

Dorn, S., and D. Relf. 1995. "Horticulture: Meeting the Needs of Special Populations." *HortTechnology* 5(2): 94–103.

EUGO. "Horticultural Therapy," Accessed April 6, 2018. https://www.eugolearning.org/learning/outcomes/horticultural-therapy.

Fung, C. Y. Y., et al. 2012. "Development of Horticultural Therapy in Hong Kong." *Acta Horticulturae* 954: 169–174.

Goodyear, S. N., and C. A. Shoemaker. 2012. "The State of Horticultural Therapy around the World." *Acta Horticulturae* 954: 159–189.

Haller, R. 2016. "Impact of a Horticultural Therapy Certificate Program on Students and Their Professional Activity," Presentation at the XIII International People Plant Symposium, Montevideo, Uruguay, November 10–12.

Haller, R. L. 2006. "The Framework," in *Horticultural Therapy Methods: Making Connections in Health Care, Human Service, and Community Programs*, edited by R. L. Haller, and C. L. Kramer. Binghamton, NY: The Haworth Press.

Haller, R. L. 2012. *Linking People and Plants Newsletter*. www.htinstitute.org.

Haller, R. L., and C. L. Capra, eds. 2018. *Horticultural Therapy Methods: Connecting People and Plants in Health Care, Human Services, and Therapeutic Programs*, 2nd ed. Boca Raton, FL: CRC Press, Taylor & Francis Group.

Haller, R. L., and C. L. Kramer, eds. 2006. *Horticultural Therapy Methods: Making Connections in Health Care, Human Service, and Community Programs*. Binghamton, NY: The Haworth Press.

Hefley, P. D. 1973. "Horticulture: A Therapeutic Tool." *Rehabilitation* 39(1): 27–29.

Horticultural Therapy Institute. 2018. "Horticultural Therapy Careers," Accessed April 10, 2018. http://www.htinstitute.org/horticultural-therapy-careers.

Kent, A. et al. 2012. "The Status of Horticultural Therapy in Canada: Practice, Research, Education." *Acta Horticulturae* 954: 165–168.

Koura, S. 2012. "The Introduction of the Japanese Horticultural Therapy Association." *Acta Horticulturae* 954: 175–177.

Kuhnert, K., J. Shoemaker, and R. H. Mattson. 1982. "Job Analysis of the Horticultural Therapy Profession," in *Defining Horticulture as a Therapeutic Modality Part 1: Profiles in Horticultural Therapy*. Manhattan, KS: Department of Horticulture, Kansas State University.

Lewis, C. 1976. *Fourth Annual Meeting of the National Council for Therapy and Rehabilitation through Horticulture*, September 6, Philadelphia, PA.

Matsuo, E. 1999. "What Is Horticultural Well-Being in Relation to 'Horticultural Therapy'?" in *Towards a New Millennium in People-Plant Relationships*, edited by M. D. Burchett, J. Tarran, and R. Wood. Sydney, Australia: University of Technology, Sydney, Printing Services, pp. 174–180.

Matsuo, E. 2012. "Developmental Features of Engei Fukushi (Horticultural Well-Being) in Japan." *Acta Horticulturae* 954: 151–154.

Mattson, R. H. 1982. "A Graphic Definition of the Horticultural Therapy Model," in *Defining Horticulture as a Therapeutic Modality Part 2: Models in Horticultural Therapy*, edited by R. Mattson, and J. Shoemaker. Manhattan, KS: Department of Horticulture, Kansas State University.

Mattson, R. H. 1992. "Prescribing Health Benefits through Horticultural Activities," in *The Role of Horticulture in Human Well-Being and Social Development*, edited by D. Relf. Portland, OR: Timber Press, pp. 161–168.

Mattson, R. H., and J. Shoemaker, eds. 1982. *Defining Horticulture as a Therapeutic Modality Part 2: Models in Horticultural Therapy*. Manhattan, KS: Department of Horticulture, Kansas State University.

NCTRH. 1981. "Hort Therapy Education." *National Council for Therapy and Rehabilitation Through Horticulture Newsletter* 8(1): 2.

Neuberger, K. 2012. "The Status of Horticultural Therapy around the World: Practice, Research, Education—Austria, Germany, Switzerland." *Acta Horticulturae* 954: 187–189.

Odom, R. 1973. "Horticulture Therapy: A New Education Program." *HortScience* 8(6): 458–460.

Olszowy, D. 1978. *Horticulture for the Disabled and Disadvantaged*. Springfield, IL: Charles C. Thomas.

Park, S. A. 2012. "Practice of Horticultural Therapy in South Korea." *Acta Horticulturae* 954: 179–185.

Point, G. 1999. "Your Future Start Here: Practitioners Determine the Way Ahead." *Growth Point* 79: 4–5.

Relf, D. 1981. "Dynamics of Horticultural Therapy." *Rehabilitation Literature* 42: 147–150.

Relf, D. 1992. "Human Issues in Horticulture." *HortTechnology* 2(2): 159–171.

Sempik, J., J. Aldridge, and S. Becker. 2003. *Social and Therapeutic Horticulture: Evidence and Messages from Research*. Loughborough, UK: Thrive and the Centre for Child and Family Research, Loughborough University.

Sempik, J., J. Aldridge, and S. Becker. 2005a. *Growing Together: A Practice Guide to Promoting Social Inclusion through Gardening and Horticulture*. Loughborough, UK: Thrive and the Centre for Child and Family Research, Loughborough University.

Sempik, J., J. Aldridge, and S. Becker. 2005b. *Health, Well-Being and Social Inclusion: Therapeutic Horticulture in the UK*. Loughborough, UK: Thrive and the Centre for Child and Family Research, Loughborough University.

Shoemaker, C. 2002. "The Profession of Horticultural Therapy Compared with Other Allied Therapies." *Journal of Therapeutic Horticulture* 13: 74–80.

Shoemaker, C. 2012. "The Practice and Profession of Horticultural Therapy in the United States." *Acta Horticulturae* 954: 161–163.

Shoemaker, J., and R. H. Mattson, eds. 1982. *Defining Horticulture as a Therapeutic Modality Part 1: Profiles in Horticultural Therapy*. Manhattan, KS: Department of Horticulture, Kansas State University.

Simson, S., and M. Straus, eds. 1998. *Horticulture as Therapy: Principles and Practice*. Binghamton, NY: Haworth Press.

Soderback, I., M. Soderstrom, and E. Schalander. 2009. "Horticultural Therapy: The 'Healing Garden' and Gardening in the Rehabilitation Measures a Danderyd Hospital Rehabilitation Clinic, Sweden." *Pediatric Rehabilitation* 7(4): 245–260.

Son, K. C., S. J. Jund, A. Y. Lee, and S. A. Park. 2016. "The Theoretical Model and Universal Definition of Horticultural Therapy." *Acta Horticulturae* 1121: 79–88.

Starling, L. A., T. M. Waliczek, R. L. Haller, B. J. Brown, K. R. Malone, and S. Mitrione. 2014. "Job Task Analysis Survey for the Horticultural Therapy Profession." *HortTechnology* 24(6): 645–654.

Thrive. 2018a. "History of Thrive," Accessed April 12, 2018. https://www.thrive.org.uk/history-of-thrive.aspx.

Thrive. 2018b. *"What is Social and Therapeutic Horticulture?"* Accessed April 12, 2018. https://www.thrive.org.uk/what-is-social-and-therapeutic-horticulture.aspx.

Watson, D., and A. Burlingame. 1960. *Therapy through Horticulture*. New York: Macmillan.

White, A. S., et al. 1972. *The Easy Path to Gardening*. London, UK: The Reader's Digest Association.

chapter two

Horticultural therapy, related people–plant programs, and other therapeutic disciplines

Rebecca L. Haller and Karen L. Kennedy

Contents

Horticultural therapy is a distinct profession. Yet it also has similarities to other therapeutic disciplines around the world that use plant-based or nature-based interactions to improve human health and well-being. Therapeutic techniques and treatment processes among disciplines are often interchangeable and depend as much on the type of program and setting in which the program occurs as the specialty of the therapist applying them. The lines between practices may be blurred, with experiences offered to the client or program participant based on principles such as working with groups, strengths-based approaches, therapeutic strategies, and processes that are found in an array of allied professions. Horticultural therapy programs may be independent, or they may be integrated into organizations that provide nature-based services such as community gardening, school gardens, urban greening, healing gardens, care farming, animal-assisted interventions, or wilderness therapy. Likewise, horticultural therapy may be offered alongside or combined with other allied therapies, such as recreation therapy, occupational therapy, mental health counseling, etc.

The overlap with each of these related practices is described in this chapter, as well as those aspects of horticultural therapy that are unique and help to define it as a profession. Further discussion is also presented regarding applications for the use of therapeutic horticulture, comparing and contrasting it with horticultural therapy.

Terminology

A wide array of treatment options and settings are based on the premise that purposeful contact with nature provides support for clients to achieve therapeutic, vocational, and wellness goals. In order to understand the elements discussed in describing these approaches, and their relationship to horticultural therapy, the terms *nature*, *therapy*, *horticulture*, and *green care* are defined in this context as follows:

- *Nature* includes outdoor environments, plants, and animals—the natural, physical world. It is used in a broad sense and includes "nearby" and "cultivated" nature such as that found in gardens and greenhouses.
- *Therapy* is a process that involves professional care or treatment of a person with assessed needs. It is intended to support individual development and improve health and well-being, but not necessarily to cure in a medical sense.
- *Horticulture* refers to the science and practice of growing ornamental and edible plants. In horticulture, nature is human influenced and affected by culture. It encompasses all aspects of gardening, including design, propagation, cultivation, harvest, etc.
- *Green care* is a broad term for "the use of agricultural farms and the biotic and abiotic elements of nature for health and therapy-promoting interventions as a base for promoting human mental and physical health, as well as quality of life" (Gallis 2013). It is also referred to as *nature-based intervention*. Another phrase to describe a similar process is *nature-assisted therapy*, which combines aspects of therapy and nature to achieve treatment goals, rehabilitation, and promotion of health. These terms are sometimes used synonymously as well as with nature-assisted therapy, or they may be considered to have distinct meanings (Bragg and Atkins 2016). For this chapter, they will be used interchangeably.

Green and plant-rich settings

One of the commonalities of horticultural therapy and other green efforts to positively affect human physical and psychological health and well-being is that they universally take place in plant-rich settings. Some of these include places where plants are cultivated, such as community gardens (allotments), school gardens, urban farms, rural farms, greenhouses, and gardens designed for health care and healing. Others are based in more natural, less human-influenced places, such as forests and wilderness. All of them offer some measure of connection with nature, which has been shown to offer intrinsic benefits of restoration (Kaplan and Kaplan 1989) and better adjustment (Louv 2005). Research has also demonstrated that contact with nature positively influences recovery from stress and illness and leads to improved health outcomes (Marcus and Sachs 2014). The settings in which these plant/nature-based therapies take place may be considered on a continuum from wilderness or natural areas to the cultivated clinical space used for horticultural therapy. Each setting offers participants some type of connection with nature.

Wilderness settings

Therapeutic programs based on interaction with natural environments that have been minimally influenced by the actions of humans are demonstrated by wilderness therapy programs. Sometimes referred to as outdoor behavioral health care, these programs typically serve adolescents and young adults, and include activities such as expeditions, camping, climbing, boating, and hiking in wilderness areas, with the aim to improve self-reliance and social skills, or heal from trauma or psychological issues. The remote physically and psychologically challenging setting is crucial to the process and offers an intimate connection to the restorative characteristics of untamed natural settings. "The more primitive the setting, the more people can experience solitude, tranquility, self-reliance, and closeness to nature" (Kaplan and Kaplan 1989). Activities at this end of the continuum of nature connection seek to leave the environment intact and undisturbed.

Forest settings

Shinrin-yoku, or "forest bathing," is a practice that occurs in an undisturbed natural place— specifically a forest—although the forest may be adjacent to an urban area, it may be one step away from wilderness on the continuum. Walking in a forest environment has been shown to boost mood, reduce stress, lower blood pressure, and be more effective in producing these outcomes than walking in an urban area (Li et al. 2011). There appears to be significance in the place of the activity. Researchers have looked not only at the restorative qualities of this environment but also at the effect of inhaling the aromatic compounds that are released by trees, which may reduce stress and blood pressure (Li et al. 2009). Forest therapy guides use mindfulness approaches to promote wellness. The practice typically serves the general population but may be integrated into psychotherapy or other practices.

Landscapes and nearby nature

For this discussion, landscapes are those horticulturally designed spaces that are intended for beauty and recreation. They may include urban parks, street trees, and residential or commercial designed outdoor spaces of various sizes and uses. Generally, the parks used for activities that bring the user closer to nature, such as walking, picnicking, or gardening, are more satisfying than those outdoor spaces used for sports. They may provide tranquility and restoration from mental fatigue, as described by Rachel and Stephen Kaplan (1989). Enhanced individual functioning, such as managing major life issues, reducing aggression, improving self-discipline, and reducing attention deficit hyperactivity symptoms have all been attributed to the presence of trees and grass in urban landscapes (Kuo 2004). Even a view of nature through a window has been shown to provide stress reduction and health benefits (Ulrich 1979). Landscapes that are purposefully designed to positively impact human health ordinarily are concerned with health promotion for the general population. They typically involve casual encounters with nature, as opposed to therapeutic interventions, because they lack the involvement of a trained therapist.

Therapeutic outdoor spaces in health care

Therapeutic landscapes or healing gardens associated with health-care facilities are places of refuge for patients, visitors, and staff members. A specialized type of landscape, they are strategically designed to offer experiences of the user's choosing, including places to sit, walk, meditate, converse, or in some way benefit from being in a garden setting. The gardens offer privacy and connection to nature to enhance well-being. Marcus and Sachs recommend that design features accommodate those whose health is compromised, and include considerations for safety, security, privacy, accessibility, physical and emotional comfort, positive distraction, and biophilia, and proper maintenance and environmental sustainability for the gardens (2014). Health-care gardens do not necessarily include therapeutic or gardening activities unless they also serve the purpose of enabling active participation in programming as described in gardens used for therapy (below).

Agricultural settings

Care farms use agricultural environments to offer services to an array of people, from those with defined medical or psychosocial needs to those who are stressed or in ill health. Professionally supervised activities may include growing crops or gardens, animal husbandry, and other farming projects. Originating in Europe, the practice takes place in nature-based settings that are even more human influenced, designed, and controlled than landscapes. Customarily they are working productive farms. Further along the continuum toward planned and managed clinical environments, care farms have actively and regularly cultivated spaces, and they may be settings for horticultural therapy programming (Sempik and Bragg 2013).

Urban gardens

At this point on the continuum, the emphasis is active, hands-on growing of plants, with a necessarily human influence on the settings. Support of community connections and social development, as well as access to healthy food, are aims of community gardens or allotments, which are found in towns and cities, on land that is set aside and shared by a group of people for cultivation. While many of these programs serve people who are excluded socially or economically, such as those living in impoverished neighborhoods, refugees, or others, they do not typically employ trained therapists or use therapeutic processes. Another example of urban garden settings includes those in schools. Schoolyard gardening has become increasingly popular, often with an emphasis on connecting to science curricula, teaching social skills, improving physical activity, and encouraging healthy eating (Waliczek and Zajicek 2016). Again, in school gardens, therapy is not necessarily conducted, yet it may be a setting for horticultural therapy, as seen in program examples in later chapters.

Gardens used for therapy

Ideal garden spaces for therapeutic use have enabling features that allow and encourage active plant care. They are carefully designed to be accessible and engaging environments that serve people who are in treatment, seeking various health and wellness outcomes. This is clinical space, whether or not it is adjoining a medical facility, and can be used to practice a therapeutic model of horticultural therapy, or it is in an environment more

Figure 2.1 Garden designed for therapy at Thrive in the United Kingdom. (Courtesy of Rebecca Haller.)

conducive to vocational or wellness programming, such as described in later chapters. Essential to the space is the capacity to grow plants and considerations for safety and accessibility as well as details that support treatment goals of the program participants. More than in the previous settings, the garden is deliberately fashioned and human-influenced to meet specific programming needs (Figure 2.1).

Enclosed environments

These gardens include indoor activity centers in hospitals, schools, residential facilities, and locked settings, such as prisons, living spaces for people with dementia, or psychiatric hospitals. Offering year-round locations for programming, they also serve those who may require a controlled and secure environment. (See Exhibit 2.4 later in this chapter for an example of this environment.) They may be spaces filled with plants and four-season growing, such as greenhouses, or have limited gardening options, such as potted plants on a windowsill or plants growing under lights. In the most restricted circumstances, where no growing space exists, therapists may bring plant materials to residents for horticultural therapy sessions. Obviously, when plants are brought to the treatment space, participants are more removed from an encounter with nature or a natural environment. Yet for someone in the most limited environment, who has no other connection to nature on a daily basis, the opportunity for multisensory stimulation combined with active plant care and nurturance benefits from the experience. Professionally and purposefully designed sessions provide a connection to the seasons, weather, and the wonders of plants and growth (Figure 2.2).

Figure 2.2 A.G. Rhodes co-treatment session with occupational, physical and horticultural therapists addresses upper extremity range of motion, fine motor skills, and attention to task in a year-round controlled and secured environment. (Courtesy of Mary Newton.)

Settings for horticultural therapy

While horticultural therapy programs have traditionally taken place in carefully designed and structured settings, there are applications of this practice in many of the settings described here. For example, *landscapes and nearby nature* are used to augment programming and provide an environment that supports therapeutic outcomes. The landscape offers stress reduction, which enables participants to be more responsive to intensive work on goals and engagement in activities. *Therapeutic outdoor spaces in health care* are typically associated with hospitals, behavioral health care, and long-term care centers. While they may not necessarily be developed to facilitate engagement in hands-on horticulture practices, they may be used for experiential sessions or for therapeutic horticulture. Exhibit 2.1 describes the use of a garden of this type. *Care farms* are commonly used in Europe and in other parts of the world to provide services for a wide range of users, including those with medical, psychological, social, physical, and vocational needs. They may incorporate horticultural therapy practices in the care or focus on providing care in a more natural, socially supported way that does not feel like therapy. In either case, horticultural therapy could take place in this setting in partnership with the farmers who not only operate the farm to produce commercially but also offer social support and a sense of normalcy to the therapy efforts. *Urban gardens*, which provide a setting for horticultural therapy, also facilitate efforts to integrate participants into the broader community. For example, programs may be conducted within the context of community gardens or allotments, and provide community integration and a chance for clients to practice social and physical skills in a more challenging environment. Programs designed to reach at-risk youth are often situated in school gardens (see Exhibit 7.4). Community-supported agriculture is another type of garden that is used for horticultural therapy. It may be located in or near an urban environment and offers an ideal situation for vocational programming. See Exhibit 13.4 for an example of a program for veterans that took place in this setting. While the types of horticultural spaces vary, Huxmann noted that a garden as a clinical setting for treatment is essential in horticultural therapy. It is an aspect of the profession that sets it apart and offers unique resources for treatment (Huxmann 2016) (Figure 2.3).

EXHIBIT 2.1

Program Example: Therapeutic Horticulture in Memory Care

The Portland Memory Garden is a unique city park in Portland, Oregon, that was specifically designed to provide a safe environment of respite and restoration for those with memory loss and their caregivers. The Friends of the Portland Memory Garden, who maintain and replenish the lush plantings for seasonal interest, have a deep commitment to providing wellness and therapeutic programs for this population.

Each summer, through a generous grant, the Friends organization hosts free horticultural therapy sessions for senior communities (organizations), especially memory care units. Every session is designed and delivered by a registered horticultural therapist (HTR) and often provides training for interns. Because the horticultural therapists have no way of knowing participants' specific needs, the emphasis is on establishing general program goals to serve the clients' overall well-being.

The psychosocial focus of each session strives to engage and enliven the participants' ability to recall past garden memories and verbally share their remembrances with others. The grounding activity called "What's in Bloom?" allows every person an opportunity to hold, experience, and delight in several selections of seasonal flowers or plants. Because the garden is intentionally planted with high-sensory plants, it provides participants many ways of exploring and experiencing the offerings. Emerging smiles are often the first indication that the plants have strong and beneficial effects on the group members.

Cognitive group goals focus on the individuals' recall and recognition of plant material presented in "What's in Bloom?" As they stroll through the garden, participants are cued to see the individual plant growing in its natural habitat and relate this to the plant they saw in the grounding activity.

Exercise and movement are goals of each session, and the wide circular pathways encourage easy and safe strolling or wheelchair maneuvering. The meandering tour includes passing by raised beds that allow all to get close to the plants for touching and smelling as well as visual enjoyment. Gross and fine motor skills are used as participants are encouraged to reach for a branch of rosemary or grasp a sprig of lavender. Many benches are strategically located around the pathway, providing places to rest or to set goals for further exploration.

Special attention is also directed to caregivers, both professionals who work for the visiting organization and family members who enjoy garden visits with their loved ones. As horticultural therapists model how to use the garden and its plantings for maximum effect, caregivers learn new skills and techniques to use in their facilities. The wheelchair-accessible bathroom allows groups to spend significant time in the garden. Additionally, the garden's design allows caregivers to close and secure the entrance gate if they should revisit the garden on their own. This allows those in their care to stroll independently in a safe and secure space.

Patricia Cassidy, MA, HTR
Contractor, Gardening for Wellness

Figure 2.3 Sensory stimulation using plants from the garden to increase ageing adults' social interactions and verbal responses at Portland Memory Garden. (Courtesy of Patty Cassidy.)

The preferred clinical space for horticultural therapy varies with the type of practice—therapeutic, vocational, or wellness—as well as program goals such as community integration, income generation, or restoration from mental fatigue. In general, the preferred settings for sessions is an outdoor garden that enables active hands-on gardening. These settings have the potential to offer the highest degree of nature connection as well as a rich source of treatment options (Hazen 2014). An indoor, plant-rich environment such as a greenhouse or plant room also offers diverse treatment options and an intimate connection with living things as well as a place for year-round horticulture. Yet it is one step removed from the outdoor experience. Next on the list of preferences is an indoor space with grow lights for plant growth and propagation, which is an environment for treatment that is the best option for programs that need to use indoor conditions for safety or security or to accommodate those who are medically fragile. They may also be used for growing in the winter months or to augment outdoor growing and treatment spaces, or they can be an option for adding growing space where budgetary or space constraints prohibit greenhouse or outdoor gardens. Another setting for treatment may be the client's residence, where the therapist can bring plants and planting equipment to the client and provide sessions in the client's room or nearby. This is found most often in medical facilities, such as hospitals or long-term care facilities, when severely limited patient mobility or cognition necessitates this approach. (See Exhibit 8.13 for an example of using a plant cart for programming.) Even in this situation, a nature connection may be facilitated through sensory stimulation, opportunities for hands-on indoor gardening, and witnessing the fascination of plants.

Note that accessing a garden, cultivated landscape, park, or an undeveloped woodland through walking, viewing, or sitting are not necessarily considered horticultural therapy. While the individual may reap many benefits from such experiences, only part of the four essential elements of horticultural therapy are in use—the person/client and the plants. Missing are goals, therapist-directed methods, and a focus on active plant engagement. Just as walking or running is not considered physical therapy, a walk in the garden is not horticultural therapy or therapeutic horticulture. Of course, passive encounters in a garden may be included in an overall program or individualized treatment plan, and such encounters can be a method that is incorporated in horticultural therapy to produce profound results. Similarly, a health-care garden or therapeutic landscape may be a setting for horticultural therapy if programming takes place there. The confusion occurs when the garden itself is considered therapy (Dorn and Relf 1995).

Nature-based therapies

Umbrella terminology for therapeutic practices and health interventions based on experiencing and interacting with nature are varied. Sempik and Bragg (2013) describe green care as the "many approaches which use nature to provide health and wellbeing services to vulnerable groups in the form of structured and organized programmes." Included in their description of green care are wilderness therapy, ecotherapy, facilitated exercise as a therapeutic intervention, care farming, horticultural therapy (and social and therapeutic horticulture), and therapy with animals. *Nature-assisted therapy* and *nature-based interventions* are other terms that are used to describe the array of human services that incorporate engagement with natural environments. Exhibit 2.2 describes green care in more detail and includes nonclinical approaches such as green exercise (Figure 2.4).

Additional types of nature-based therapies are presented in the following sections to provide a basis for understanding similarities and contrasts with horticultural therapy. Each practice overlaps with one or more of the essential elements of horticultural therapy—client, goals, therapist, and plant.

Figure 2.4 YMCA Green Care Program. (Courtesy of Kristen Greenwald.)

EXHIBIT 2.2

Perspective: Green Care

Green care is an umbrella term for complex interventions that utilize nature and natural environments in a wide range of active techniques addressing mental health, physical health, social rehabilitation, and employment opportunities for vulnerable populations. Care, in this context, encompasses both therapy and wellness programs (Sempik et al. 2010). Foundational theories and various research studies provide a framework for which necessary components of a program must be met in order for it to maintain therapeutic value. Therapeutic programs that are currently considered green care programs include animal-assisted therapy, horticultural therapy, wilderness therapy, care farming, and green exercise (Sempik et al. 2010).

Within the green care framework, programs are either structured therapy programs with clearly defined and documented client-centered goals, such as animal-assisted therapy, wilderness therapy, and horticultural therapy, or such programs are designed to provide broader benefits, such as green exercise and care farming (Sempik et al. 2010). A care farm, for instance, may consist of programming for those with mental health challenges, developmental disabilities, cognitive disabilities (neurodiverse functioning), or physical disabilities with the goal of providing benefits such as meaningful employment and social opportunities. Green exercise, or exercise in natural landscapes, is supported by research to provide mental and physical human health benefits, but it is often not facilitated by a therapist to obtain a specific documented treatment goal. Conversely, an animal-assisted program provides clinical interventions in which a trained professional and animal(s) partner to co-treat specific mental health treatment or educational goals. Similarly, wilderness therapy programs utilize natural settings and outdoor activities to obtain documented mental health and physical treatment goals as exemplified by adaptive outdoor programs, and therapeutic mountaineering programs.

A key theory that underlies the green care framework is the biophilia concept developed by E. O. Wilson. This theory postulates that humans are inherently connected to nature and have an "innate tendency to focus on life and lifelike processes" (Wilson 1984). Furthermore, Wilson defines *biophilia* as "the innately emotional affiliation of human beings to other living things." Wilson observes how the human tendency to focus on life and lifelike processes is a biologically based need and integral to development as individuals and as a species (Kellert and Wilson 1993). Today, children may have limited exposure or few positive experiences in nature and with animals (Louv 2005; Melson 2001). Therefore, human dependence on nature extends far beyond material and physical sustenance to include a human craving for aesthetic, intellectual, cognitive, spiritual meaning, and satisfaction (Kellert and Wilson 1993).

Another foundational theory that supports the green care framework has been developed by Stephen and Rachel Kaplan. Their theory of attention restoration asserts that mental fatigue is reduced after spending time in nature or viewing a picture of a natural setting (Kaplan 1995). Excessive visual stimuli—noise, movement, and visual complexity—lead to damaging levels of psychological and physiological arousal.

(Continued)

EXHIBIT 2.2 (Continued)

Perspective: Green Care

Nature inspires "involuntary attention," which does not require "directed attention," and creates experiences of "being away" and "soft fascination" with elements that exist in nature (Kaplan and Kaplan 1989). Natural environments have less complexity—noise, movement, and intensity—and have patterns that reduce arousal and subsequent feelings of stress (Simson and Straus 1998). Restoration occurs when stimulus overload is reduced, and attentional capacity is restored, through exposure to nature.

These theories have led to numerous research studies examining the possibility that humans may be naturally connected to nature. They have also inspired research related to the human health benefits of human-nature connection. A primary finding from various research studies is that contact with nature improves psychological health by reducing stress levels and enhancing mood (Ulrich 1981, 1979; Kaplan and Kaplan 1989; Kaplan 1995; Hartig et al. 1991, 2003; Nakamura and Fujii 1990, 1992). Research also supports the existence of a direct link between the amount of accessible local green space and psychological health (Takano et al. 2002; De Vries et al. 2003). These findings support the therapeutic programs that fall under green care. Furthermore, research that informs the therapeutic efficacy of green care program settings indicates that attentional restorative effects occur in a variety of nature-based settings, such as wilderness areas (Hartig et al. 1991), prairies (Miles et al. 1998), forests (Park et al. 2010), and community parks (Cimprich 1993). These findings suggest promising benefits if horticultural therapists coordinate programming with social and environmental justice initiatives that give equal access to people and plant connections.

Philip Tedeschi, LCSW
Clinical Professor and Executive Director, Institute for Human-Animal Connection, Graduate School of Social Work, University of Denver

Kristen Greenwald, MSW, LSW
Child and Family Therapist, Garden Resource Outreach

Wilderness therapy

As previously described, the practice of wilderness therapy applies traditional therapy practices in natural remote places, which are away from cities and towns. Applying experiential treatment to groups or individuals, they focus on personal resilience and empowerment, and they may serve people who seek addiction or mental health recovery, improved social skills, self-reliance, and behavioral care. Programs vary in their scope and methods and range from an emphasis on adventure to more self-reflective or contemplative approaches. Participants do not act upon the environment in which the therapy takes place but rather seek to leave no trace of their presence there.

Ecotherapy

Ecopsychology recognizes and describes essential connections between humans and other forms of nature, such as plants, animals, rocks, oceans, etc. Ecotherapy addresses this fundamental need for nature connections and is used by mental health professionals to improve mental health and foster healthy interactions with the earth (Buzzell and Chalquist 2009). It is often practiced in outdoor settings, including a range of settings from wilderness to gardens.

Care farming

Offering contact with green nature, gardens, fields, and farm animals, care farming uses structured programming to address social, cognitive, physical, psychological, or educational needs. It may be only one part of a working farm, or it may be applied throughout the farm's operation. Working relationships with the farmers, staff members, and clients are cultivated with purposeful, meaningful tasks and experiences. Active engagement and manipulation of the environment is regularly practiced through raising and caring for animals, gardening, woodlot management, cultivating field crops, and land conservation.

Animal-assisted therapy

Used in nursing homes, mental health settings, prisons, and hospitals, animal-assisted therapy provides links among the therapist, the client, the animal, and nature in general. Those links assist the client to relax and be more receptive to therapy. Most commonly applied with dogs or horses, animal-assisted therapy is used to address the full range of treatment domains, such as cognitive, psychosocial, and physical.

Comparisons to horticultural therapy

Horticultural therapy is both parallel and dissimilar to other nature-based therapies. Widely accepted elements of horticultural therapy include a client with specified and documented *goals*, and a *trained therapist* who applies the use of horticultural activities (plants) to promote goal achievement (see Chapter 1 for definitions of horticultural therapy). It may focus on outcomes related to recovery from illness or injury, vocational skills, or placement, and/or it may address overall health and wellness. Nature-based therapies include the following elements: a *client*, a *trained professional*, *documented goals*, and *contact with nature*. Gardens are a type of so-called nearby nature and offer a connection with the natural world through plants, insects, birds, weather, seasons, and soil. Indoor and outdoor gardens provide plant-rich spaces that encourage this connection (Kaplan and Kaplan 1989). The key distinction of horticultural therapy is that, through the act of gardening, participants (clients) act on nature. Gardens are human-influenced spaces, in harmony with nature, yet cultivated, nurtured, and managed. The act of raising plants offers a myriad of opportunities for relationships, physical and mental self-improvement, and a nature connection that is increasingly seen as essential for human health. The opportunity to care for another living entity is a hallmark feature of this type of therapy, providing a rich source of analogies and life lessons as well as motivation and inspiration.

In practice, the therapeutic use of horticulture may also be combined with other nature-based therapies to offer a full range of treatment and wellness options. Exhibit 2.3 describes a university-based application of this concept.

EXHIBIT 2.3

Program Example: Nature-Based Therapeutics in Public Horticulture

Nature-based therapeutics (NBT) is an emerging field that includes, but isn't limited to, the following modalities: therapeutic horticulture, restorative environments, facilitated green exercise, therapeutic landscapes, care farming, healing gardens, and animal-assisted interventions. The focus of NBT is on the purposeful healing power of nature—through plants, animals, and natural landscapes—to provide measurable beneficial outcomes in human health and well-being.

The University of Minnesota Nature-Based Therapeutic Services is a partnership between the Minnesota Landscape Arboretum and the Earl E. Bakken Center for Spirituality and Healing. The partnership provides a broad range of opportunities, understanding, and a wide skill base to ensure access to the most current research and best practices in the field. The partnership recognizes the strengths and expertise to make the best use of resources from both integrative medicine and nature-based science. NBT services are comprised of five areas: direct program delivery, professional training, academic graduate courses, research, and outreach.

NBT services operate from a model of well-being created by Dr. Mary Jo Kreitzer, director of the Earl E. Bakken Center for Spirituality and Healing at the University of Minnesota. The Well-being Model is based on extensive work and research around integrative health and identifies six dimensions that contribute to human well-being: health, relationships, security, purpose, community, and environment.

NBT direct programs are delivered to a variety of people from diverse cultural backgrounds and with varied diagnoses and abilities, such as for people with Parkinson's disease and other movement disorders, children and elders together intergenerationally, adults with persistent mental illness, children on the autism spectrum or with co-occurring severe emotional or behavior disorders, high school students with chemical health issues, children and adults with eating disorders, adults with developmental and cognitive differences, and veterans.

NBT direct program goals are organized in five domains, with outcomes such as the following:

- Physical—relaxation, sensory awareness, restoration, self-regulation, body awareness.
- Social—collaboration, relational, outward perspective, team cohesion, communication.
- Psychological—resilience, motivation, self-confidence, self-efficacy, self-awareness.
- Cognitive—human impact on the environment, phenology, citizen science, concentration.
- Spiritual—relational and interactional, intrapersonally, interpersonally, and with nature.

The outcomes of NBT programs are based on whole-system healing as a way of integrating the health and well-being of individuals, communities, organizations, societies, and the environment into all living systems. In other words, it is essential

(Continued)

EXHIBIT 2.3 (Continued)

Program Example: Nature-Based Therapeutics in Public Horticulture

for NBT direct program goals to consider the implications of how NBT and other service interventions (e.g., physical therapy, psychotherapy, occupational therapy) will affect the person and their relationships with family, friends, community, work, and the environment.

The facilities used to implement NBT direct programs are varied and nimble due to the wide range of users, locations, finances, and accessibility. For example, some NBT programs are set in a larger therapeutic landscape designed by local professionals with fully accessible raised-bed gardens, automatic watering system, equipment shed, comfortable and flexible seating, and shelter. Others may be less defined and more organic, with ground beds, a nylon tent for shelter, and hoses connected to a nearby water source.

All of the NBT direct programs are financially sustained by yearly contracts with the organization served.

Jean M. Larson, PhD, CTRS, HTR
Manager, Nature-Based Therapeutic Services, Assistant Professor, Nature-Based Therapy Studies Earl E. Bakken Center for Spirituality and Healing, Minnesota Landscape Arboretum, University of Minnesota

Horticultural therapy and other allied therapies

As previously noted, horticultural therapy takes place in a variety of institutional and community-based environments, and programs can be designed to address almost any type of situation facing the human condition. The distinguishing characteristic that sets horticultural therapy apart from other allied disciplines is the interaction with plants. This interaction includes all aspects of plant care, from growing house plants on a small scale indoors to farms and greenhouse production. It also includes harvesting and using the product of the harvest (e.g., drying and preserving produce, cooking meals, arranging flowers, crafting items such as wreaths, or using preserved materials). The inherent reaction to the green growing environment and the response to a growing, living entity that survives and thrives under the care of the participant are important aspects of horticultural therapy, adding an element unlike other allied disciplines (Lewis 1996).

There is common ground between horticultural therapy and other allied disciplines, including occupational therapy, speech therapy, therapeutic recreation, art therapy, music therapy, mental health counseling, special education, and vocational rehabilitation. The commonalities can be summarized as follows:

- Participant needs are identified through assessment.
- Intervention is planned based on identified goals and objectives.
- Goals and objectives are written and measured in behavioral terms, which include modifications based on abilities.
- Intervention is activity based.
- Documentation is an essential part of the process to ensure that both progress and ongoing needs are communicated.

Figure 2.5 Rehab patient practicing dynamic balance and standing endurance, with physical therapist providing close supervision, during group horticultural therapy session. (Courtesy of Einstein Marketing.)

Horticultural therapists often use a holistic approach and, as such, treatment goals and objectives may be addressed in multiple disciplines. As a result, co-treatment opportunities maximize the benefits of horticultural therapy sessions in some facilities. For example, working alongside a patient in a rehabilitation hospital, an occupational therapist has the opportunity to work on bilateral hand use and left-side neglect with a person in treatment following a stroke, while the horticultural therapist focuses on psychosocial coping skills. As noted in Exhibit 2.4 this strategy can be useful for a number of disciplines (Figure 2.5).

EXHIBIT 2.4

Program Example: Co-treatment in Rehabilitation

The Alice and Herbert Sachs Therapeutic Conservatory is a dedicated space for horticultural therapy at MossRehab in Pennsylvania. It was designed to serve patients recovering from spinal cord injuries, brain injuries, strokes, amputations, or other complex neurological conditions. The conservatory is a warm, plant-filled, natural environment, removed from the noise, bright lights, and clinical atmosphere of the hospital. Divided into three sections, the conservatory includes a greenhouse with an indoor garden and seating for respite and relaxation, and a grow room and workroom for plant and nature-based activities. Under the guidance of a registered horticultural therapist (HTR), who collaborates and co-treats with the clinical team, patients are able to meet physical, occupational, speech, recreational, and psychosocial goals outside the traditional therapy gym.

Patients may visit the conservatory seven days a week for de-stressing, socializing, and rejuvenating. Participation in clinical horticultural therapy sessions gives patients an opportunity to engage in functional leisure activities while addressing a wide range of treatment goals. The physical setting and the modality

(Continued)

EXHIBIT 2.4 (Continued)

Program Example: Co-treatment in Rehabilitation

of horticulture afford patients the benefit of connecting with nature in an intimate way. Therapists have observed that patients are often able to reach goals not previously realized. For example, they may achieve longer endurance for standing and ambulating; better focus and concentration on the task; and, in some cases, speaking for the first time or with more clarity than in previous sessions.

Horticultural therapy sessions consist of groups, ranging from three to ten patients, and individual sessions. A patient, family member, therapist, physician, nurse, or another practitioner can make referrals to horticultural therapy. For co-treats, a treating therapist, such as a physical therapist, occupational therapist, speech and language pathologist, or recreational therapist, contacts the horticultural therapist with information about the patient and relates specific goals for the session. The horticultural therapist creates an activity that serves the patient's needs and interests. If the co-treat is to be within a group session, the horticultural therapist adapts the project accordingly.

Example 1 Physical Therapist Co-treat with Horticultural Therapy Group

Individual physical therapy goal: To increase independent standing tolerance and static balance, patient will engage in horticultural therapy activity while standing for ten-minute intervals, three times with no more than two seated rest breaks.

Group horticultural therapy goal: To improve divided attention strategies, patient will follow multistep directions, perform sequential planting steps, and engage in social communication with peers during task engagement.

Horticultural therapy task: Repotting a four-inch houseplant into a five-inch pot.

Adaptation for physical therapy co-treat: An elevated table, adjoining the group table, allows patient to stand and engage with the group while the physical therapist provides distant supervision.

Example 2 Speech Language Pathologist Individual Co-treat with HTR

Individual speech goal: To improve swallow function, patient will utilize compensatory strategies with minimum assistance.

Individual horticultural therapy goal: Patient will articulate one fact about herbs and fill, seal, and package herbs that are grown in the program for tea.

Horticultural therapy task: Sample four warm herbal teas. Make tea bags by scooping herbs and sealing bags.

Co-treat technique: The speech language pathologist cues the patient on swallowing strategies. The horticultural therapist engages the patient in cognitive and fine motor activities.

Abby Jaroslow, HTR,
Certified Horticulturist, Horticultural Therapist, The Alice and Herbert Sachs Therapeutic Conservatory, MossRehab, Einstein Health care Network

Distinguishing horticultural therapy from therapeutic horticulture

The definitions of horticultural therapy and therapeutic horticulture have evolved as the specifics of these and other similar practices have continued to advance. Due to the similarity of language and the diversity of program styles and types, the terminology can sometimes be confused. The distinction can be viewed as a spectrum. (See Figure 2.6 for a spectrum of the use of horticulture as therapy.) On one end, a therapeutic horticulture experience can be a gently directed passive encounter for an individual or group with a horticultural therapist in a plant-rich environment, without specific goals or documentation yet with purpose and meaning. This experience may be the best level of intervention to meet client needs. Horticultural therapy begins at a moderate level of intervention, with a horticultural therapist documenting individual progress on group goals. This level of intervention is appropriate when the goals are the same for all participating in the session. Documentation that reflects an individual's progress on objectives specific to them is the highest level of intervention. The need for uniform terminology for using horticulture in care and health promotion has been discussed, yet the terminology still varies in countries around the world (Bragg 2016). In addition, in some instances, the term *therapeutic horticulture* may be synonymous with *horticultural therapy*, or the full spectrum may be described as "social and therapeutic horticulture" (Annerstedt and Währborg 2011). See Chapter 1 for more discussion on definitions.

The factors that differentiate one end of the spectrum from the other are:

- Clients seeking to improve overall health and well-being to clients with defined treatment goals and objectives.
- Level of client engagement with growing plants from individual sensory engagement to active gardening.
- Therapist interaction with the client supporting an independent experience with the general intention for improved well-being to therapist-directed sessions focusing on goals/objectives.
- The use of goals and objectives to define and document client progress on identified needs ranging from the desire for improved general health and well-being to group goals for all clients with no documentation, to individual goals and objectives with documented outcomes for each client.

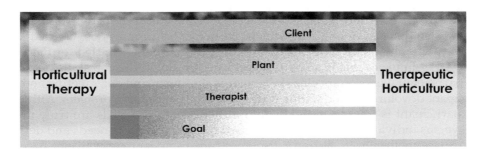

Figure 2.6 The Therapeutic Use of Horticulture Spectrum. Model by Karen Kennedy and Rebecca Haller. (Design by Moss Cremer.)

Programs may fall at different places within these variables, depending on the circumstances. All intervention levels can be important and meaningful to the clients served in the right situation. However, the addition of client goals and objectives adds additional value. Goals and the steps to achieve them (objectives) serve to inform the therapist of the client's specific needs and drives the techniques selected to support client achievement. Goals and objectives also motivate the client by breaking the goal into manageable steps while providing a vision of the progress to come. Written documentation provides the communication tool of the outcomes and the plan for the future for the client and care team. Sometimes a therapeutic horticulture program can be transformed into a horticultural therapy program with the additions of client goals and documentation of progress or change. This may result in the therapy being held in higher regard by other professionals and employers.

The horticultural therapist uses discretion about employing the most fitting level of intervention to best suit the clients, desired outcomes, and type of program. Exhibit 2.5 describes the range of horticultural therapy and therapeutic horticulture intervention strategies used within a single program setting. The flexibility of employing multiple program types maximizes the range of client needs served as well as the use of a therapeutic garden space. For example, the Gathering Place, a nonprofit providing support to cancer patients and their families, recognizes the importance of enabling opportunities for solace, contemplation, camaraderie, and developing friendships in the garden as well as nutrition education, therapeutic horticulture and horticultural therapy programs. Their garden serves multiple purposes, much like a room within a building. Exhibit 2.6 provides more details about how their therapeutic garden is integrated into their healing community for people touched by cancer.

EXHIBIT 2.5

Program Example: Horticultural Therapy Models

At the Greenhouse at Wilmot Gardens, we use different therapeutic models depending on the group we are working with. The majority of our groups follow the therapeutic horticulture model. In other words, the groups are led by a registered horticultural therapist, but individual goals are not set for each participant. Instead, group goals are defined based on the common needs of the group. With our movement disorders group, for example, our goals are to increase or maintain fine and gross motor skills, range of movement, and eye-hand coordination. Although we don't individually document a participant's progress and report back to a treatment team, we do observe individual skills and deficits and adjust the activity as needed.

We also have a model, in which participants' progress is charted and documented individually, but everyone in the group is being charted for the same goals. An example of this is an autism spectrum group; in this job skills program, each participant works on specific skills such as signing in and out, greeting another participant or volunteer, and letting staff know if he or she has a problem. Although each participant is tracked individually for these goals, the goals are truly group goals versus individually tailored goals based on an individual's needs and desires.

We have another group, also an autism spectrum disorder group, with which we use a horticultural therapy model. Participants in this group undergo an intake assessment and then a program-oriented assessment. After attending a few sessions, each participant

(Continued)

EXHIBIT 2.5 (Continued)

Program Example: Horticultural Therapy Models

meets with the horticultural therapist to explore and identify specific goals that he or she would like to work on to increase both quality of life and job readiness. Those goals are individual and participant specific, and they are charted on a regular basis.

All of these therapeutic models are valuable; one is not more important than another because they all reach and help people in different settings. What is important is that the horticultural therapy practitioner is using professional skills to effect positive change within the individual or group.

Leigh Diehl, HTM, FLA
Director of Therapeutic Horticulture, Wilmot Gardens at the University of Florida

EXHIBIT 2.6

Program Example: Meeting the Needs of People Touched by Cancer

People touched by cancer include immediate and extended family members, best friends, coworkers, and other support people for the individual diagnosed with cancer. Coping with the implications of the cancer diagnosis, and short- and long-term treatment regimens have an impact on the patient and their support network for long periods of time. Everyone involved has a variety of needs that may shift over time and that can be addressed through experiences along the people–plant continuum. Gardens are a nonjudgmental space for restoration, exploration, and self-expression, and they support multidisciplinary therapeutic activity (Sempik et al. 2003). As such, the same garden space can provide inspiration for a variety of intervention strategies, including educational classes, therapeutic horticulture experiences, and horticultural therapy utilizing wellness or therapeutic models for sessions. People–plant experiences can meet people wherever they are along the cancer journey and are each valuable toward the overall goal of creating positive strategies for coping with cancer.

While not all organizations serve the whole range of people touched by cancer, the Gathering Place, a community-based organization in Beachwood, Ohio, illustrates the range of programming possible. The Gathering Place works to meet the psychosocial, spiritual, and physical—but not medical—needs of everyone in the diagnosed person's immediate care network. Norma's Garden, designed by Virginia Burt of Virginia Burt Designs, Canada, is a thoughtfully designed garden space arranged in a series of rooms to support diverse experiences. This garden is only one-third of an acre, yet it has a large impact. It is not only a visually pleasing feature for the organization but also serves as a place for:

- Learning how to grow and harvest fresh vegetables, herbs, and flowers; how to use a plant-rich environment to incorporate stress management techniques; and how to change eating habits to include a plant-rich diet.

(Continued)

EXHIBIT 2.6 (Continued)

Program Example: Meeting the Needs of People Touched by Cancer

- Therapeutic intervention such as counseling, support groups, art therapy, therapeutic horticulture, and horticultural therapy.
- Individual respite and spiritual growth, such as journaling, meditation, walking the labyrinth, or relaxation.
- Physical recovery, including yoga, tai chi, and working in the garden.
- Self-expression, including participating in a drumming circle, gardening, flower arranging, and art therapy.
- Play, especially in the children's spaces.

The garden is a destination for connecting with friends and colleagues over lunch or a cup of tea too. It is also a welcoming location for events for donors or visiting professional groups. It truly is an extension of the building in terms of providing usable space for all types of programming.

Karen L. Kennedy, HTR
The Gathering Place, Contract Program Staff

The value the garden contributes to an organization often goes beyond its role on the therapeutic horticulture–horticultural therapy spectrum. As noted in this comment from a family member of a patient at Greensboro Hospice and Palliative Care of Greensboro, "She looked towards the rosebushes, smiled and said, 'this is my idea of peace and tranquility.' Then she started to cry and said, 'my husband is passing, and all this beauty just washes away the pain'" (Sally Cobb, email message to authors, February 20, 2018). Though beyond the scope of a therapeutic horticulture or a horticultural therapy program, the needs of family members seeking respite and healing can still be found in the garden.

Conclusion

This chapter explored the distinctions and similarities between horticultural therapy and other allied professions and green care. Similarities include the application of standard treatment processes, the quest for similar outcomes for those served, and the utilization of human-nature connections to improve well-being. Horticultural therapy combines all of these, yet it is distinct due to the active participation in nature that revolves around the care of plants, in a professionally directed, goal-driven treatment process.

References

Annerstedt, M., and P. Währborg. 2011. "Nature-Assisted Therapy: Systematic Review of Controlled and Observational Studies." *Scandinavian Journal of Public Health*, 39: 371–388.

Bragg, R., and G. Atkins. 2016. *A Review of Nature-Based Interventions for Mental Health Care*. Natural England Commissioned Reports, Number 204.

Buzzell, L., and C. Chalquist, eds. 2009. *Ecotherapy: Healing with Nature in Mind*. San Francisco, CA: Sierra Club Books.

Cimprich, B. 1993. "Development of an Intervention to Restore Attention in Cancer Patients." *Cancer Nursing*, 16: 18–32.

De Vries, S., R. A. Verheij, P. P. Groenewegen, and P. Spreeuwenberg. 2003. "Natural Environments: Healthy Environments? An Exploratory Analysis of the Relationship Between Greenspace and Health." *Environment and Planning A*, 35: 1717–1731.

Dorn, S., and D. Relf. 1995. "Horticulture: Meeting the Needs of Special Populations." *HortTechnology*, 5(2): 94–103.

Gallis, C., ed. 2013. *Green Care for Human Therapy, Social Innovation, Rural Economy, and Education*, Vol. 12. New York: Nova Science Publishers.

Hartig, T., G. Evans, L. Jamner, D. Davis, and T. Garling. 2003. "Tracking Restoration in Natural and Urban Field Settings." *Journal of Environmental. Psychology*, 23(2): 109–123.

Hartig, T., M. Mang, and G. W. Evans. 1991. "Restorative Effects of Natural Environment Experiences." *Environment and Behavior*, 23: 3–26.

Hazen, T. 2014. "Horticultural Therapy and Healthcare Garden Design." In *Therapeutic Landscapes: An Evidence-Based Approach to Designing Healing Gardens and Restorative Outdoor Spaces*, ed. C. C. Marcus, and N. A. Sachs. Hoboken, NJ: Wiley. 253.

Huxmann, N. J. 2016. "The Garden as Setting for Horticultural Therapy." *Acta Horticulturae*, 1121: 39–45.

Kaplan, R., and S. Kaplan. 1989. *The Experience of Nature: A Psychological Perspective*, New York: Cambridge University Press.

Kaplan, S. 1995. "The Restorative Benefits of Nature: Toward an Integrative Framework." *Journal of Environmental Psychology*, 15: 169–182.

Kellert, S. R., and E. O. Wilson, eds. 1993. *The Biophilia Hypothesis*. Washington, DC: Island Press.

Kuo, F. E. 2004. "Horticulture, Well-Being, and Mental Health: From Intuitions to Evidence." *Acta Horticulturae*, 639: 27–34.

Lewis, C. A. 1996. *Green Nature Human Nature: The Meaning of Plants in Our Lives*. Urbana, IL: University of Illinois Press.

Li, Q., et al. 2009. "Effect of Phytoncide from Trees on Human Natural Killer Cell Function." *International Journal Immunopathol Pharmacol*, 4: 951–959.

Li, Q., et al. 2011. "Acute Effects of Walking in Forest Environments on Cardiovascular and Metabolic Parameter." *European Journal of Applied Physiology*, 11: 2845–2843.

Louv, R. 2005. *Last Child in the Woods: Saving Our Children from Nature-deficit Disorder*. Chapel Hill, NC: Algonquin Books.

Marcus, C. C., and N. A. Sachs. 2014. *Therapeutic Landscapes: An Evidence-based Approach to Designing Healing Gardens and Restorative Outdoor Spaces*. Hoboken, NJ: John Wiley & Sons, 56–58.

Melson, G. F. 2001. *Why the Wild Things Are*. Cambridge, MA: Harvard University Press.

Miles, I., W. Sullivan, and F. Kuo. 1998. "Prairie Restoration Volunteers: The Benefits of Participation." *Urban Ecosystems*, 2: 27–41.

Nakamura, R., and E. Fujii. 1990. "Studies of the Characteristics of the Electroencephalogram When Observing Potted Plants: Pelargonium Hortorum 'Sprinter red' and Begonia Evansiana." *Technical Bulletin of Faculty of Horticulture, Chiba University*, 43: 177–183.

Nakamura, R., and E. Fujii. 1992. "A Comparative Study of the Characteristics of the Electroencephalogram When Observing a Hedge and a Concrete Block Fence." *Journal of the Japanese Institute of Landscape Architecture*, 55, 139–144.

Park, B. J., Y. Tsunetsugu, T. Kasetani, T. Kaqawa, and Y. Miyazaki. 2010. "Physiological Effects of *Shinrin-yoku* (Taking in the Atmosphere of the Forest): Evidence from 24 Forests Across Japan." *Environmental Health and Preventative Medicine*, 15(1): 18–26.

Sempik, J., J. Aldridge, and S. Becker. 2003. *Social and Therapeutic Horticulture: Evidence and Messages from Research*. Loughborough, UK: Thrive, in association with the Centre for Child and Family Research, Loughborough University.

Sempik, J., and R. Bragg. 2013. "Green Care: Origins and Activities." In *Green Care for Human Therapy, Social Innovation, Rural Economy, and Education*, ed. C. Gallis. New York: Nova Science Publishers, 20–21.

Sempik, J., R. Hine, and D. Wilcox, eds. 2010. *Green Care: A Conceptual Framework, A Report of the Working Group on the Health Benefits of Green Care, COST Action 866, Green Care in Agriculture*. Loughborough, UK: Centre for Child and Family Research, Loughborough University.

Simson, S., and M. Straus. 1998. *Horticulture as Therapy: Principles and Practice.* New York: Food Products Press.

Takano, T., K. Nakamura, and M. Watanabe. 2002. "Urban Residential Environments and Senior Citizens' Longevity in Megacity Areas: The Importance of Walkable Green Spaces." *Journal of Epidemiological Community Health,* 56: 913–918.

Ulrich, R. S. 1979. "Visual Landscapes and Psychological Well-Being". *Landscape Research,* 4(1): 7–23.

Ulrich, R. S. 1981. "Natural Versus Urban Scenes Some Psychological Effects." *Environment and Behavior,* 13: 523–556.

Waliczek, T., and J. Zajicek. 2016. *Urban Horticulture.* Boca Raton, FL: CRC Press, Taylor & Francis Group, 8–9.

Wilson, E. O. 1984. *Biophilia.* Cambridge, MA: Harvard University Press.

chapter three

The therapist–client relationship

Jay Stone Rice

Contents

This chapter considers elements of the therapeutic relationship that may assist horticultural therapists working in the diverse spectrum of horticultural therapy programs. The therapeutic process is a rich mosaic that draws from deep wells of inquiry into the nature of human well-being. Horticultural therapeutic relationships will be discussed from the perspectives of archetypal images, cultural narratives, treatment models, and treatment approaches to assist horticultural therapists in deepening their work with people and plants.

Definition of horticultural therapy

Horticultural therapy differentiates itself from other therapeutic modalities because it includes plants, plant care, and cultivation in the therapeutic process. Horticultural therapy takes many forms and may occur in individual or group settings. The goals of treatment might be to support physical rehabilitation, emotional health, wellness, and/or vocational rehabilitation. Horticultural therapy programs are further differentiated by their frequency. Clients may participate in horticultural therapy programs on a daily, weekly, or less frequent basis. For example, in a short-term hospital physical rehabilitation program, a client may be seen only once, while in vocational programs, the client and therapist typically interact for several hours on a daily basis.

What then is the definition of *therapy* that applies to all of these expressions of horticultural therapy? The American Heritage dictionary defines therapy as (1) the treatment of disability or illness, (2) a healing power or quality (1981). It is derived from the Greek: *therapeia*, "to be of service," and *therapeuein*, "to be an attendant." These definitions suggest a nuanced relationship between treating, a goal-directed active process, and attending to or serving, which emphasizes the quality of the relationship.

The current bias in health care is toward active treatment. Time is scarce and monetized to maximize productivity and cost-effectiveness. Goals are established, such as increasing mobility or reducing psychological distress, and the means to accomplish these goals are emphasized over the quality of the therapeutic relationship. In the haste to reduce suffering or increase efficiency, treatment often fails to address the client's inner struggles with the difficulties she or he faces (Sacks 1984).

Horticultural therapy is an activity-based treatment modality. Frequently plants are considered a means or a tool for accomplishing a goal in much the same way that a biofeedback device or an exercise machine may be used for treatment. However, there is an ancient root meaning of *horticultural therapy* in the Hebrew Bible. *Ezekiel 47:12* asserts "the fruit thereof shall be for food and the leaf thereof for healing" (Jewish Publication Society 1985). It would be easy to construe this as referencing herbal or plant medicine. In fact, many pharmaceuticals are created from synthetic compounds that mimic medicinal plants. The biblical Hebrew word, *Refuah*, used in this passage to connote healing or therapy, means to "stitch together." Gershon Winkler (2003), a religious scholar and translator of early Jewish texts notes that the Jews were initially a tribal people with an indigenous understanding of the interrelationship of all life. What, then, does the relationship with plants stitch together? Native Americans regard plants as ancestors to humans for they created the atmosphere necessary for human life on Earth (Buhner 2006). Breathing embodies the interdependent relationship between humans and plants, which is critical for mutual survival. Indeed, in oral earth-based cultures, it was understood that human health was a function of balance with Earth (Mehl-Madrona 2007).

Earth-based wisdom and archetypes

Horticultural therapy is often viewed as an adjunctive therapy to supplement medical, psychological, physical, and vocational therapies. This perception is based on a particular worldview that has held sway for many centuries. This outlook obscures understanding of horticultural therapy's critical contribution toward bridging the perceived separation of humans and nature. Carl Jung's conceptualization of archetypes can extend horticultural therapists' understanding and appreciation of the importance of their work.

Jung (1875–1961), was a Swiss psychiatrist and analyst who made significant contributions to the development of modern psychology. As a descendent of generations of farmers and ministers, he recognized that nature nourished the human soul. In comparing modern life with older earth-based indigenous tribes in East Africa, he observed "it is as if our consciousness had somehow slipped from its natural foundations and no longer knew how to get along on nature's timing" (Sabini 2002).

In 1925, Jung traveled to Taos, New Mexico, in the four corners region of the American Southwest to meet with Hopi elders. Jung later observed how Americans had particular psychological challenges to address that were the result of being a largely immigrant population coming from other continents. Jung believed that it was essential for Americans to connect with the wisdom of these indigenous earth-based inhabitants for guidance in living emotionally grounded and physically healthy lives (Sabini 2002).

Jung believed that the natural human mind developed over three million years in conjunction with the cycles of Earth. The inherited wisdom of this collective evolutionary journey is stored as symbolic images in the unconscious mind. Jung called these symbolic images archetypes (Jung 1933). For example, seed pods have been found in Neanderthal graves (Nadel et al. 2013). This indicates that the vital relationship between humans and plants extends back to archaic human roots. These seed pods may have been placed in the grave so that the deceased would have sustenance in the afterlife. They may also suggest these early human ancestors sought solace in facing death by recognizing their link to natural cycles represented by the flowers.

Cultivating plants stimulates connection to this deeper archetypal understanding of human nature. The older wisdom that plants convey is that human experience is best understood within a context of ever-changing cycles that lead from seeds to germination, growth, fruiting, seeding, and to dying away, regeneration and new starts. Horticultural therapists working in vocational, physical rehabilitation, and mental health group or individual programs facilitate this essential relationship between humans and plants (Figure 3.1).

Figure 3.1 Cultivating people–plant relationships. (Courtesy of Pacific Quest.)

Archetypes and therapy

For Jung, the two major archetypes that form the modern cultural image of doctors and therapists are the Hero and the Wounded Healer (Whitmont 1969). The Hero is one who vanquishes illness and disease. Jonas Salk and Marie Curie represent this archetype in their development of tools and means to understand and combat human illness and disease. Aspiring to help others is a motive present in almost all horticultural therapists. If the therapist seeks to vanquish a client's suffering or misfortune, the Hero archetype is mobilized. To a large extent, the Hero's journey necessarily leads the therapist to the limits of the archetype. Humans must inevitably accede that life and nature are seldom under their control or direction. How each person addresses the challenges of being human is the domain of the Wounded Healer archetype. From this perspective, therapeutic alliances are partially shaped by the common human experience of facing loss, wounding, and vulnerability.

Rarely is any human activity solely under the influence of only one archetype. In particular moments many different archetypes are drawn upon. Becoming aware of what is stirring within helps the therapist adapt and adjust to the needs of her or his client. For example, a horticultural therapist working in a vocational program develops comprehensive program plans for each session or series of sessions. The therapist may start the day with a coherent map of what she or he hopes to achieve. However, a particular member of the group may be struggling with self-confidence. The therapist encourages the client to slow down and guides that client through the task at hand. Nevertheless, the client cannot seem to make use of what the therapist is offering. Within the Hero archetype, the therapist may begin to feel like she or he is failing and may become anxious because she or he had not envisioned this circumstance. Horticultural therapists in this situation may find that they are getting angry or irritated with their client for making it so difficult to feel successful. In this moment, both therapist and client might be feeling like failures! If the therapist is able to access the Wounded Healer archetype, an interesting dynamic arises. She or he may begin to empathize with the client's feelings of doubt or discouragement. Respecting the mutual challenges each face may lead to a sensitive, in-the-moment adaptation that brings the experience of success to their interaction. For this shift to occur, the therapist must release her or his original plan in order to come into the present where an achievable goal appears, one that incorporates the client's current needs.

A horticultural therapist working in a wellness program has a program goal of teaching people how to relieve distress by cultivating plants. If the Hero archetype is in charge, the therapist might look for immediate confirmation that the program is helping clients feel better. Timing is a primary teacher when working within the garden. The gardener knows that it takes time for a seed to sprout, put down roots, gather in nutrients, and move toward flowering. Along the way, challenges from weather, insects, and other plants may have to be addressed. Gardeners know that they tend to their garden rather than control it. Most gardeners have been humbled by the challenges nature brings. In a horticultural therapy wellness program, the therapist cultivates the seeds of self-care that grow over time in their clients, according to each client's individual pace. This is the understanding of the Wounded Healer archetype.

Therapeutic relationship

There are two therapeutic relationships within horticultural therapy: the people–plant relationship and the therapist–client relationship. The relative emphasis or impact of each of these relationships differs between vocational, therapeutic, and wellness horticultural therapy programs. Understanding the therapist–client relationship assists the horticultural therapist in each facet of this work.

The client's treatment goals are defined by the nature of the program. In a vocational horticulture program, the treatment goals may be preparing for job readiness; in a wellness program, treatment goals may be a healthier lifestyle; in a therapeutic program focused on physical outcomes, the goal may be supporting the client's physical recovery or functioning. Often therapy is thought of as the curing or fixing of a person's illness, suffering, or life situation. The *American Heritage Dictionary* defines *cure* as "restoring health" (American Heritage Dictionary 1981). *In Care of the Soul*, Thomas Moore (1992) notes that the Latin root *Cura* contains additional meanings including "attention," "devotion," and "husbandry." Here the goal is not to fix people as much as to be with them, helping them manage their condition or circumstance.

The philosopher Martin Buber discusses qualities of relationship in his seminal book, *I and Thou* (Buber 1971). Buber wrote in German, and his preeminent translator into English was Maurice Friedman. Friedman noted that the German word that has been translated as "Thou" is better interpreted as the more intimate "You." Buber proposed that when inhabiting roles such as doctor and patient or therapist and client, the relationship form may obstruct true meeting and connection. Buber called this an "I-It" relationship. For Buber the "I-You" relationship is one where both the therapist and client encounter a unique other, and new learning occurs that is potentially meaningful to each one.

In horticultural therapy training, considerable and necessary attention is placed on the ethical boundaries required in professional therapeutic work. Buber would not dispute this. This quality of authentic meeting does not require sharing personal information. Rather, Buber suggests that if the therapist stays connected to her or his own human uncertainties and vulnerabilities when within the professional role, a deeper relationship becomes possible where the client feels less isolated, different, or disempowered by her or his challenges. Exhibit 3.1 describes an "I-You" interaction between a horticultural therapist and a young child.

EXHIBIT 3.1

Program Example: Preschool, Texas

From a therapeutic standpoint, the goal of the Bloom and Grow Wildlife Garden we installed at the back of the property might have been a lofty one. "Envision a beautiful future." I heard the phrase from Richard Louv, author of *Last Child in the Woods*, and chair emeritus of the Children and Nature Network. I wanted the preschool children of Vogel Alcove in Dallas, Texas, all of whom are affected by homelessness, to be exposed to the cycles of nature and, even if they didn't completely comprehend the symbolism, to grow up knowing that there are periods of dormancy in nature, as in life, when things look bleak, but life is resilient and spring returns to awaken the beauty.

With that goal in mind, the children enjoyed working the soil and planting annual seeds and tender perennials. That summer, zinnias enlivened the garden with bright

(Continued)

EXHIBIT 3.1 (Continued)

Program Example: Preschool, Texas

colors while Gregg's mistflower and salvia attracted monarchs, hummingbirds, and bees. The garden teemed with life in all of its stages, and the children romped and played among it.

In autumn, the garden provided the backdrop for a story-telling scarecrow that sat atop straw bales surrounded by decorative gourds and pumpkins that were brought in for the occasion. It was about this time when Jenni, then four years old, revealed to me that her father had recently been imprisoned, and she didn't know when she would see him again. Jenni's mom was working long hours and attending school to earn her belated diploma, and the aunt that collected her in the late afternoon was frequently inattentive to the soulful eyes searching for reassurance. Jenni was anxious. We rubbed our hands on the Russian sage and took a calming breath. I gave Jenni a vase and she cut some orange zinnias for her classroom. I would have loved to say, "Everything will be okay." But there was no certainty. So instead I said, "This is really hard. But you are strong."

Winter that year was harsher than usual. Dallas always exists in extremes, but that year's lows were a record-breaking 15°F. We lost a number of plants that are typically well adapted to the zone. Jenni told me her sister was fighting in school and getting in trouble. Her mom was mad. She wanted more flowers, but we searched and found none. It was too cold to plant now, but I assured her there would be more when it warmed up. Each time I came to the classroom, she asked me again, "Is it time yet?"

In early March, as temperatures began to warm, we went back to the Bloom and Grow Garden. We pulled back long, brown, dormant stems to discover some bright white pansies with yellow centers that we planted in November but thought we had lost. "Look!" cried Jenni. "They made it! They are strong!" My skin tingled, and tears welled in my eyes. "Yes, Jenni," I said. "They are resilient. Just like you."

Kathryn E. Grimes, MAT, HTR

Therapeutic narrative and focus

The horticultural therapist's task is to design activities that aid her or his clients. In order to accomplish this, therapists strive to understand their clients' needs. Therapists must remain cognizant that their thoughts, perceptions, and observations of their clients are influenced by their subjective experience or narrative. A narrative is the story or understanding that a person or culture lives within. Personal narratives may be shaped by family, culture, religion, experience, and professional training. How horticultural therapists views their professional identity, their clients, and their work becomes part of a narrative that helps organize and make meaning in their lives. Narratives may limit understanding. As Exhibit 3.2 portrays, a narrative may connect certain dots of experience and leave others unnoticed.

Lewis Mehl-Madrona is a psychiatrist, physician, and author who weaves Native American healing practices with conventional western medicine. In his book, *Narrative Medicine*, Mehl-Madrona (2007) maintains that understanding the stories people embody is

EXHIBIT 3.2

Perspective: Horticultural Therapy, an Ecological Narrative

When I approached the dean of my doctoral program in clinical psychology with my plan to do my dissertation on a horticultural therapy program, I encountered a clash of narratives immediately. She responded, What does gardening have to do with psychology? Instantly, I recognized that part of my work was to understand and respond to this query. It is interesting to note that psychology as an academic discipline is not considered part of the natural sciences. This separation of humans from the natural world is a narrative that limits human health and has grave consequences for the well-being of our biosphere. Horticultural therapy implicitly addresses this separation from our true human nature.

Jay Stone Rice, PhD LMFT

a key to healing work. He questions the modern narrative that "experts" can cure a person from the outside without sufficiently getting to know the story the client lives within. He suggests this expert narrative does not include sufficient respect for what people already know or have learned through their life experiences and challenges. If horticultural therapists listen or observe well, they can help their clients better understand and utilize what they have already learned but may not sufficiently recognize.

For example, Rosalyn struggled with self-confidence. After having a meaningful career, she switched directions and devoted herself to raising a family of four children. Rosalyn had a questioning mind and frequently identified areas of her life where she felt she could do better. She welcomed advice and ideas from others and often became confused and overwhelmed by the task of determining what was right for her. Sometimes she would begin a new direction and then become consumed with self-doubt. She was given the suggestion to consider her ideas and new directions as seeds she was planting. She paused for a moment and shared the following story. Her grandfather emigrated from southern Europe to Idaho. He loved roses and worked hard to grow them in a challenging climate. His tenacity left a deep impression on Rosalyn. In fact, all of her children have first or middle names that are connected to roses. She recalled one particular autumn when she was tormented with doubt while preparing dinner for her husband's business associates. Sitting in her garden, she looked at her roses, which included some she had brought from her grandfather's house at the time of his passing. She reflected on his patience and dedication and understood that she could apply this to her own life path.

Mehl-Madrona emphatically states that cause-and-effect narratives are too simplistic for understanding one's life. A native elder once told him "If you think you know what is going on, you're wrong. If you know what is going on, it's trivial. It is always much more complicated than you could ever understand" (Mehl-Madrona 2007). This does not imply that horticultural therapists cannot gain proficiency through education and experience. Rather, when well-trained horticultural therapists become comfortable with complexity and the inability to fully understand themselves or others, they are better able to observe, listen, and learn from each client. Exhibit 3.3 provides an example of this dynamic.

EXHIBIT 3.3

Perspective: Taking Time to Be

As good as someone is at horticultural therapy projects and as important as it is to address motor skills, cognitive abilities, sensory, and more, there is an essential element that is often missed. That element is the gift of passive yet attentive time in nature. If programs have a focus on doing, the benefits of being are often missed. While active programming is valuable and can address specific goals, it is important to create space for being in the garden and the benefits derived from such planning. My horticultural therapy sessions for many years were very active either physically and/or cognitively. What was offered in the program expanded after the following experience:

Harold (not his actual name) who had been diagnosed with schizophrenia, was marginally involved in the horticultural therapy group at the adult day center he attended. He would typically sit at a table away from the main group and would often be speaking stream of consciousness thoughts, seemingly unable to respond in context to questions. Only occasionally would Harold carry out a project when given supplies. One day in the garden, toward the end of a session in which the group had been tending the plants, we took time to sit, relax, and have an experience of just being in the space. After a period of quiet reflection, participants were invited to express what thoughts came to mind. All but Harold took time to speak. Most commented on the beauty of the surroundings, the colors of the flowers, or the blue sky. When asked if he had something to share, Harold slowly took a breath, looked at the group and the garden and then said, "We are one with all."

That experience of "oneness" Harold spoke of is something we all strive for. As horticultural therapists, we can facilitate opportunities that support a greater sense of connection to the outer and inner worlds. As a result, the participants and leaders can potentially experience greater well-being, not just during a session but throughout day-to-day life.

Pamela A. Catlin, HTR
Horticultural Therapy Consultant, Faculty, Horticultural Therapy Institute

Models of treatment

The attempt to understand human experience and to provide therapeutic help to others has contributed to the development of many models of treatment. While each model arises to advance the ongoing narrative of how best to understand human experience, therapists can become confused if they seek to determine which model is the right one. Barry Lopez (1986) suggests that each human being is on a journey to understand the complexity of their lives and their world. It is less important that people come to a shared unified understanding than to respectfully share what each comes to understand. The horticultural therapist may find more resonance with one or another of the models presented here. The therapist may find that different models are more helpful in a particular instance with a client or group. Fluency across models facilitates the ability to customize the therapeutic approach and thus meet the individual client's goals within each program setting.

Biopsychosocial model

Modern medicine has developed significantly through a process of specialization both in research and treatment. While specialization has stimulated great advances, it has created an imbalance that can hinder the development of integrative treatment approaches. These approaches seek to include larger contextual issues that also influence the course of wellness or disease. The biopsychosocial model developed as clinical thinking grew beyond simple cause-and-effect orientations (Engel 1977). In this view, human conditions are frequently shaped by multiple interactions between biological, psychological, and social factors. These contribute to a broad range of variation within a particular group. For example, a horticultural therapist working with a group of veterans experiencing post-traumatic stress disorder (PTSD) will often observe that her or his clients' differences shape how this disorder affects each of them.

Social-ecological model

The social-ecological model was developed by Urie Bronfenbrenner (1979), a research psychologist who studied the intersecting social systems that shape human development. He found that individual experience occurs within a social context that includes the family, neighborhood, community institutions, social policy, cultural beliefs, and physical environment. The social-ecological context influences how successful a treatment intervention will be with a given population.

In developing the Garden Project for the San Francisco Sheriff's Department, Catherine Sneed recognized that her students in the jail faced many issues that were not of their making. Her program emphasized individual responsibility for life choices as well as the provision of social supports that were frequently absent in the inner-city communities from which the inmates came. Understanding the social-ecological context of the population the horticultural therapist serves may deepen empathy and enhance the design of horticultural therapy sessions.

Disease model

It is natural to focus on the issue that brings clients to a horticultural therapy program. The disease model helps the therapist understand the challenges faced by his or her clients. The disease model helps clarify and focus treatment goals. It also may reduce a client's anxiety by providing an explanation for his or her difficulties. However, this model may have unintended adverse effects. Often people who have a disease or disorder can feel as if they are defined by deficit. They may struggle with a loss of identity as people begin to relate to them solely as their condition.

Strength-based model

The strength-based model focuses on the whole person rather than their particular reason for seeking a therapy program. Almost everyone experiences challenges at some point in life. A strength-based model looks for inherent resiliency that may support adapting and growing toward wholeness and wellness. Oliver Sacks, the clinical neurologist and author, suggests that the therapeutic clients he served broadened his understanding of human experience. Recognizing their adaptive and creative physiological, neurological, and psychological responses to disease can enhance appreciation for human complexity

(Sacks 1995). Horticultural therapy focuses on supporting the strengths within each client. The therapist remains interested and curious about how each client meets the challenges life has presented rather than solely focusing on somehow fixing them. By doing this, the therapist helps the client discover what grows through loss and what remains in the midst of life's inevitable changes.

A principle of traditional Chinese medicine notes that when a person who is ill cares for another, the healing Qi—life energy—will flow through the person who is ill to the person she or he seeks to help, providing benefits to both (Reichstein 1998). The horticultural therapist helps clients grow stronger through the care of plants. Preliminary studies of horticultural therapy programs for older populations have demonstrated pain and stress reduction, improved attention, decreased falling, and reduction in medication use (Detweiler et al. 2012).

Positive psychology model

The positive psychological approach to treatment was developed by Martin Seligman and others at Stanford University (Seligman 2002). These researchers noted how psychology frequently presented life as a problem-focused narrative. They recognized that treatment seemed to naturally focus on what was wrong in their patients or their lives. Seligman began exploring what qualities enhanced the experience of a good or positive life. He noted how a good life was not defined by the absence of illness, aging, or unexpected change. Rather he observed how character values, interests, relationships, and a person's authenticity seemed to determine wellness more than circumstance.

Mihaly Csikszentmihalyi (1990) observed how the concept of *flow* influences the positive quality of life experiences. *Flow* is defined as "complete absorption in what one is doing." Flow is differentiated from control. Control represents judging an experience by whether it goes according to one's plan, idea, or expectation. Flow represents total engagement with the present moment. If change arises in the moment, then the focus shifts to how to move with this change. Gardeners often experience flow when gardening if they are able to resist the desire to get everything done as quickly as possible. A horticultural therapist may also observe flow in group gardening activities. At the Horticultural Therapy Institute, students are given a group assignment to silently clear an area in preparation for spring. As their work unfolds they wordlessly become absorbed in their tasks and locate the flow that harmonizes their individual and collective efforts.

Forgiveness is another determinant of positive psychology. During a recent interview, Marjorie Walters, a family therapist and researcher focusing on how high-conflict divorce affects children, noted that imperfection is an element of all human relationships. Having the capacity to forgive is an essential characteristic for children to thrive after a divorce (Burr 2013). Forgiveness spans emotions and behaviors (Exline and Baumeister 2000). Forgiveness is generally viewed as benefitting the person who is forgiven. However, it also has significant consequences for the health of the person who forgives. Witvliet et al. (2001) observed in their study of forgiveness at Hope College in New York that subjects who practiced forgiveness had lower stress levels, fewer negative emotions, fewer cardiovascular problems, and a stronger immune system.

Gratitude is another essential element in positive psychology. When facing a substantial life challenge, it is easy to sink into despair and lose a sense of any goodness in life. Exhibit 3.4 depicts how recognizing the gifts that life continues to bring can provide some comfort at these times.

EXHIBIT 3.4

Perspective: Gratitude

My father, Charles Rice, of blessed memory, conveyed awareness of gratitude in a phone conversation I had with him one night during the last years of his life. He had lost his vision due to macular degeneration. He shared with me how he would go to get his mail late at night so that people in his apartment building would not see him fumbling to open his mailbox. I remarked how hard it was to lose something that had been so integral to functioning in his life. He responded that "it could be worse." I asked what was he thinking of? "We are talking on the phone," he said, "and I know who you are, and I know who I am. I am thankful for this."

Jay Stone Rice, PhD

Positive psychology suggests that a person's perspective on life shapes her or his narrative. This narrative has an impact on that person's emotional and physical functioning. The horticultural therapist does not directly teach someone how to be positive about life. Rather the horticultural therapist can utilize the garden and people–plant interactions to convey these principles. Forgiveness can be practiced when a client mistakenly steps on a plant while weeding the garden. Flow can be identified when a client remarks that the time went quickly because she or he was completely absorbed in the task. Gratitude for the beauty and sustenance that plants provide can be expressed by the horticultural therapist as an integral part of the group activities.

Therapeutic approaches

The history of modern therapy has seen the development of many therapeutic techniques and approaches toward treatment. These interventions have been designed for individual and groups. The goals of treatment may be improved school functioning, vocational development, and emotional well-being. Therapies such as horticultural, equine, and animal therapy utilize nonhuman relationships to support improved functioning. Psychology has contributed an understanding of how the relationship between therapist and client can support growth and understanding. These insights have relevance across treatment modalities.

Psychodynamic

The psychodynamic approach was initially developed by Sigmund Freud (1999). It suggests that clients are always communicating information about themselves through their words, actions, and interactions with others. For example, a horticultural therapist may ask a group to clear dead plants from the garden. An anxious participant may inadvertently remove a living plant. Another might continuously ask for guidance on what should be removed because the client fears making a mistake. Remembering that their behavior is communicating something important about them will help the horticultural therapist intervene in an empathic way. The observant therapist will notice how much each participant expresses her or his personality or emotional state in the way she or he works. The horticultural therapist who is working with people over a period of time will naturally learn more about their clients than a horticultural therapist who sees their clients infrequently. However, all horticultural therapists know that each type of plant needs some individualized attention. Even the same type of plant may have different needs depending on its location in the garden. This type of knowledge also applies to working with people: Therapists bring their experience and understanding to each new therapeutic relationship.

The psychodynamic model underscores the importance of the horticultural therapist's awareness of self. The therapist's choice of words, tone of voice, and body posture all communicate information to the client as well. Often, how a client responds to the therapist can be understood more clearly when these factors are considered. Horticultural therapists working with groups must be alert to the feelings they may have about different group members. In particular it is important to pay attention to both people they favor working with and people they wish to avoid. The psychodynamic model recommends that therapists seek support and guidance as needed to further their work with clients.

Client centered

This approach was developed by Carl Rogers (1951), an early pioneer in humanistic psychology. Its primary orientation is positive regard for each individual and her or his inherent capacity to learn through experience. The horticultural therapist acknowledges each client's strengths and capacity to determine goals for treatment. The therapist develops individualized treatment for their clients based on their expressed goals and interests. Therapeutic interventions assist clients in accomplishing their goals. This model is essential when working with clients on a short-term basis because the therapist necessarily draws more from the client's self-knowledge. Exhibit 3.5 describes a client-centered approach to working with youth who have an autism spectrum disorder (Figure 3.2).

EXHIBIT 3.5

Program Example: Therapist–Client Relationship

Designed to teach viable, marketable job skills that help people secure employment in the garden, farm, or nursery business, a community-supported agriculture (CSA) garden and greenhouse program for young people diagnosed with autism spectrum disorder (ASD) can include many activities. Pruning trees, bushes, or vines; mulching; composting; planting; watering and weeding; harvesting; packaging and delivering vegetables and produce; as well as indoor greenhouse work all involve skills that can be demonstrated, taught, monitored, and measured for goal-oriented achievement. In order to provide training and support, verbal, visual, physical, and hand-over-hand assistance may be required of the therapist to ensure that the client can complete the task safely and effectively. Data collected regarding the specific number of prompts or types of assistance required by the participant, and the progress made can be examined for modifications of goals and to notice changes in the client's functioning.

(Continued)

EXHIBIT 3.5 (Continued)

Program Example: Therapist–Client Relationship

In order to provide assistance and training supports effectively, the relationship between the client and the therapist is critical. That relationship requires contemplation and thought to be effective and efficient for the participant.

Clinicians supporting individuals with developmental differences are given many opportunities to work with and assist people; to share information with them; and to challenge them to use what they learn to help improve confidence, self-esteem, and quality of life. In the development of programming for young people diagnosed with ASD, it is important to focus on several key components of not only the program-related activities but also the therapeutic working relationship with the participant.

Therapists must realize that each person's cognitive level of function is as unique and complex as she or he is. In order to teach, a supporting staff member must first be able to reach their participants. Having an in-depth knowledge and understanding of who each person is and how each person learns, works, acts, and reacts helps foster a relationship that encourages trust and respect. Getting to know the client as a person is the first and most important step in this process and is at the very core of this connection. Effective ways to initiate this process include reading historical documentation about the person; researching that person's likes and dislikes; talking with other people who know and have worked with the individual; and, most important, *spending time with the person*.

In a therapeutic/vocational program, it is most effective and appropriate to take a person-centered approach to all aspects of the overall planning. Helping participants understand and identify what they want to learn or do is vital to ensuring that they will be successful in their endeavors. Concentrating on what is important to and for a participant will help determine the what and how of their overall therapeutic experience.

True therapeutic intervention should also incorporate empowerment of the participant. By empowering, a facilitator motivates participants to learn how to learn—not just to satisfy the requirements of a chore or task at hand but also to encourage their thoughts; work ethic; and desire to obtain, contribute to, and/or overcome something they have been challenged to accomplish. Giving participants a say, choice, and voice in their program process empowers them to want to learn and to seek that greater knowledge and understanding.

John Fields
Certificate in Horticultural Therapy,
Director of Operations GHA Autism Supports

Figure 3.2 Using a client-centered approach. (Courtesy of John Murphy.)

Relationship and attachment

Attachment therapy is based on the work of John Bowlby (1988), who studied the attachment relationship between parents and young children. He noted how infants have a range of inherent purposeful behaviors to keep their primary caregivers engaged in caring for them. Evolution favors successful attachment because human babies require significant care after birth in order to survive. Bowlby, along with researcher Mary Ainsworth, observed three different attachment relationships that developed based on the quality of early relationships with caregivers: secure, anxious, and avoidant. Researcher Mary Main later added disorganized attachment (Karen 1998). The quality of early attachment relationships may significantly shape expectations regarding relationships throughout one's life.

The relevance of attachment for horticultural therapy is twofold. While the quality of the relationship that the client forms with the therapist will reflect their attachment assumptions, a calm, nonjudgmental, supportive therapeutic approach will benefit them. This is equally true for therapists working with clients on a short-term basis or over time. Attachment also has relevance for therapeutic horticulture and vocational horticultural programs. In "A Domain of Sorts," Robert Coles (1978) observes the attachment of children to their surroundings in a hollow in eastern Kentucky. He eloquently observes the unfolding of the relationship with the land, plants, water, and animals, which convey secure attachment. Many of the difficulties that humans face are caused by other humans. Animal therapy, as well as horticultural therapy, highlights nonhuman interventions that are also inherently integral to our experience of well-being and connection with life. Plants do not judge humans and are therefore usually experienced as emotionally safe (Figure 3.3).

Self-psychology

Self-psychology, developed by Heinz Kohut (1984), proposes that people generally think of themselves as a self. The quality of self-experience ranges from cohesiveness to fragmentation. Kohut hypothesized that, as social beings, humans acquire a sense

Figure 3.3 Use of plants to form therapist–client relationship at Anchor Center for Blind Children. (Courtesy of Janet Schoniger.)

of self through significant relationships. Kohut suggested when a person is accurately seen and understood, they feel whole and cohesive. While Kohut focused primarily on human relationships, his model illuminated the growth that chemically dependent county jail inmates experienced through their participation in the San Francisco Sheriff's Department's Garden Project. Many of the Garden Project's students came from fragmented families and communities. They had endured multiple traumas and family disruptions and were provided with little support or understanding for how these experiences had an impact on their sense of self. Kohut noted that drugs are often used to numb pain and provide a temporary experience of cohesion (Kohut and Wolf 1978). Working in the garden provided these students with a new coherent identity as part of nature. Their capacity to co-create life emerged as the seeds they planted grew into food and flowers, which were given to those in need within their communities. Students who participated in the Garden Project exhibited increased hope and the desire to change: key factors in determining the readiness for addiction treatment (Rice and Remy 1998).

Techniques for cultivating a therapeutic relationship

Horticultural therapy students aspire to become efficacious horticultural therapists. Developing a therapeutic style of relationship is akin to learning a craft (Needleman 1993). There are specific techniques to this endeavor, yet each therapist expresses these differently depending on her or his temperament, personality, and particular focus within horticultural therapy. The craft of therapy is best learned through coursework, personal development, and work with a mentor/supervisor. Developing craft takes time and becomes more individualized and natural with experience. Grounding, intention, mindfulness, and empathy are essential therapeutic techniques.

Grounding

Just as a garden begins with the quality of soil, a horticultural therapist's work with others is based on their own connection to the ground. Unlike plants, humans are constantly in motion and can easily be thrown out of balance. Before working with another, it is helpful to have a ritual that connects the therapist with the earth. This can be as simple as spending a few moments in the garden feeling the energetic contact with the earth. The therapist may take a few deep breaths and observe how the breath is a vehicle for joining all of life.

Intention

Horticultural therapists often come to their work from other activities, commitments, and challenges that each life entails. It is helpful to take a moment, after connecting with an awareness of ground, to recall why they aspired to be a horticultural therapist and then set an intention for the work to be done that day. This intention is not the same as an activity or session plan. Rather, it represents a value such as the commitment to supporting others with care.

Mindfulness

Mindfulness is a Buddhist meditative awareness practice that focuses attention on the quality of one's thoughts and experiences. In his books, *The Mindful Brain* and *The Mindful Therapist*, Daniel Siegel (2007, 2010) integrates current research in neurophysiology with this 2,500-year-old practice. He describes mindfulness as a perspective on how the brain reacts with experience. Siegel suggests that a mindful therapist integrates attention to her or his own experience with attention to the client. Cheryl Wilfong (2010) weaves mindfulness and gardening in her book *The Meditative Gardener* (2010). Exhibit 3.6 describes a horticultural therapy program that incorporates mindfulness. Mindfulness requires training and practice. Horticultural therapists who are interested in this approach can find classes and retreats offered in educational, medical, and meditation settings (Figure 3.4).

EXHIBIT 3.6

Program Example: Horticultural Therapy and Mindfulness, Canada

A program combining horticultural therapy and mindfulness practice called Green Mindfulness was developed and implemented at Memorial University as a wellness-related activity for students (Quinlan et al. 2016). Horticultural therapy is a treatment modality focused on the use of natural work, more specifically horticulture, to meet goals based on a formally diagnosed issue or problem (Haller and Kramer 2006; Simpson and Straus 1998; Ulrich and Parsons 1992). Mindfulness is a practice that involves being fully present in the here and now and being aware of whatever is happening in the moment, free from the lens of judgment; it involves developing an awareness of mind and body (Birnie et al. 2010; Bishop et al. 2004). The success of the Green Mindfulness student program led developers to consider the application of the program to other populations, namely, prison populations. The team was inspired by work on Riker's Island, a jail in the East River in New York City, which offers a

(Continued)

EXHIBIT 3.6 (Continued)

Program Example: Horticultural Therapy and Mindfulness, Canada

horticultural vocational program for inmates (Jiler 2006). Her Majesty's Penitentiary (HMP) in St. John's, Newfoundland, the primary facility for housing medium- and maximum-security male prisoners, where the general term of incarceration is two years less a day, was identified as a target population. The proposal for programming was welcomed by the manager and social worker for institutional programming because it was something never offered before. A five-week program based on the Green Mindfulness experience of students was developed.

Program participants were identified by the social worker and included men living with addiction and/or other mental health issues, and ranging in age from youth (19) to senior (65+). The program had several goals. First, the program was to introduce the inmates to mindfulness, with a focus on mindful practice in a noisy environment, with the hope of aiding in their recovery and rehabilitation. Other program goals included using horticultural therapy activities to ground the practice, develop a new vocabulary associated with each activity, and create positive interpersonal interactions and learning. Participants were housed in different areas of the prison, meaning that they may have met for the first time in the program or were meeting again after not being in contact for a long time. Beginning each session with a mindfulness check-in and mindful breath practice helped establish a positive group dynamic at the start of the session. This was followed weekly by mindful practice through the plant-based activity and guided meditation using Tibetan bells (*tingsha*) and/or a singing bowl.

Five unique sessions were developed. In the first session, the men introduced to mindful practice and how mindfulness would be used during plant-based activities each week. The first activity involved using various growing media and learning plant propagation techniques. The mindfulness component of this session encouraged the men to focus on using their senses (e.g., focusing on tactile sensations) to ground them in the activity. The olfactory sense of the peat evoked strong memories for some of the men; discussion of memories and reminiscence was encouraged, creating a positive, nostalgic atmosphere. The second session involved the propagation, division, and rooted cuttings of mint and lemon balm. The men used leaves to make tea and participated in a mindful practice focused on touch, olfaction, and taste.

The third session focused on propagating with seed and culinary herbs and on participating in a mindful practice focused on touch (seed and potting media) and olfaction (herb plants). The fourth session was a creative printmaking activity, with directed plant samples as the inspiration for sketching the image to be created. Mindful practice was used to focus the men's attention on the activity through observing textures, pressure, colors, etc. The activity for the final session, chosen by the men, involved potting herbs for the facility kitchen and creating air plants on fiber pots.

Feedback from the men highlighted their engagement in the meditative practice and application of learned mindfulness practice throughout their week. The men expressed concern about bringing their plants back to their cells due to frequent cell tosses. It was

(Continued)

EXHIBIT 3.6 (Continued)

Program Example: Horticultural Therapy and Mindfulness, Canada

important to the men and facilitators that the men have their plants in their cells because the plants evoked strong memories, and caring for them and watching them grow gave the men a sense of hope. Later permission was granted for the men to have their plants, which they propagated and shared with other men in their area of the prison. In a place that is stark and barren, the presence of a small plant had such a powerful, positive effect.

Heather Quinlan, PsyD
Grace Centre, Eastern Health

Norman Goodyear, PhD, PAg HTR Retired

Figure 3.4 Mindfulness practice in the garden at Skyland Trail, Atlanta. (Courtesy of Libba Shortridge.)

Siegel utilizes the acronym PART to describe the elements that comprise the mindful therapy process: presence, attunement, resonance, and trust (Siegel 2010). *Presence* is shaped by receptivity to one's own experience as well as the experience of others. Presence conveys openness and fosters the neurological integration necessary for new learning. *Attunement* is the process of whole body listening to oneself and another. The therapist listens to the content of the words the client is speaking as well as to the emotional tone. The therapist may determine a client seems tense, agitated, or fearful by observing how she or he moves, breathes, or appears physically. Active listening, described in Exhibit 3.7, is a therapeutic technique that details the attunement process. Attunement draws on the neurological

EXHIBIT 3.7

Technique: Active Listening

The act of listening takes into account a combination of verbal and nonverbal cues on both the part of the horticultural therapist and the client. Nonverbal cues include facial expressions, body posture and positioning, gestures, eye contact, head nodding, dress and appearance, and more, while verbal cues involve spoken words and syntax, audible sounds, volume and tone, silence, and interruptions, among others (Stumbo and Navar 2011). In paying attention to these cues, the horticultural therapist can adjust appropriately. Active listening techniques are enhanced when the horticultural therapist focuses on delivering culturally sensitive, client-centered services. Here are some considerations for actionable techniques for active listening.

Guidelines for Active Listening

- *Provide undivided attention;* set aside any distractions and face the person directly. Display positive body language by making eye contact and keeping an appropriate, comfortable distance from the other person.
- *Offer encouraging verbal and nonverbal prompts,* such as "I see," or nodding, to indicate to the person that you are listening. Validate what the person is saying by restating the person's words, either by repeating or summarizing what you just heard. Acknowledge the person's stated feelings.
- *Ask open-ended questions* rather than yes or no questions to encourage more elaborate, detailed responses and to encourage dialogue.
- *Nonverbal behaviors* are important means of communication for all participants, particularly for individuals who are nonverbal or have speech that is difficult to understand. Gestures and other physical movements, posture, and level of engagement in the activity may provide information to the horticultural therapist about an individual's confidence level, interest in the activity, or present emotional and physical state. Additionally, the horticultural therapist should be mindful of his or her own nonverbal cues and how they may affect individuals' participation in activities.
- *Use pauses and silence effectively,* allowing for time between words for the person to gather his or her thoughts or to reflect. Too much chatter may discourage communication by the participant and be a detriment to concentration.
- *Reflect and provide feedback* on the statements and behaviors of the other person. Does it appear that she or he wants advice or feedback on how to handle a certain situation? Or might she or he be seeking assistance with a certain task? Do the person want to share her or his experience or simply vent?
- For a client with perseverating behaviors or ruminating thoughts, *redirect* to the task at hand.
- *Identify follow-up action steps* for participants and check back with them at the next session or as needed (Grohol 2017; Hoppe 2006).

Susan Conlon Morgan, MS
Certificate in Horticultural Therapy
The Horticultural Link

capacity to feel another's experience. *Resonance* is a neurological process whereby the nervous system of one person forms a relationship with the nervous system of another. In music, two notes come into resonance through sympathetic vibrations. This is the biological basis of empathy. *Trust* emerges when presence, attunement, and resonance are felt by clients. The ability to let another person help or guide them is predicated on the experience of trust. Often trustworthiness is considered a static quality or attribute. Within a therapeutic interaction, trust arises in the present moment as a possibility for opening and growth.

Empathy and empathic failures

Empathy is often characterized as the ability to enter into the experience of another. Empathy is perhaps better described as the ability to feel within oneself another person's experience. The capacity for empathy, as noted above, draws on an inherent aspect of the nervous system. Awareness of one's own experience while being aware of another person's experience is sometimes challenging. Therapists may think they understand another when in fact they might be confusing their own experience with the client's experience. Kohut (1984) described this as a failure of empathy. Therapists can often catch if an empathic failure has occurred if they pay attention to the client's reactions or responses to her or his words, actions, tone of voice, or body language.

It is best to anticipate that failures will occur. Fortunately, clients do not need the therapist to be perfect. In fact, Kohut believed empathic failures may serve to deepen the therapeutic relationship. For example, a horticultural therapist may suggest to a client in a senior care facility that she or he can use a particular adaptive tool. When the client seemingly resists the therapist's suggestion, the therapist may mistakenly believe the client lacks confidence or sufficient information. The therapist might then provide more detailed instructions. At some point the therapist may notice a distressed look on the client's face. If the therapist responds by saying, "I seemed to have misunderstood what you were trying to tell me. Would you mind explaining to me why this tool does not work for you?" she or he will deepen the connection to the client. Kohut observed that people generally do not require perfection; they require the therapist's genuine interest in getting to know and understand them.

Therapist's personal history

Horticultural therapy integrates knowledge of plants with the needs of specific human populations. To be effective, horticultural therapists also must understand how their own life experiences shape how they view the people they serve.

If the therapist's parent or partner suffered a traumatic brain injury in the military, this may influence her or his work with veterans. If the therapist grew up on a farm, she or he will have had experiences that would be foreign to someone growing up in a metropolitan area. If the therapist is a child of immigrant parents, this may influence her or his work with a refugee group.

The horticultural therapist's philosophy of treatment may be predicated on a particular worldview or cultural belief. For example, if self-reliance was emphasized in the therapist's family or community, she or he may sometimes judge others for showing vulnerability or dependency. If the therapist's culture of origin views people who have professional careers as being more valuable than people who work directly with the earth, she or he may inadvertently minimize the value that horticultural therapy brings to clients. If the therapist comes from a family or community that did not welcome diversity of human experience, she or her may find moments where working as a therapist will necessitate expanding her or his worldview.

Horticultural therapists' experiences of gardening and nature profoundly influences their work. If therapists fight with their garden to get the result they envision, they may do the same with their clients. However, if therapists relish the ongoing interaction with the living garden, this will likely influence their horticultural therapy work. Horticultural therapists can foster their own growth through gardening, personal therapy, supervision, and consultation with respected peers. There may be difficult patches in the work of horticultural therapists, where the soil of their compacted past experiences will need aeration, moisture, compost, and added nutrients in order to grow. It is important to remember that the plants and people therapists work with are the ecosystem that will support their own seed's unfolding.

This chapter discussed how archetypes, narratives, and subjective experience influence the development of a therapeutic relationship. Therapeutic models and approaches were examined for their applicability to the broad range of horticultural therapy programs. While much of the focus of this chapter has been on the therapist–client relationship, it is important to remember that the use of plants is horticultural therapy's unique contribution to the care of others. Jack Kornfield, author and teacher, has been instrumental in bringing mindfulness practices to the cultivation of well-being. He has noted that he keeps pictures of revered teachers and saints, such as the Buddha, the Dalai Lama, and Mother Teresa, in his office as reminders that "I am only middle management in the firm" (Talk given at Spirit Rock Meditation Center). Considering the intelligence and wisdom that plants embody can similarly provide important perspective to the horticultural therapist.

Key terms

Archetypes According to Carl Jung, primordial images reflecting universal themes or basic patterns of human experience residing in the unconscious mind.

Adjunctive therapy An additional treatment that assists the primary therapy.

Flow according to Mihaly Csikszentmihalyi, compete absorption in an activity where energy, focus, and pleasure are intertwined.

Neurophysiology Study of the nervous system.

Psychodynamic A perspective that suggests human emotions, behaviors, and feelings are shaped by psychological dynamics that underlie our conscious awareness.

References

The American Heritage Desk Dictionary. 1981. Boston: Houghton Mifflin.

Birnie, K., M. Speca, and L. E. Carlson. 2010. "Exploring Self-Compassion and Empathy in the Context of Mindfulness-Based Stress Reduction (MBSR)." *Stress and Health*, 26(5): 359–371. doi:10.1002/smi.1305.

Bishop, S. R., et al. 2006. "Mindfulness: A Proposed Operational Definition." *Clinical Psychology: Science and Practice*, 11(3): 230–241. doi:10.1093/clipsy/bph077.

Bowlby, J. 1988. *A Secure Base: Parent-Child Attachment and Healthy Human Development*. New York: Basic Books.

Bronfenbrenner, U. 1979. *The Ecology of Human Development: Experiments by Nature and Design*. Cambridge, MA: Harvard University Press.

Buber, M. 1971. *I and Thou*. Translated by Walter Kaufmann. New York: Touchstone Press.

Buhner, S. H. 2006. *Sacred Plant Medicine: The Wisdom in Native American Herbalism.* Rochester, VT: Bear & Company.

Burr, G. 2013, September 23. "Impact of Divorce on Children and Their Relationships." *Divorce Issues Today.*http://divorcetalkradiolive.com/category/divorce-issues-today/page/102/.

Coles, R. 1978. "A Domain of Sorts." In *Humanscapes: Environments for People,* ed. S. Kaplan, and R. Kaplan. North Scituate, MA: Duxbury Press, 91–93. (Reprinted from *Harper's,* November 1971, 116–117).

Csikszentmihalyi, M. 1990. *Flow: The Psychology of Optimal Experience.* New York: Harper & Row.

Detweiler, M. B., et al. 2012. "What Is the Evidence to Support the Use of Therapeutic Gardens for the Elderly?" *Psychiatry Investigation,* 9(2): 100–110.

Engel, G. L. 1977. "The Need for a New Medical Model: A Challenge for Biomedicine." *Science,* 196: 129–136.

Exline, J. J., and R. Baumeister. 2000. "Expressing Forgiveness and Repentance: Benefits and Barriers." In *Forgiveness: Theory, Research, and Practice,* ed. M. E. McCullough, K. I. Pargament, and C. E. Thoresen. New York: Guilford Press, pp. 133–155.

Freud, S. 1999. *The Interpretation of Dreams* (Oxford World Classics). Oxford, UK: Oxford University Press.

Grohol, J. 2017. "Become a Better Listener: Active Listening." https://psychcentral.com/lib/become-a-better-listener-active-listening/ (accessed August 9, 2017).

Haller, R. L., and C. L. Kramer, eds. 2006. *Horticultural Therapy Methods: Making Connections in Health Care, Human Service, and Community Programs.* New York: Haworth Press.

Hoppe, M. H. 2006. *Active Listening: Improve Your Ability to Listen and Lead.* Greensboro, NC: Center for Creative Leadership.

Jewish Publication Society, ed. 1985. *The Jewish Bible: Tanach: The Holy Scriptures.* Philadelphia, PA: Jewish Publication Society.

Jiler, J. 2006. *Doing Time in the Garden: Life Lessons Through Prison Horticulture.* Oakland, CA: New Village Press.

Jung, C. G. 1933. *Modern Man in Search of a Soul.* New York: Houghton Mifflin Harcourt.

Karen, R. 1998. *Becoming Attached: First Relationships and How They Shape Our Capacity to Love.* Oxford, UK: Oxford University Press.

Kohut, H. 1984. *How Does Analysis Cure?* Chicago, IL: University of Chicago Press.

Kohut, H., and E. S. Wolf. 1978. "The Disorders of the Self and Their Treatment: An Outline." *International Journal of Psycho-Analysis,* 59: 413–425.

Lopez, B. H. 1986. *Arctic Dreams.* New York: Charles Scribner's Sons.

Mehl-Madrona, L. 2007. *Narrative Medicine: The Use of History and Story in the Healing Process.* Rochester, VT: Bear & Company.

Moore, T. 1992. *Care of the Soul: A Guide for Cultivating Depth and Sacredness in Everyday Life.* New York: Harper Collins.

Nadel, D., et al. 2013. "Earliest Floral Grave Lining from 13,700-11,700-Year-Old Natufian Burials at Raqefet Cave, Mt. Carmel, Israel." *Proceedings of the National Academy of Sciences of the United States of America,* 110(29): 1174–1178.

Needleman, C. 1993. *The Work of Craft: An Inquiry into the Nature of Craft and Craftsmanship.* New York: Kodansha USA.

Quinlan, H., P. Cornish, N. Jenkins, and S. N. Goodyear. 2016. "Green Mindfulness: A Novel Approach to Student Wellness." Paper presented at the *XIII International People Plant Symposium: Plants, Cultures and Healthy Communities/Plantas, Culturas Healthy Comunidades Saludables,* November 10–12, Montevideo, Uruguay.

Reichstein, G. 1998. *Wood Becomes Water: Chinese Medicine in Everyday Life.* New York: Kodansha USA.

Rice, J. S., and L. L. Remy. 1998. "Impact of Horticultural Therapy on Psychosocial Functioning Among Urban Jail Inmates." *Journal of Offender Rehabilitation,* 26(3/4): 169–191.

Rogers, C. R. 1951. *Client-Centered Therapy: Its Current Practice, Implications, and Theory.* New York: Houghton, Mifflin Company.

Sabini, M., ed. 2002. *The Earth Has a Soul: The Nature Writings of C.G. Jung.* Berkeley, CA: North Atlantic Books.

Sacks, O. 1984. *A Leg to Stand On*. New York: Touchstone Books.

Sacks, O. 1995. *An Anthropologist on Mars*. New York: Vintage Books.

Seligman, M. 2002. *Authentic Happiness: Using the New Positive Psychology to Realize Your Potential for Lasting Fulfillment*. New York: The Free Press.

Siegel, D. J. 2007. *The Mindful Brain: Reflection and Attunement in the Cultivation of Well-Being*. New York: Norton.

Siegel, D. J. 2010. *The Mindful Therapist: A Clinician's Guide to Mindsight and Neural Integration*. New York: Norton.

Simson, S., and M. Straus, eds. 1998. *Horticulture as Therapy: Principles and Practice*. Binghamton, NY: Haworth Press.

Stumbo, N. J., and N. H. Navar. 2011. Communication techniques. In *Facilitation of Therapeutic Recreation Services: An Evidence-based and Best Practice Approach to Techniques and Processes*, ed. N. J. Stumbo, and B. Wardlaw. State College, PA: Venture Publishing, pp. 67–86.

Ulrich, R. S., and R. Parsons. 1992. "Influences of Passive Experiences with Plants on Individual Well-Being and Health." *The Role of Horticulture in Human Well-Being and Social Development*, ed. D. Relf. Portland, OR: Timber Press.

Whitmont, E. C. 1969. *The Symbolic Quest: Basic Concepts of Analytical Psychology*. Princeton, NJ: Princeton University Press.

Wilfong, C. 2010. *The Meditative Gardener: Cultivating Mindfulness of Body, Feelings, and Mind*. Putney, VT: Heart Path Press.

Winkler, G. 2003. *Magic of the Ordinary: Recovering the Shamanic in Judaism*. Berkeley, CA: North Atlantic Books.

Witvliet, C. V., T. E. Ludwig, and K. L. Vander Laan. 2001. "Granting Forgiveness or Harboring Grudges: Implications for Emotion, Physiology, and Health." *Psychological Science*, 12: 117–123.

chapter four

Development of the profession
Assets and issues

Rebecca L. Haller and K. René Malone

Contents

The term *horticultural therapy* was first used in the United States in the late 1940s, following World War II (Olszowy 1978), when the practice began to emerge as a distinct profession. The use of horticultural therapy has developed internationally since that time, and it is useful for students, practitioners, educators, and administrators to understand the assets and issues of this interdisciplinary field. Horticultural therapy offers unique benefits to clients who receive treatment as well as to therapists and organizations who employ it as a therapeutic modality or program. This chapter reveals many of those unique benefits, explores some of the challenges of the profession, and provides recommendations to take advantage of assets and navigate current issues successfully. Issues are categorized as those that have an impact on employment in horticultural therapy and those that influence program sustainability.

Assets

There are many unique benefits offered by horticultural therapy. This chapter portrays the beneficial use of plants and horticulture activities to pursue goal-oriented, specific, health-related outcomes for participants. Using a popular outdoor hobby to motivate and engage clients, horticultural therapy may use indoor and outdoor garden space to further support healing and immerse participants, visitors, and staff members with the passive, but beneficial, benefits greenspaces bring (Marcus and Sachs 2014). Tools, garden structures, plants, and garden areas are easily adaptable and bring gardening to all age groups, cultures, backgrounds, and abilities. The horticultural therapist, limited only by agreeable growing conditions (both indoor and out) when using live plants, can tap into numerous options for programming. Plants intrinsically offer an endless variety of sensory input, provide metaphorical opportunities, and obviously possess attributes of beauty and diversity (Lewis 1996).

Patient satisfaction

Research on the benefits of gardens and natural environments offers insight into why therapeutic gardens offer restoration (Kaplan and Kaplan 1989). (See Chapter 5 for an in-depth discussion of people–plant research and theories.) Vast anecdotal support for horticultural therapy comes from both patient testimonies, family stories, and staff interactions and observations. Charles Lewis (1996) shares a story of a resident in a geriatric center who spent a few days on a home visit with family and when asked to extend her visit, she stated, "I have all those plants in my room. They need me! I cannot be away from my babies." This shift of focus extrinsically rewards individuals who often feel a loss of control in health-care settings. It provides a sense of purpose, an opportunity to give as well as receive care, and may encourage more sustained and active engagement in an activity (Jarrott and Gigliotti 2010). When given a choice, this innovative approach to therapy can reconnect those whose previous life stages were engaged in positive gardening activities and may often be sought after by family members, or the individuals themselves, when choosing their health-care facility. Deadheading marigolds in an outdoor patio filled with interesting colors, scents, and textures, for example, may provide more intrinsic motivation to pursue fine-motor therapy versus picking up wooden blocks inside a physical therapy room crowded with other patients. Often horticultural therapy interventions do not feel like therapy to the patient or client and can in fact seem like a break from other intensive therapies.

With a trend toward increased opportunities for choice in health care, it is beneficial to understand why horticultural therapy is chosen. A study conducted at Rusk Institute for Rehabilitation Medicine in New York City looked at patients' choices for voluntary participation in horticultural therapy and found that those being treated for nervous system impairments who also had plant preferences—indicating prior experience with plants—were the most likely to choose horticultural therapy as an intervention (Orr et al. 2004). Further surveys and research to discern the influences on patient choice are needed to better understand how and why people participate in horticultural therapy programming.

Offers purposeful activity

An exceptional benefit of using hands-on gardening in programming is that it provides an activity that has meaning and is purposeful. Clients can see tangible results of their efforts and are often motivated to engage in the therapy as a result. A skillful horticultural therapist uses this aspect to encourage participation by clients to achieve their goals. At times, however due to the very nature of working with plants, clients may be unaware they are involved in actual therapy. Exhibit 4.1 describes how a therapist used potential campus beautification outcomes to engage a group of high school students to embrace tasks that would also ultimately benefit their own personal skill development. Exhibit 8.10 (in Chapter 8) describes how a patient in rehabilitation for a brain injury was prompted to actively work on speech and language goals through a horticulture-related project during co-treatment with a horticultural therapist.

EXHIBIT 4.1

Techniques: Purposeful Goals, Special Education

Imagine how it would feel to be attending high school knowing that you were not going to graduate, that you were different from the majority of students who changed classes all day long. Whether because of an IQ that was a hindrance to your learning or due to environmental reasons, or perhaps a combination of both, imagine you were in a class sometimes referred to as a "special class."

The year that I was made aware of these students happened to be the year that I had agreed to become the Chair of the Beautification Committee of the high school that my daughter attended. It was also the year that I had studied horticultural therapy through courses at the Horticultural Therapy Institute in Denver, Colorado.

With a background in public school teaching, training in horticultural therapy, and a desire to instill self-esteem and confidence in students while teaching useful vocational skills, it seemed the perfect opportunity was available to apply my services. Preparation involved pitching the idea of a once-a-week, two-hour horticultural therapy class with these "special" students to their teacher, the principal, and administrative staff in the county office. All parties agreed to give the horticultural therapy class a chance. Resources were secured through budgeted funds as well as grants and donations from a variety of school and community resources.

(Continued)

EXHIBIT 4.1 (Continued)

Techniques: Purposeful Goals, Special Education

Now, the biggest challenge of all: how to hook the students! They had experienced year after year of failure and isolation from the mainstream of students, and they did not feel that school was a worthwhile pastime. Using my knowledge of the extensive research on the positive effects of plants on humans was the key to engaging them.

The day came when I met with the students to explain the program. I had placed a vase on each of the nine desks in the classroom. Some of the vases held colorful cut flowers, some held dried-up leaves and sticks, and the rest were empty. The students came into class and took a seat at their desks. I began by explaining that each of them had a vase before them on their desk and that I could tell by looking at their faces which students had the bright, colorful flowers. Then they switched vases and began to describe how they felt when they were viewing the bright flowers compared to the brown, dried leaves.

I then shared this with them: "Sometimes when we look at the landscape, we view a scene like the dried-up leaves and broken sticks, and personally, that scene causes me to make an unpleasant face—it makes me feel uncomfortable. Other times, we may see a landscape that we hardly notice, such as the empty vase, which doesn't have much of an effect on us. At other times, we see a scene like the bright, colorful flowers and when we do, we may end up smiling, staring, and trying to take in all of the beauty. If we pay attention to how that makes us feel, we notice we feel happy and peaceful. There is an area on your campus on a busy corner with lots of people passing by each day that reminds me of the dried-up leaves and sticks—each time I drive by I feel uncomfortable with what I see. In this class we are going to be able to replace the unattractive areas like that with beautiful plants, which you will be planting and taking care of. You will be able to change the way that people feel when they drive by the campus. Changing the way people feel is a very powerful thing!"

Instilling hope in the students—hope that they had the power to create a meaningful change—was the key to success. Because the students were able to experience the positive feelings themselves that come from nature's beauty, they were able to invest themselves in the program with trust.

Sally Cobb, BA, HTR
Horticultural Therapist, Hospice and Palliative Care of Greensboro

Reaching diverse audiences

Horticultural therapy employs gardening activities to engage participants for treatment, work skill enhancement, and quality of life. It has a distinct advantage because of the plant and gardening modality. Not only is gardening a popular pursuit and familiar to many clients but it also offers a wide assortment of possibilities for treatment sessions and activities, providing opportunities to connect therapeutically with clients from many different backgrounds.

Cultural diversity

Another significant factor that horticultural therapy can address is the increasing cultural diversity among clients served in any type of program. For example, in the United States, the 2010 Census Bureau examined the nation's changing racial and ethnic diversity and stated that "while the non-Hispanic white alone population is still numerically and proportionally the largest major race and ethnic group in the United States, it is also growing at the slowest rate. Conversely, the Hispanic and Asian populations have grown considerably, in part because of relatively higher levels of immigration" (US Census Bureau 2011).

Cultural competency in the health-care system is defined as "the broad concept used to describe a variety of interventions that aim to improve the accessibility and effectiveness of health care services for people from racial/ethnic minorities ... developing largely in response to the recognition that cultural and linguistic barriers between health care providers and patients could affect the quality of health care delivery" (Truong et al. 2014). Horticultural therapists can employ strategies and interventions using food types, cultural methods of preparing traditional foods, ceremonial plant use, and the like, to connect with clients.

Accessible to a wide variety of participants

In addition to reaching culturally diverse groups of people, horticultural therapy offers a universally accessible activity. Gardening tasks may be accomplished at virtually any age and with any ability. Horticultural therapists are adept at working with people across the lifespan and with people with varied illnesses, injuries, life circumstances, and abilities. With appropriate design, tool use, or adaptations and modifications, many gardening tasks may be successfully performed. (See Chapter 11 for more details on these skills.) In addition, gardening itself offers diverse opportunities, from digging the soil to garden planning, indoor plant cultivation, cooking, planting seeds, creating botanical arrangements, etc. A skilled practitioner can offer active plant cultivation in some form to an exceedingly diverse clientele. Furthermore, plants respond to care no matter whether the client is verbal or nonverbal, ambulatory or nonambulatory, calm or stressed, affluent or impoverished, etc. With support, structure, and careful selection of activities, the horticultural therapist enables success for diverse participants (Figure 4.1).

Figure 4.1 The serene environment of a greenhouse enables a patient with dementia to focus and decreases restlessness. (Courtesy of Beth Bruno.)

Public relations

Horticultural therapy offers ample opportunities for positive public and media relations for the organization employing it. It is an asset and an innovative approach to treatment, has visuals that are notable for their beauty and human interest, and is ripe with affirmative accounts of personal successes. The general public as well as family members and potential customers of the organization can easily relate to images of clients or residents actively engaged in horticulture. Savvy horticultural therapists and administrators use public relations opportunities to build financial and community support and to increase demand for programming. Similarly, they value the programming not only for its efficacy in achieving treatment goals but also for its role in a quality public image for the organization. The public relations potential, along with how other benefits to the organization positively affect that organization's finances, are described further in Exhibit 4.2.

EXHIBIT 4.2

Perspective: Horticultural Therapy and the Bottom Line

When presented with a proposal for a horticultural therapy program, inevitably the first question asked by a facility, program, or hospital administrator is, "What is the bottom line?" Unless the horticultural therapist plans to volunteer his or her services and pay for the program out of goodwill, funding the program is a critical concern.

Funding is dictated, of course, by the model of the facility, whether it is for-profit or nonprofit, private pay or government funded, acute care, residential, vocational, or even a prison. A for-profit facility may wish to offer horticultural therapy, typically viewed as a unique service to clients, but must be willing to allocate funding toward it. In this situation, a horticultural therapist may be contracted for services to provide programming for an hourly fee rather than taking a position on staff. Horticultural therapy programs offered by a nonprofit facility may need to rely on grants, fundraisers, or private donations. The facility may also pay for contracted hourly services or commit to a part- or full-time horticultural therapy program. In many cases, a nonprofit may have a salaried horticultural therapist on staff while greenhouses, gardens, plants, and supplies are provided by grants and other fundraisers.

Third-party reimbursement, or lack thereof, has long been the Achilles heel of horticultural therapy programs. Physical therapy (PT), occupational therapy (OT), and speech and language pathology (SLP) represent most rehabilitation services because those therapies are able to bill for treatment and acquire third-party reimbursement for the hospital or subacute facility. Services are rendered to a client, and payment is provided by Medicare or a private insurance carrier. This ability to obtain reimbursement pays for PT, OT, and SLP positions and provides a major source of income for the facility. Since horticultural therapy is not reimbursed in this manner, it is often viewed as adjunctive. In other words, it may be a nice addition to a rehabilitation program yet a detraction from "the bottom line" or balanced budget.

How can the addition of a horticultural therapy program positively affect this proverbial bottom line? There is more to the bottom line than the heavily leaned upon and most obvious third-party reimbursement. Whether a facility agrees to

(Continued)

EXHIBIT 4.2 (Continued)

Perspective: Horticultural Therapy and the Bottom Line

fund a program fully or relies on other means, a horticultural therapy program can positively affect finances, far exceeding the cost of the program.

For a nonprofit facility, a well-executed horticultural therapy program can be an effective draw for grant funding and private donations. Working closely with the company's grant writer or development office allows the facility to raise funding for greenhouses, gardens, and other capital projects that improve the facility and the services it offers. Improved patient outcomes increase discharge rate and income potential for the facility. Gardens associated with horticultural therapy programs enhance curb appeal and visibility. The uniqueness of horticultural therapy in many cases provides wonderful visuals for marketing opportunity and materials. Community outreach and tours draw potential clients who may not have been aware of the facility's services. Studies show that garden spaces increase employee retention. This saves cost in human resources by decreasing employee turnover. Most important, horticultural therapy programming provides excellent customer service and patient satisfaction as families and clients enjoy the beautiful spaces. All of these aspects can positively affect the financial health of the organization that includes horticultural therapy as part of its business.

Kirk W. Hines, HTR
Horticultural Therapist, A.G. Rhodes Health and Rehab

Linking with nature's impact on well-being

Public awareness of the benefits of nature on health and well-being has grown worldwide in recent years, with recognition of the importance of plant-rich spaces to increasingly urban populations. Gardening and gardens may be seen as antidotes to the stress of urban environments, modern lifestyles, and the prevalence of electronic devices (Louv 2005; Williams 2017). The psychosocial welfare of urban and rural dwellers seems to be inextricably linked to access to nature, including gardens, with researchers, planners, and residents gaining awareness of this important link (Waliczek et al. 2016).

Gardening activities as well as passive experiences in verdant or wooded landscapes are gaining recognition as a research-based approach to improving health from common and universal diseases found in adults and children (Li 2011; Soga et al. 2016; Williams 2017). Heart disease, stress, diabetes, obesity, and depression are top threats to human well-being around the world; however, the popular news media and governmental agencies are sharing an increasing number of reports on the benefits of nature to everyday health (US Department of Agriculture 2018). Horticultural therapists may use the research and public awareness on the health benefits of contacts with nature to inform practice and garner support for programming.

Reaching beyond the clients served

The gardens produced by and for horticultural therapy programs include clinical hands-on spaces as well as productive vegetable gardens, therapeutic gardens, and healing environments. Gardens may have multiple designs to achieve therapeutic value for clients and nonclients alike. By their very nature, horticultural therapy gardens can have a positive impact on those who come into contact with them, including other staff members, family members, and community residents. Everyone who walks by or has access to the garden can benefit from the beauty and restorative qualities of a plant-rich environment. Many programs use greenhouses as essential spaces for programming. According to Kirk Hines, "They are the ultimate tool, and allow the therapist to have everything they need readily on hand, and are invaluable from a marketing standpoint" (Kirk Hines, March 7, 2018). Greenhouses provide growing spaces as well as a nice setting that may be showcased, particularly in winter, when outdoor gardens may be dormant. Of course, the horticultural therapist benefits personally from working in this environment as well. Regular access to plants and outdoor gardens is often a huge attraction as a career path and can be restorative from the stresses of human service work. While designed for therapeutic programming, horticultural therapy gardens actually have a far wider influence. Wagenfeld, an occupational therapist, and Winterbottom, a professor of landscape architecture, addressed critical landscape and design elements for therapeutic gardens, breaking down gardens for movement and physical rehabilitation, solace and comfort, learning, and sensory and community use (Winterbottom and Wagonfeld 2015). Recognizing the wide impact of a garden in health care and human service facilities and applying a multi-user approach for designing a therapeutic garden (AHTA 2015a and 2015b, Marcus and Barnes 1999) can lead to vast appeal for visitors and staff members in addition to clients. Recognition of this allows the facility to offer multiple benefits for the user while promoting itself to the public as being aware of the value of outdoor spaces. Lovely garden images frequently grace the facility brochures, offering opportunities to promote awareness of horticultural therapy programs and funding options in a very visual way (Figure 4.2).

International interest and use

Recognition of horticultural therapy as a distinct profession is growing and already exists in many countries around the world. In some areas, this was spurred by the creation of the National Council for Therapy and Rehabilitation through Horticulture in 1973. In other parts of the world, a cultural and spiritual component of connecting with the earth itself naturally evolved into horticulture as therapy. In some areas organizational roots parallel the emergence of the profession in the United States, beginning in the late 1900s. Many countries have seen rapid professional growth through public awareness, academic curriculum development, and subsequent increases in the number of practitioners (Park 2012). Horticultural therapy is spreading and gaining awareness and support particularly with programs in Europe, Asia, and the Americas. Professionals meet internationally at conferences and symposia and publish in journals across borders, but no international organization specifically focused on horticultural therapy has yet been formed. There is, however, an organization that leads the effort to gather and disseminate knowledge in the broader topic of people and plant relationships called the International People Plant Council. Every two years, the council holds an international symposium on people and plant interactions, including horticultural therapy. It also publishes the proceedings of these meetings in *Acta Horticulturae*, as part of the International Society for Horticultural Science, based in Belgium. The organization's international efforts facilitate communication across borders and disciplines, offer discussion of many topics related to horticultural

Figure 4.2 Gardens at Craig Hospital provide positive benefits beyond the patients served. (Courtesy of Rebecca Haller.)

therapy, and help to raise the profile of professionals in these areas. Future international collaboration on research, evidence-based practice, and uniform terminology specific to horticultural therapy will be important steps for the global advancement of the profession.

Issues

All human service professions have encountered challenges, and the field of horticultural therapy is no exception. Horticultural therapy has been hampered by lack of public awareness, funding, and evidenced-based research as well as credibility among medical professions and other allied therapies. These challenges are common in any young profession; however, many, such as music, art, recreation, and dance therapies, have emerged at a quicker pace with a firmer backing in evidenced-based research, solid health outcomes, and name recognition by the public, which in turn has created demand for services.

Overcoming these issues has been a professional goal in the United States, a goal led by the American Horticultural Therapy Association (AHTA). Worldwide organizations, horticultural therapy practitioners, educators, and researchers, as well as other allied professionals who use horticultural therapy interventions, have also worked to define and promote the profession (Fung 2012; Kent 2012; Koura 2012; Neuberger 2012; Park 2012; Shoemaker and Diehl 2012). In some countries, the practice of horticultural therapy is incorporated as part of green care or other nature-based therapies (Gallis 2013; Bragg and Atkins 2016). This section identifies challenges and offers insights for working within the emerging profession, with a focus on employment and the closely related issue of program sustainability.

Employment and compensation

Overall, perhaps the number one issue for horticultural therapists is that of employment. Securing work that is well compensated requires a variety of proficiencies and actions on the part of the therapist in an emerging field. This section on employment discusses considerations and perspectives, including employment types, distinguishing characteristics of the professional horticultural therapist, image, and recommended actions.

Employment models

As described in Chapter 13, there are three primary models of paid employment in the United States. The horticultural therapist may work as an employee, a contractor, or a consultant, or as any combination of these. Briefly, an employee is paid a salary or hourly wage, with taxes and any benefits paid by the employer, who has considerable control over the hours and place of work as well as job tasks to be performed. A contractor delivers a product and works with much more flexibility. The employer contracts with the horticultural therapist for specific services—typically a series of regular sessions and sometimes garden oversight. A consultant provides advice—that is, consults with an organization to produce various results.

Katcher and Snyder (2010) summarize key job factors to compare working as an employee, contractor, or consultant. As with most entrepreneurial ventures, motives for running a consultant business include controlling one's schedule and workload, thus allowing freedom from standard time restraints. When contracting horticultural therapy services, the facility typically dictates the scheduling of activities or sessions but does not have control over the place and hours worked beyond that. Conversely, an employee follows the schedule, location, and timing determined by the employer. Other financial and administrative considerations for contractors and consultants include the need to furnish one's own health, auto, and professional liability insurance as well as maintaining detailed records for tax purposes. Katcher and Snyder note additional positive factors for an employee, including higher job security than a contractor; a contractor must renew agreements with those who engage his or her services on a regular basis and also must do more marketing of services. An employee may also have more stimulation from others in the workplace and receive benefits such as health insurance and retirement.

For ongoing practice and program delivery, horticultural therapists most often work as employees or contractors, with consulting usually provided outside regular employment by experienced professionals who give advice on programming, design, and development. Horticultural therapy consultants may solely offer advice, typically through some form of on-site assessment, although phone conversations or other indirect forms of communication may occur, both indoors and out, if applicable. This form of business is only limited by the entrepreneur's imagination, and opportunities vary widely, including advising health-care facilities on programming, evaluating garden sites and offering design suggestions, and assisting individuals or department staff members with their own businesses or programs. Care must be taken not to give away valuable services, which should be provided by horticultural therapists. Instead, it should be considered a chance to educate those requesting consultation of the differences between gardening activities and horticultural therapy interventions. Many horticultural therapists may consider this avenue, wishing to impart their expertise in this area and/or in hopes of landing employment through the connections made. Exhibit 4.3 describes how a professional therapist combines skills in horticultural therapy and garden design for consulting work in the United States.

A survey of those who completed coursework in horticultural therapy over a span of 15 years at the Horticultural Therapy Institute and Colorado State University in the United States indicated that people were more often working as employees than contractors (Haller 2016), although contracting was a model used by 35% of respondents. Those who enter the field of horticultural therapy as a second career often choose contracting for the flexibility it provides (Christine Capra, Rebecca Haller March 2, 2018). Advantages and disadvantages of contracting are described by Pam Catlin in Exhibit 4.4. Note that contractors are sometimes denied access to confidential patient information or fail to negotiate time to record information in medical records. In order to provide high-quality professional services to clients, it is recommended that the horticultural therapy contractor builds access to this information into any written agreement with an organization.

EXHIBIT 4.3

Program Example: Horticultural Therapy Consulting

Melanie Hammer, MSW, HTR is the owner of Therapeutic Gardens by Design, LLC in Oconomowoc, Wisconsin. She consults with health-care facilities to design horticultural therapy gardens (Melanie Hammer, personal communication, September 15, 2017). Hammer said that her job entails evaluating obstacles that keep people from coming outside, giving recommendations to the chief executive officer and department heads, and combining results with a landscape designer, using Hazen's three, one-hour therapeutic garden design meeting format (Hazen 2014). The design meetings gather considerable input from staff. Prioritizing client safety and goals, Hammer is also the general liaison between behavioral health-care staff and the landscape architect at Rogers Hospital, a psychiatric facility. In that role, she seeks to ensure that effective communication occurs between the two for the best design and program outcomes. As a horticultural therapist, she also trains staff to utilize these spaces successfully as part of treatment with clients.

Jonathan Irish, MA, LPC, HTR
Coordinator of Horticultural Therapy, Rogers Memorial Hospital

EXHIBIT 4.4

Perspective: Advantages and Disadvantages of the Contracting Model of Employment

As in all things, there are advantages and disadvantages to working as a private contractor. Advantages include, but are not limited to:

- The ability to create a career in towns where opportunities for full-time horticultural therapy employment are limited or nonexistent.
- Increased control of wages.
- Increased control over one's own schedule.
- The means to work with a variety of populations and organization, if desired.
- Flexibility to offer a variety of services: consultation, programming, design/implementation, etc.
- The potential for contracted work to develop into a full-time, salaried position.

Disadvantages include, but are not limited to:

- No long-term commitment to a program at sites, with horticultural therapy considered an addition or project rather than part of the core business of the organization.
- The need for off-site storage space for materials and supplies.
- Low sense of ownership of the program by the organization, especially if the program is funded by individuals, grants, or donations.
- Research and financial planning must be done personally to obtain benefits such as medical, dental, etc.
- Sometimes organizations will not share personal information of clients with contractors due to their understanding of privacy regulations, possibly resulting in limited information to the therapist that would be beneficial in providing consistent person-centered horticultural therapy.
- Programs might be less fully developed in a contract than it would be with a full-time, salaried horticultural therapist on site, due to time and financial constraints.

Pamela A. Catlin, HTR
Horticultural Therapy Consultant, Faculty, Horticultural Therapy Institute

Additional aspects of contracting that influence employment and workplace relationships are considered in Exhibit 4.5. Note that flexibility, the skills to serve a broad array of participants, high professional standards, and the ability to market the service to employers are all traits that contribute to a successful career as a contractor in horticultural therapy. The discussion of employment models applies to systems found in the United States. In other parts of the world, employment models vary and depend on the systems in place

EXHIBIT 4.5

Perspective: Aspects of Private Contracting that Influence Practice and the Experiences and Relationships of the Horticultural Therapist

Often, private contractors need to adapt the horticultural therapy services offered to meet the needs of the community in which they work. Demographics, population size of the area served, and average income levels are just three factors that affect the development of horticultural therapy contract services.

Demographics play a role in determining what populations to serve. An area that has a high number of retirement-age residents would potentially support more horticultural therapy programs for older adults than for youth. A community known for being a hub for recovery programs would have a different horticultural therapy focus. It is beneficial, however, to develop skills in working with more than one population in order to maximize the number of potential contract sites.

The population size of the area being served and the average income for the area influence the horticultural therapy contractor in several ways. A large city offers more potential contract sites and supports higher contract fees. Smaller towns may require more flexibility in terms of services offered, and average pay tends to be lower. Other factors in determining what horticultural therapy services to provide and what level of pay will be supported are the type of sites and the title of the horticultural therapist. Clinical settings such as hospitals, physical rehabilitation centers, and psychiatric hospitals tend to support higher pay than assisted and skilled nursing communities. A job title of horticultural therapy program specialist would likely bring in a higher wage than horticultural activities contractor. Becoming acquainted with the language and culture of potential sites is a step in determining the most effective title to use.

The person desiring to be a successful horticultural therapy contractor needs to:

- Be a self-starter and be highly motivated.
- Hold to high standards regardless of the work site's staffing standards, as in some cases hourly wage earners have less motivation and attention to detail than a person engaged in a career as a contractor.
- Develop the ability to work successfully with a variety of personalities and methods of operating.

There are challenges to becoming a private contractor. Attention to the following can help ensure greater success:

- *Time and money:* It takes time to develop a private contracting career in horticultural therapy. One should have a means of additional financial support during the years it takes to become an established, secure business.
- *Family impact:* Discuss the pros and cons of becoming a private contractor with those around you who will be affected. Changes in income, benefits, and travel requirements are just three areas that can have an effect on family relations.

(Continued)

EXHIBIT 4.5 (Continued)

Perspective: Aspects of Private Contracting that Influence Practice and the Experiences and Relationships of the Horticultural Therapist

- *Marketing:* Never quit marketing. Even when the business is fully established, it is important to continue promotional activities. A waiting list for those desiring horticultural therapy services is essential for those times when changes in administrations, grant funding, and other circumstance can result in loss of contracts.

Pamela A. Catlin, HTR
HT Consultant, Faculty, Horticultural Therapy Institute

in each area of governance. In the United States and other countries, horticultural therapists may work in health-care environments, care farms, community-based settings, and/or horticulture businesses. Varied types of programs and program sites influence the employment model used, and how clinical the program approach may be.

Making the case for the professional

As described in Chapter 1, the roots of horticultural therapy began in the United States with an emphasis on activity that was offered by untrained volunteers who combined their knowledge of horticulture with a desire to help others. While it was a noble start, the casual nature of these types of programs still influence the reputation and employment of horticultural therapists today. Since the 1970s, a more professional approach to using horticulture as a goal-focused treatment modality emerged. Even so, it continues to be necessary for horticultural therapists to inform others of the distinction between a professionally delivered program and that of a layperson who uses horticulture for a social service, or someone who just feels better after weeding a flower bed in his or her garden (Figure 4.3).

Following are some of the distinguishing factors of delivery of services by horticultural therapists versus laypersons who may provide excellent activities but are without professional training. The most critical concern in this situation is safety to the client. Plant choices, specific client precautions, and lack of knowledge of disease and medically related illnesses are all avenues that may harm, intentionally or not, a client if the deliverer of services is uneducated. Options for garden accessibility, and adaptation or modification of horticultural therapy activities are skills typically deficient in untrained individuals. Pervasive beliefs that all gardening is good, and the consequences of these unintended perceptions, is described by Leigh Diehl (2012).

Many people think of horticultural therapy as an extension of gardening, believing that if they are knowledgeable in gardening, they can be a practitioner. While people with that belief have good intentions—they are trying to combine something they love with helping people—they are actually doing a disservice to the profession by not recognizing the importance of training and experience. True horticultural therapy and therapeutic horticulture programs are led by professionals trained in the use of horticulture as a means toward human well-being. Horticultural therapy is much more than a love of gardening and the desire to help people; it involves significant education and training in sensorimotor, cognitive, and psychosocial functioning as well as special populations and diagnoses. Horticultural therapists also study therapeutic methods

Figure 4.3 Working in the garden and/or gathering a bouquet may address physical, cognitive, social, or emotional goals. (Courtesy of Beth Bruno.)

and frameworks, program planning and evaluation, and methods for creating treatment plans and conducting activity analyses. Horticulture and its application to therapy is, of course, another major component of the curriculum. While many people can be effective in volunteering or assisting in a horticultural therapy program, the trained horticultural therapy practitioner has the background to ensure the quality and success of activities and the overall horticultural therapy program.

As described in Chapter 1, horticultural therapy is an *interdisciplinary* field; knowledge in horticulture, human sciences, *and* horticultural therapy is essential. Furthermore, a professional delivering health or human services such as horticultural therapy is expected to learn and exemplify a high standard of professional behaviors. In *Developing Professional Behaviors*, Kasar and Nelson list ten behaviors that build on one another and address the following areas:

- Dependability.
- Professional presentation.
- Initiative.
- Empathy.
- Cooperation.
- Organization.
- Clinical reasoning.
- Supervisory process.
- Verbal communication.
- Written communication (Kasar and Clark 2000).

Many seasoned, allied health-care professionals have progressed through these developmental stages during their career and transition to the field of horticultural therapy effortlessly, requiring only skills needed to deliver horticulture as an effective intervention for the client. While a mature adult who has worked in other fields outside the health-care arena or a young, enthusiastic, intelligent student may exemplify strengths in the first six listed professional behaviors, the similarities typically stop there. Focused education, internships, and experience allow horticultural therapists to distance themselves from expert gardeners who lead groups for individuals to meet some form of overarching goals even if those goals address some health-related need. Upon hiring a horticultural therapist, an employer can expect the professional to have had some exposure to clinical reasoning, the supervisory process, and advanced written and verbal skills that are nonexistent in a layperson who has had no basic framework to address these areas when working with clients. For example, Kasar and Nelson, expanding on each of the ten behaviors, describes clinical reasoning. Differing from the nontechnical skill called common sense, they cite Hall's definition of clinical reasoning as "the process by which clinical judgment generates clinical decisions that define and determine the procedures chosen," which gives rise to a "best rather than a right answer" (Kasar and Clark 2000). This reasoning, when developed along with evidence-based practice, is based on these best outcomes, and the activity developed by the horticultural therapist ensures the best success focused solely on the client needs. (See Chapter 14 for more on evidence-based practice.) Understanding the need to adapt and/or reevaluate clients when the need arises requires a deeper form of clinical reasoning and is typically gained through continued work alongside other care team professionals or experience gained in practice (Figure 4.4).

Communication, both verbal and written, is expected to be on a higher level than that of a layperson. While horticultural therapy practitioners may use general language to communicate with clients, family members, and members of the public, it is critical that communication among other health-care peers, such as those on an interdisciplinary team, is understood and understandable by the horticultural

Figure 4.4 Internship experiences enhance clinical reasoning skills. (Courtesy of John Murphy.)

therapist. The phases of interviewing and assessing the client, and evaluating and implementing the activity, require skill in facilitation techniques and an understanding of multiple delivery models. Receiving information from the client involves a fundamental understanding of the various techniques of both speaking to and questioning the client and using verbal skills such as probing, reflecting, interpreting, confronting, informing, summarizing, and listening techniques to gather necessary and specific information (Austin 2013). Formalizing the input received from the client into specific treatment goals and outcomes, as well as co-treatment with other professionals, further distinguishes the professional horticultural therapist from others offering horticultural activities such as those from master gardener volunteers. Professional processes used in horticultural therapy, including treatment, activity, and session planning, are described in *Horticultural Therapy Methods* (Catlin 2017; Haller 2017; Kennedy 2017). It is expected that a professional horticultural therapist can deliver quality programming to clients served, as well as develop and fully utilize horticulture resources for programming.

The horticultural therapist must be able to communicate and distinguish professional practice from laypersons' use of horticulture at all stages of programming, both within the organization as well as in external communications. This includes program proposals, solicitations for funding and donations, interdisciplinary communications with team members, and descriptions of benefits and treatment goals, and also communication with administrators and other decision makers. Careful and accurate descriptions and framing of the professional approach to all audiences is a necessary action for success in horticultural therapy. Exhibit 4.6 illustrates a list of treatment goals for students with "significant physical, medical, behavioral, and intellectual disabilities" included in a proposal for a new horticultural therapy program. By articulating potential treatment goals, the author demonstrates the professional approach to horticultural therapy practice. Additional descriptions of various communication strategies for other aspects of program management may be found in Chapter 13.

EXHIBIT 4.6

Technique: Communicating a Professional Approach in a Program Proposal

Note: Exhibit 4.6 is an excerpt from a written proposal to establish a new horticultural therapy program.

Developing a horticultural therapy program for the five- to twenty-one-year-old students of Morrison Center School in Scarborough, Maine, will enhance the school's therapeutic environment, support classroom academics, and offer prevocational opportunities to students with significant physical, medical, behavioral, and intellectual disabilities. Students will be referred to the program by their classroom teachers and therapists. This program will function as a closed group in an effort to work with each student's specific therapeutic goals. Selection of each student group will be carefully considered by therapeutic staff members and teachers to ensure a cohesive and productive work environment.

(Continued)

EXHIBIT 4.6 (Continued)

Technique: Communicating a Professional Approach in a Program Proposal

The following are examples of treatment goals, which may be addressed in horticultural therapy.

Cognitive:

- Develop skills in sequencing and memory recall.
- Acquire compare and contrast skill.
- Develop recognition/identification skill.
- Practice good decision making.
- Become aware of possible career options.

Physical:

- Stand or sit for the duration of an activity as needed.
- Improve coordination.
- Increase muscle tone.

Emotional:

- Show empathy toward fellow participants.
- Practice patience and wait for your turn.
- Develop positive behaviors.
- Improve self-esteem.

Social:

- Develop a sense of purpose.
- Collaborate with the group.
- Speak in a respectful manner.
- Communicate feelings in a positive manner.

Colleen Griffin, HTR
Dempsey Center, Lewiston, Maine

The use of gardening activities by volunteers

As described in Chapter 1, volunteers in the United States began bringing gardening to veterans of World War II, as well as those who were ill or incarcerated, as early as the 1940s. Most of these efforts were organized by garden clubs, with the desire to provide care and relief to a variety of populations. They delivered programs typically referred to as garden therapy. Since that time, volunteers have continued to offer services to a wide array of people in hospitals, prisons, and other settings. Master gardeners routinely provide such services in many communities across the United States. They may be entirely managed by volunteers, or they may support professionally led programs. The volunteers typically have not received training in horticultural therapy practices. Due to the preponderance of such volunteer programs, horticultural therapists often find that organizations expect them to donate their time to programs and may resist paying them at a professional level. It is necessary to communicate to potential employers the distinctions between professional programs and those delivered by laypersons. It is also important to encourage volunteer roles among garden club members and master gardeners that support professionalism rather than supersede it (Shoemaker 2016).

The use of gardening activities by allied professionals

What are potential considerations for the use of gardening activities by a professional with training and credentials in a related health care or human service field? Is that different than horticultural therapy? It depends on how it is used, what skills are used, and its application. For example, does the therapist use established practices of treatment and documentation, such as those described in Chapter 12? Does the therapist have the interdisciplinary training required for success? Similar to a gardener who wants to help people, a therapist who does some gardening with clients, but who is not trained in horticultural therapy techniques, is not necessarily equipped to provide quality services. Horticulture knowledge is essential in order for the practitioner to lead successful sessions and to have beautiful thriving gardens that motivate and inspire participants. Similarly, knowledge of horticultural therapy practices is needed to maximize positive outcomes for the clients served through gardening. A love of gardening is not enough.

On the other hand, therapists with credentials in health care, occupational therapy, therapeutic recreation, counseling, and other fields who acquire the training, skills, and knowledge in horticultural therapy may have opportunities for funding and employment that are challenging for someone without professional credentials to achieve. They may also offer solid counseling or other skills that enhance outcomes for program participants. However, many allied professionals do not seek horticultural therapy training and registration due to the time and financial commitments. It is important to emphasize that, overall, the key to successful programming is for any practitioner to acquire interdisciplinary training and experience. For example, a therapist in an allied profession who completes coursework in horticulture and horticultural therapy and who does an internship supervised by a proficient horticultural therapist will likely be knowledgeable and skilled in horticultural therapy practice. They may even help to raise the professional image of the field by their dual credentials and level of practice.

Professional image

The image of horticultural therapy has been shaped by historic activity-based work of laypersons, as previously described, and by the public perception of gardening as a leisure pursuit that is ordinary and familiar. Only with additional information about the use of horticulture for treatment outcomes does that image change. Horticultural therapists need to reach administrators, other professionals, and members of the public in order to positively affect employment opportunities and compensation levels. It is also important to have standards and credentials, a solid body of research, clear definitions, and connections among closely related practices.

Employers

Current and potential employers must be informed about the professional approach and the potential for horticultural therapy to have a positive impact on the people served as well as the organization that uses it. As previously stated, an employer can expect the professional to have had some exposure to clinical reasoning, the supervisory process, and advanced written and verbal skills. But many employers are unaware of this skill set in horticultural therapists. Typically, the therapist has a much higher expectation of the processes and outcomes than does the administrator. One indication of this is the prevalence of employment situations where the horticultural therapist is not required to document outcomes or even be part of treatment teams. It is advisable to keep records, as described in Chapter 12, regardless of whether or not it is mandatory to do so and to find ways to share this information with teams while adhering to privacy laws and ethics. Practice with professionalism is one important step to heighten the image of the profession.

Another factor to consider with employers is related to program placement. Where is the program in the organizational structure? Is horticultural therapy considered part of activities, or is it considered a clinical treatment program? Clinical programs are often represented on treatment teams, have more expectations for working on client goals, and are usually paid at a higher rate. Exhibit 4.7 provides further advice from two experienced horticultural therapists on the steps to take to be paid a fair wage.

EXHIBIT 4.7

Technique: Making a Case for Commensurate Pay

How might a horticultural therapist communicate to employers, or potential employers, the professional nature of horticultural therapy practice in order to be paid wages that are commensurate with that level of service? Here are how two experienced professionals respond.

Kirk Hines, HTR:

1. Provide a solid scope of what your services would entail. Be clear that you bring *professional* services. For example, your services may include seeing patients as a credentialed therapist, co-treating with other allied professional and working alongside them, managing the gardens and greenhouses that are used as therapy and treatment spaces, and possibly working with a development department to raise funds for facilities, etc. You will be pretty busy with all that.
2. You may need to start at a lower salary than you desire and work to build credibility. That may take some time to establish. Then ask for a salary survey to check for what a commensurate wage would be in your region and type of business. You may need to provide data from colleagues in horticultural therapy in order for the employer to have enough information for this process. This is different than requesting a raise; rather, it is seeking adjustment to a higher wage category. (Editor's note: You may also emphasize the management role that you fill in horticultural therapy, as described in Chapter 13.)

(Continued)

EXHIBIT 4.7 (Continued)

Technique: Making a Case for Commensurate Pay

Pamela Catlin, HTR:

Particularly when proposing new horticultural therapy programs, people often may believe it is a nice volunteer opportunity and question why there should be a charge. What has been the most helpful has been to inform administrators in three particular areas:

- Education requirements to be a registered horticultural therapist. This is usually an eye opener, showing administrators that the horticultural therapist isn't just anyone they might meet at the garden center or plant nursery.
- How the use of horticultural therapy is not busy work or just activity. Rather, it is a modality that, when used professionally, assists each person in reaching his or her specific goals. This is persuasive when approaching sites that practice person- or patient-centered care.
- Describing that we have a national organization, the American Horticultural Therapy Association, that promotes the profession practiced internationally, with many populations receiving the benefits of horticultural therapy.

Being informed about approximately what fees other adjunct therapists in your geographic area are charging is helpful. Keep in mind the level of materials and physical work that go into horticultural therapy versus music or art therapy and charge accordingly.

Rebecca L. Haller, HTM

Other professionals

Horticultural therapy is still widely unknown or misunderstood by other professionals, largely due to a lack of information on what it is that horticultural therapists do. There are many parallels in practice among horticultural therapists and those in allied health care, including treatment planning, goal setting, and documentation. The processes are quite alike across types of therapy and human services, and additional professional similarities exist. For example, horticultural therapists in the United States have adopted revised standards of practice and code of ethics (AHTA 2015a, 2015b) and have had a professional registration system in place since 1975 (Shoemaker and Diehl 2012; AHTA 2017). The most important actions to take to improve the image of horticultural therapy by other professionals are to conduct oneself professionally, seek and accept roles on care/treatment teams, provide interdisciplinary communication, and report positive outcomes seen with the clients served.

Members of the public

Working on the public profile of horticultural therapy is challenging and exciting. One challenge is to overcome the views that it is just gardening and/or that it is anytime someone with a disability engages in gardening. These images are bolstered by feel-good stories in the press that are intended to convey the heartfelt job done by those who garden with people who are disabled, disadvantaged, or incarcerated, or who are recovering from illness or injury. Only rarely do popular media include any mention of the more clinical or goal-driven aspects of the work, as that is not the norm for the popular press. Generally, they want to make the story accessible to the general public. Even so, opportunities are abundant for getting press coverage of horticultural therapy programs, and getting such coverage is a distinct asset brought to an organization by this field. Creating a fact sheet with basic information about the program and the positive outcomes for clients is advised to provide the writer or reporter with accurate details. To prepare for interviews and talks, it is also useful to identify key points and practice articulating them succinctly and verbally in lay terms. National and regional associations also play a key role in conveying an accurate and positive image of the profession through websites, information services, press releases, etc.

Credentials

Simply put, credentials affect professional image. They indicate that a certain level of professional education and training has been achieved and ongoing learning opportunities and education are pursued to ensure the consumer of qualifications. In many fields credentialing includes testing for competency. In the United States, the system of professional registration does not require an exam but rather is based on education and a supervised internship (AHTA 2017; Haller and Capra 2017). This system has also been replicated and modified in other countries, as mentioned in Chapter 1. Steps have been taken by AHTA in recent years to better position the field for the eventual creation of a competency exam by a credentialing body. However, distinct challenges exist due to the relatively small number of practitioners who would be eligible to take an exam and the high expense of creating one. Some interim steps that may be taken would seek to dramatically increase the numbers of those who are professionally registered (Horticultural Therapist Registered [HTR]). This might be accomplished through encouraging new professionals to seek the HTR credential and by communicating the benefits of the credential to employers. An exam and requirements for annual maintenance and recertification would transform credentialing to include assurances of competency, rather than relying on education as the sole determinant of qualification, and thus elevate the image of the profession and professionals who practice it.

Another reason to seek the HTR credential involves the skills that a therapist gains in the process of completing coursework and an internship in order to become professionally registered. These skills may be applied in many ways, including horticultural therapy services, therapeutic horticulture, or social and therapeutic horticulture. One individual who currently practices therapeutic horticulture states that, due to the nature of her programming, outcomes will unlikely change when she receives professional registration. Her reasons for pursuing HTR are to improve program evaluation for current programming, increase skills at matching specific activities for specific populations, as well as the ability to offer horticultural therapy internship opportunities through her program (Rachel Deffenbaugh, e-mail with R. Malone, September 27, 2017). In other words, the program itself may not change, but her professional competence will be positively affected by the training that is required for professional registration.

Research

Health-care systems are demanding more specific health outcomes in order to pay for the coverage of services. Horticultural therapy practitioners must contribute evidence-based outcomes in order to encourage inclusion of horticultural therapy services for consideration into current health-care legislation and public policy.

While the body of knowledge in horticultural therapy continues to grow worldwide, there is still a great need for scientifically sound research that provides data about the clear connection between horticulture and health outcomes (Frumkin 2004). Literature searches for evidence on the effects of social and therapeutic horticulture reveal a growing body of literature, but most reports did not provide hard evidence or include experimental results (Soga 2017). Sempik et al. (2003) point to a need for a more scientifically rigorous approach and advised publishing in "'mainstream' medical, scientific, social science and horticultural journals" in order to lead to more funding and support for this work. Horticultural therapists believe that working with plants is good for people. They regularly witness positive change in the clients they serve. They document those changes in case notes and charts. They report on the effects to colleagues at work, at conferences, and in journals. However, most practitioners work with a relatively small number of clients and are often not well trained in research methods. Therefore, the research tends to have small sample sizes, with statistically insignificant results. Furthermore, the effects of horticultural therapy are often difficult to separate from those of the plant-rich environment, those connected to activity, and many other factors for which the researcher must provide control groups. It is advised that research in horticultural therapy be conducted by people versed in social science research, in cooperation with those whose expertise involves horticulture and its therapeutic application. Qualitative studies may be conducted and used not only to inform practice but also to inform others of the efficacy of the therapeutic use of horticulture.

In order to improve confidence, credibility, and employability practitioners must use research and other literature. Building a body of knowledge requires using the literature that exists and publishing what is learned (Relf 2016). Each type of horticultural therapy program demands an array of literature to inform, support, and convey its worth to employers. It is recommended that universities team with associations worldwide to identify research priorities and also train and encourage interdisciplinary researchers. Horticultural therapy researchers could benefit from the following suggestions made by respected professors in the field of therapeutic recreation. Buettner and Richeson (2012) propose useful categories for creating consistent, clinical research, citing the need to address the optimal attributes of the intervention and the dosage (amount and length of time) of the intervention as well as looking at the effects of co-treatment. In addition, they prioritize the need for focused health outcomes, listing depression, pain, and anxiety as the top three areas. For horticultural therapy, Relf outlines theoretical models and elements for developing research foci based on a model of horticultural therapy that includes the four elements of a trained professional, a client in treatment, treatment goals, and horticultural therapy activities. She recommends that research is needed to better understand each of these elements and the interrelationships among them (Relf 2012). Additional recommendations on the types of research needed for any profession suggests at least two types of research—that which "contributes to the knowledge base" in an academic way and that which "contributes to service innovation and service development" for practitioners. Research is also important to inform policy (McKee 2004).

Documentation tools

Horticultural therapists regularly document the many steps of treatment and the outcomes of services, including client assessment, treatment planning, and charting progress on group or individualized goals and objectives. (See Chapter 12 for more information on documentation.) In many professions, standardized tools have been created and adopted for these types of uses, but only a few documentation tools designed for horticultural therapy have been published to date (Melwood 1980; Chambers et al. 1992; Im et al. 2018). In addition to using tools provided by the organization in which horticultural therapy is practiced, it is common for therapists to produce specific tools that work for them, even developing unique charting systems for individual clients. Due to the diverse models of treatment, populations served, and the nature of program sites, it is very unlikely that any scale or form could be applicable to all situations. Yet basic tools that could be adapted for an array of programs would be beneficial to aid practitioners in documenting the benefits of horticultural therapy. Exhibit 4.8 illustrates this need.

Therapeutic horticulture

How does the use of therapeutic horticulture affect the image, and therefore the employment, of horticultural therapists? Chapter 2 describes the distinctions between therapeutic horticulture and horticultural therapy. Generally therapeutic horticulture does not demand that the therapist address or document specific treatment goals, and so the field may be considered less clinical. Employers are very likely to be unable to differentiate between the two approaches and therefore may not expect that the horticultural therapist has the skills to address treatment professionally nor expect to pay for that level of care. Trained horticultural therapists have the knowledge to skillfully run therapeutic horticulture programs in those situations where that approach is preferred. It is up to the therapist to convey the scope, potential, and treatment procedures used in horticultural therapy.

EXHIBIT 4.8

Perspective: The Need for Standardized Measurement Tools

There is a need for standardized measurement tools for use by horticultural therapy professionals. Significant time and cost savings could be realized if a basic measurement toolkit were developed that provided easy-to-use templates of measurement tools, such as surveys and checklists. Ideally it would include projected outcomes for a variety of populations and be applicable for a broad range of program goals and objectives. It should be developed as a collaborative effort involving horticultural therapists and allied professionals.

The need for a basic measurement toolkit in horticultural therapy became evident when we, as contractors, were asked to use a standardized measure to document the success of a program and show the benefits and effectiveness of the horticultural therapy program in order to sustain funding from outside sources. The lack of a standardized tool directly affected the ability to demonstrate the program outcomes to funders and resulted in the program being on hold, risking its future.

The creation of a credible measurement toolkit would be beneficial to horticultural therapists, the organizations who employ them, and funders. The basic toolkit could provide a basis for further development of customized measurement tools and contribute to the effective development and management of horticultural therapy programs.

Pat Czarnecki, HTR, MBA
President, GreEn'ergy, LLC

Nature connection

Employers who are interested in therapeutic gardens and incorporating nature into their settings and programming may not understand that horticultural therapy is related to, yet distinct from these endeavors and that it involves active participation in gardening activities designed to meet treatment goals. As interest in nature connections continues to grow internationally, however, it behooves the horticultural therapy profession and its organizations to build relationships with this movement. Chapter 2 describes nature-based therapies and differentiates them from horticultural therapy. Cooperating with those efforts for research and programming is in the best interest of all. For example, horticultural therapists may collaborate with those working for community-wide access to nature, school gardens in every school, urban greening and farming, ecotherapy, and others. Perhaps the horticultural therapy contribution is to effectively reach people with disabilities, social disadvantages, and others to promote recovery, vocational training, and/ or wellness. A challenge for achieving professional horticultural therapy employment is to clearly distinguish horticultural therapy from related efforts and to promote and maintain a professional image.

Program sustainability

It has been a challenge historically to establish horticultural therapy programs that out-last the influence of those who initiate or develop them. The issue is that programs often cease operation when horticultural therapists or others who champion the program leave their positions. Similarly, when initial funding ends, programs may not be sustained by the organization. Changes in health-care systems, administrators, and facilities are also factors that may lead to the closing of a program. Many of these types of changes are inevitable and require flexibility and resourcefulness on the part of a program manager. Chapter 13 outlines an array of best practices for managing programs, including those that foster excellence and sustainability. This section of Chapter 4 explores considerations that may influence program endurance as well as suggested actions that may lead to more long-lived programs.

Funding

Sustained funding for horticultural therapy is a key factor in program longevity. The various sources of money for horticultural therapy are related to the type of program, the institution or company employing it, and the structure and placement of the program within the organization. Whether the organization is a for-profit or nonprofit business, governmental or nongovernmental, solid and reliable funding is essential.

Program models

The program model employed affects costs and potential income and funding. For example, production of a substantial amount of plants and/or produce of high quality is a natural outcome of vocational horticultural therapy programs. In fact, it is usually a necessary component in the provision of a workplace environment that offers effective vocational training and placement. The sale of those products provides program as well as funding opportunities. The clients benefit from the high standards of quality production, community integration through sales and delivery of products, wages earned, and the self-esteem and satisfaction that comes from creating items of beauty and value. Sales of items may contribute a substantial portion of the budget to operate the program—again ultimately benefiting the client with sustainable programming. Additional funding may come from governmental and nongovernmental sources to provide vocational services, as described in Chapter 9. For example, an organization may bill a government agency to work with individuals with developmental disabilities in a supported employment situation.

In therapeutic or wellness horticultural therapy programs, some funds may also come from plant sales run by the patients or participants. However, this is likely to be only a very small part of the budget. Still, many programs rely on sales of plants and products to fund or supplement the budget for supplies and materials. Most funding must be procured from other sources. In therapeutic programs, those sources may include the organization itself, government funding, direct private pay from clients, insurance money, grants, donations, or any combination of these. The strategies used are as varied as the programs themselves. Exhibit 4.9 describes a multifaceted approach to funding horticultural therapy services at a university public garden in the United States.

EXHIBIT 4.9

Technique: Funding Horticultural Therapy

Grants are highly competitive, and funders want to ensure that their funds have an impact. After ten years of grant writing, I have found that if you can write winning grants, you will always have a job. A successful grant writer receives funding for one in ten grants. For example, in the past few years, I have received 31 grants totaling $300,000 but have 18 unawarded grants totaling $740,000. Successful grant writing requires time and effort. Internet search engines can help find local foundations. Programs like Foundation Directory Online, Pivot, and Grants.gov are also helpful. Check with your organization or local library to gain access to any fee-based grant search engines.

The following are key factors for successful grant writing:

- Use keywords. Match your program to foundations with similar interests including population and the goals of your program. Foundations supporting specific medical conditions, gardens, horticulture, and nature may fund you. Use keywords such as *disability, education, youth, therapy, garden, horticulture,* and *vocational training.*
- Know the foundation's grant size, process, and time line.
- Look at the foundation's past awards to get an idea of what it may fund.
- Follow all directions for the application; incomplete applications will be rejected.
- Include relevant research and best practices.
- Be clear, outlining the need for your program and its impact.
- Include assessments, evaluations, and measurable outcomes. Include outcome data you have. For new programs, reference assessments and results of similar programs.

(Continued)

EXHIBIT 4.9 (Continued)

Technique: Funding Horticultural Therapy

- Find a colleague to review your application to catch errors and rephrase items that may be confusing.
- Relationships count. Supporters of your organization may have connections to foundations that can open communication for your program. When working for the Amputee Coalition, we received several grants through connections of board members or volunteers.

Don't get discouraged if you don't get funded. Ask the foundation for feedback so you can improve on your next grant proposal.

Developing *fee-for-service* programs requires an understanding of all costs of your program. Consider the following material costs when setting up your program fees:

- Containers.
- Plants: self-propagated or purchased from a nursery.
- Growing medium.
- Other supplies.
- Fertilizers.
- Greenhouse costs.

You will also need to consider the following when determining staffing cost:

- Program length.
- Preparation and documentation time.
- Off-site programs, include mileage and travel time.
- Salary and benefits.

Keep in mind that salary and benefits often cost an organization as much as an employee salary. If you are paid $25 an hour, factor in another $25 for your benefits for a total of $50 per hour. Check with your human resources office to determine benefit costs. These factors will give you a total cost for the program.

At the University of Tennessee Gardens, we have offered programs to local organizations for five years. We were able to subsidize our rates because of a grant, which partially funded the programming. As those funds ended, we have increased our fees, and the organizations we work with are willing to pay the increase because administrators have seen the positive benefits to their clients. We currently work with assisted living facilities, memory care facilities, and mental health organizations.

Derrick R. Stowell, MS, HTR, CTRS
Education Director, University of Tennessee Gardens

Sustained and integrated program versus short-term project

The ideal for sustainability is that the horticultural therapy program is integrated into the business of the organization. It is built into operations, staffing, budget, and other working processes. Horticultural therapy is an established part of treatment, and practitioners work with interdisciplinary teams to help clients meet therapeutic goals of a clinical and/or nonclinical nature. It may be adopted as an innovative approach by an established agency, or it may be the basis for a new company whose focus is the use of horticulture for therapy and human service. In either case, the key concept is that it is essential to the work being done. This integrated approach and attitude contrast with a gardening project (Majuri 2002) that is intended to be a temporary enrichment or interesting diversion for participants. Those projects are usually much less sustainable, and typically end when key personnel and start-up funding are gone. Furthermore, a gardening project may be difficult to distinguish from the gardening done by laypersons or those allied therapists who have an interest in gardening. They may in fact employ professional horticultural therapy techniques and trained personnel, but they may not have full credibility or be integrated into the systems of the agency. Therefore, program longevity is continuously tenuous. In proposing or developing new horticultural therapy programs, it behooves the therapist to consider the advantages of a more integrated approach, communicate that potential to administrators and all stakeholders, and build in strategies for endurance in all aspects of program design (Majuri 2002).

Funding sources

Contrasts exist between programs and projects, and funding sources may thus be categorized between those that are included in regular organizational budgets (internal funding) and those who rely on outside funding sources (external funding). Of key importance to horticultural therapy program development, longevity, and well-paid employment are adequate and sustained funding. When a program is integrated into the operations of an organization, that funding is more stable. When administrators, a board of directors, and/or an owner make a financial and institutional commitment to this type of programming, horticultural therapy may flourish and offer substantial positive outcomes.

Some strategies and sources for outside funding include payments for services by outside parties, individual payments, grants, endowments, and donations. In the United States, for an organization to receive grants or donations, it must be a nonprofit organization. Schools and other government-related entities also are eligible for these sources of support. National and regional laws and regulations vary considerably around the world. Payments by outside parties include fees for services such as those described in the discussion about vocational programs. They also may include billing an insurance company or government agency for co-treatment with an allied professional, such as occupational therapy, or for filling a particular niche in the social service system, such as working with a disabled or disadvantaged clientele or providing services as part of a broader effort to combat a regional problem. Identifying the needs and concerns of communities, foundations, and/or individuals is the first step toward program development that is to be funded by these sources. Horticultural therapy offers a distinctly unique approach to meeting those needs and may be articulated as such. The fact that it also fits into many clinical and nonclinical approaches and may dovetail with nature-based and green initiatives are all assets for gaining funding from outside sources.

Professional practice

Important to program success, credibility, funding, and longevity is that the horticultural therapist meets high standards of professional practice. Maintaining communications, meeting program goals, providing effective treatment, working collaboratively, managing staff members and volunteers, and evaluating program processes and outcomes are all essential skills described in further detail in Chapter 13. The therapist must also meet ethical standards of the profession and the organization in which they work. Because horticultural therapy treatment is based in gardening, the quality and maintenance of indoor and outdoor gardens are direct reflections on the skills and proficiency of the therapist as well as on the field of horticultural therapy itself. This may present both opportunities as well as challenges. As previously described, beautiful gardens invite positive public relations and offer restorative experiences to all who encounter them. Conversely, gardens that are poorly managed present an unprofessional image and may undermine efforts to establish sustained and enduring programs. Depending upon the type of clientele and program, gardening with clients may be insufficient to maintain attractive gardens. Horticulture professionals or landscape maintenance crews may be required, and should be considered in budgeting.

Communication

Essential for longevity is that therapists continuously communicate and market their programs. It is not enough to convey horticultural therapy assets, processes, and outcomes only when proposing or developing a new program. Why? Personnel, administrators, board members, clients, families, funders, and other stakeholders change over time. It cannot be assumed that the same individuals are still working when information was first conveyed. It is crucial to conduct regular program evaluations and provide the data collected to key decision makers, as well as internal and external funding sources, in order to document the benefits of the program. (See Exhibit 4.10 and Chapter 13 for further discussion about program evaluation.) In-house guidance is necessary for those who work with horticultural

---

EXHIBIT 4.10

Perspective: Program Evaluation

How many current horticultural therapy programs are routinely evaluated for performance and value? While that question is unstudied, horticultural therapists need to adopt the practice of gathering data to methodically understand a program's benefits. Program evaluation offers an opportunity not only to find out what works and what doesn't and to methodically assess the processes, inputs, and outcomes, but it also can be used to communicate the worth of a program. In order to facilitate this communication, it is advantageous to identify to whom the evaluation will be given before designing the scope and details of an evaluation process. Who needs to know the results of the evaluation process? Strategically gathering information that matters to the intended reader and decision maker may serve to advance the status of the program and contribute to the likelihood of its continuation.

Rebecca Haller, HTR

therapy, including related clinical staff members, facility maintenance personnel, support staff members, and volunteers. Seeking and developing relationships with those who champion the program can be beneficial to bridge the span of personnel and administrative turnover. Champions may include those who manage an organization, such as board members, as well as devoted program volunteers. Those who see a horticultural therapy program in action directly are often inspired and become advocates for it because directly witnessing the positive effects is powerful. Depending on the setting and the need for confidentiality, it may be appropriate and possible to invite people to observe or participate in horticultural therapy sessions. If not, or in addition, photographs and video recordings may be used to present horticultural therapy in action. Of course, witnessing a session only shows the outer work that is done. More information about the skills and processes used by the therapist, including assessment, treatment planning, activity modification, effective garden use, documentation, and other skills is necessary in order to understand some of what is required to sustain successful client outcomes.

As an emerging profession, horticultural therapy is relatively unknown by those involved in health care and human services, as well as by the general public, so they need information in order to support uninterrupted programs. Another reason to communicate is that program successes are more significant when they are shared with others. It is always advantageous to chronical positive outcomes—both in stories and anecdotes as well as documentation of treatment outcomes.

Historically, the distinction between horticultural therapy programs and gardening activities has not always been clear to those who secure staff members for and fund those efforts. Nor is the dissimilarity of a professionally led program and that of an avid gardener volunteer program always understood by administrators. It takes frequent communication to be sure the practice and the need for a trained professional are clear. This becomes particularly critical when the horticultural therapist leaves an organization. For program continuity, ideally another trained and qualified horticultural therapist is hired to replace that person. This is often the crux of an enduring program, so it is advised that all decision makers be kept apprised of the positive consequences of continuing professional services in horticultural therapy. More details on communication as it relates to program management is found in Chapter 13.

Research

The importance of research to the advancement of the profession of horticultural therapy cannot be overstated, as was described in the discussion about professional image earlier in this chapter. It is essential for professional credibility, evidence-based practice, funding, and even future policy making. It affects program sustainability because a solid knowledge base is fundamental for any profession. It is probable that horticultural therapy programs that are viewed as offering evidence-based treatments are more likely to be integrated into the business of an organization, with the therapist paid at a professional wage level. Health-care systems are demanding more specific health outcomes in order to pay for the coverage of services. Horticultural therapy practitioners must support claims of efficacy with peer-reviewed research in order for consideration in current health-care legislation and public policy. It behooves the therapist to be familiar with research, communicate it to stakeholders, use it to inform practice, and seek opportunities to contribute to the body of knowledge of horticultural therapy through research of any type that is feasible within the scope of his or her work and connections (Relf 2016).

References and research that are applicable to horticultural therapy practice appear throughout this book. Chapter 14 also describes the use of research in evidence-based practice.

Growing the profession

The effective use of horticulture for therapy, rehabilitation, and wellness has great potential to grow and thrive for decades to come. Approaches and practices vary widely across program types, populations served, service delivery systems, and countries. This diversity contributes to the strength of the profession and its ability to respond to future needs and possibilities. Some of the actions that may help the profession to grow include broadening the populations, both underserved and unrealized, as well as working to improve employment opportunities.

Broadening the populations served

One way to grow the profession is to expand the audiences that are served by horticultural therapy. This may happen organically as individual practitioners propose and establish new and innovative programs to meet the needs of currently underserved populations. For example, many people are displaced annually due to catastrophic weather events, war, famine, and political actions. In developed countries, obesity rates continue to climb. Many people are left out of the mainstream of societies due to class, ethnicity, income, abilities, and age. Starling (2016) suggests that additional opportunities might exist related to broader community concerns, such as food sustainability or unemployment. Urbanization creates an imperative for the provision of purposeful natural connections. It is possible for horticultural therapy to serve in any or all of these areas. To date, horticultural therapy has worked rather quietly through individual programs that offer important services to targeted clients and communities. International initiatives could spur the creation of large-scale responses that include horticultural therapy. It may be advantageous to think globally and devise strategies for using horticultural therapy on a broad scale while also promoting programming at the local level.

Connecting with related green initiatives and movements will also broaden program types and people served. Urban horticulture, green roofs, measures that connect children and nature (such as school gardens), growing and eating local food, nature-based therapies (such as care farms), as well as health-care design that offers access to gardens are all approaches with which horticultural therapy may collaborate. Horticultural therapy has the flexibility to apply professional practices and standards in a wide variety of settings and circumstances; thus, it does not jeopardize its professional image by broadening its scope of practice.

Another way to expand the audiences reached includes continuing the efforts to improve the credibility of horticultural therapy and the techniques of the practice itself. As previously discussed, research is essential for reputation: It provides theoretical support and informs evidence-based practice. It is imperative that the profession promotes and raises the level of support for high-quality interdisciplinary research and its dissemination internationally. Communicating those findings is essential for improving the effectiveness and outcomes of programming.

Employment and pay

Broadening the audiences served is important in order to increase opportunities for employment and professional pay levels, buy equally essential for the development of any human service or health-care profession is influencing those who pay for services. This requires strategic interactions with insurance companies and government agencies in order to gain broad financial support for horticultural therapy. Having a solid body of knowledge through research is key to gaining that support. Organizations may cooperate with practitioners as well as insurance companies and agencies to identify research needs and priorities, seek funding to support that research, and use it to lobby for institutionalized financial support for horticultural therapy programs.

Summary

This chapter identified many of the universal assets and issues that can affect the future of horticultural therapy and offered some considerations and actions for the practitioner. The intention is to encourage the horticultural therapist to take advantage of the beneficial aspects to propose, develop, fund, deliver, sustain, and communicate the benefits of well-regarded programs. Issues and obstacles are common in any profession, particularly during the developmental stages. Many of those matters directly affect employment of horticultural therapists and the lifespan of programs. Awareness of these issues will help the professional to plan deliberately and strategically to overcome them. As horticultural therapy continues to spread around the world and expand in those areas where it is now practiced, additional hurdles will be inevitable, but opportunities will also be presented because whenever and wherever there are challenges, there are also possibilities for professional reflection, collaboration, and eventual improvements. In the future, as people seek more natural environments for healing, working, and playing, horticultural therapy will play an increasingly important role in human health and well-being.

References

American Horticultural Therapy Association. 2015a. "Code of Professional Ethics for Horticultural Therapists." Last modified March 13, 2015. https://ahta.memberclicks.net/assets/docs/ahta%20code%20of%20ethics.pdf.

American Horticultural Therapy Association. 2015b. "Standards of Practice for Horticultural Therapy." Last modified March 13, 2015. https://ahta.memberclicks.net/assets/docs/ahta_standards_of_practice.pdf.

American Horticultural Therapy Association. 2017. "AHTA Professional Registration Policies and Procedures." https://ahta.memberclicks.net/assets/docs/Professional_Registration/2017_ahta_professional_registration_applicant_policies_and_procedures.pdf.

Austin, D. R. 2013. *Therapeutic Recreation Processes and Techniques: Evidenced-Based Recreational Therapy*, 7th ed. Urbana, IL: Sagamore Publishing.

Bragg, R., and G. Atkins. 2016. "A Review of Nature-Based Intervention for Mental Health Care." Natural England Commissioned Reports, no. 204.

Buettner, L., and N. Richeson. 2012. "ATRA Research Agenda Summary," *Annual in Therapeutic Recreation* 20: 68–60.

Catlin, P. 2017. "Activity Planning: Developing Horticultural Therapy Activities and Tasks." In *Horticultural Therapy Methods: Connecting People and Plants in Health Care, Human Services, and Therapeutic Programs*, ed. R. L. Haller, and C. L. Capra. Boca Raton, FL: CRC Press, Taylor & Francis Group, pp. 37–62.

Chambers, N., and P. N. Williams. 1992. "Developing a New Computer-Accessed Data Base for Horticultural Therapy Research at Rusk Institute," In *The Role of Horticulture in Human Well-Being and Social Development*, ed. D. Relf. Portland, OR: Timber Press.

Diehl, E. R. M. 2012. "Insights from the Editor of the Journal of Therapeutic Horticulture: My Top Ten List," *Acta Horticulturae* 954: 33–49.

Frumkin, H. 2004. "White Coats, Green Plants: Clinical Epidemiology Meets Horticulture," *Acta Horticulturae* 639: 15–25.

Fung, C. Y. Y., and E. T. Y. Shum. 2012. "Development of Horticultural Therapy in Hong Kong," *Acta Horticulturae* 954: 169–173.

Gallis, C., ed. 2013. *Green Care for Human Therapy, Social Innovation, Rural Economy, and Education.* New York: Nova Science Publishers.

Haller, R. L. 2016. "Impact of a Horticultural Therapy Certificate Program on Students and Their Professional Activity." Presentation at the XIII International People Plant Symposium, Montevideo, Uruguay, November 10–12.

Haller, R. L. 2017. "Goals and Treatment Planning: The Process." In *Horticultural Therapy Methods: Connecting People and Plants in Health Care, Human Services, and Therapeutic Programs*, edited by R. Haller, and C. Capra. Boca Raton, FL: CRC Press, Taylor & Francis Group, pp. 37–62.

Haller, R. L., and C. L. Capra, eds. 2017. *Horticultural Therapy Methods: Connecting People and Plants in Health Care, Human Services, and Therapeutic Programs.* Boca Raton, FL: CRC Press, Taylor & Francis Group.

Hazen, T. 2014. "Horticultural Therapy and Healthcare Garden Design." In *Therapeutic Landscapes: An Evidence-Based Approach to Designing Healing Gardens and Restorative Outdoor Spaces*, ed. C. C. Marcus, and N. A. Sachs. Hoboken, NJ: John Wiley & Sons.

Im, E.-A., S.-A. Park, and K.-C. Son. 2018. "Developing Evaluation Scales for Horticultural Therapy," *Complementary Therapies in Medicine* 37: 29–36.

Jarrott, S., and C. Gigliotti. 2010. "Comparing Responses to Horticulture-Based and Traditional Activities in Dementia Care Programs," *American Journal of Alzheimer's Disease and Other Dementias* 25(8): 657–665.

Kaplan, R., and S. Kaplan. 1989. *The Experience of Nature: A Psychological Perspective.* New York: Cambridge University Press.

Kasar, J., and N. Clark. 2000. *Developing Professional Behaviors.* Thorofare, NJ: Slack Incorporated.

Katcher, B. L., and A. Snyder. 2010. *An Insider's Guide to Building a Successful Consulting Practice.* New York: AMACOM.

Kennedy, K. 2017. "Planning Horticultural Therapy Treatment Sessions." In *Horticultural Therapy Methods: Connecting People and Plants in Health Care, Human Services, and Therapeutic Programs*, ed. R. Haller, and C. Capra. Boca Raton, FL: CRC Press, Taylor & Francis Group, pp. 37–62.

Kent, A., T. Ruffini, and S. N. Goodyear. 2012. "The Status of Horticultural Therapy in Canada: Practice, Research, Education," *Acta Horticulturae* 954: 165–168.

Koura, S. 2012. "The Introduction of the Japanese Horticultural Therapy Association," *Acta Horticulturae* 954: 175–177.

Lewis, C. A. 1996. *Green Nature Human Nature: The Meaning of Plants in Our Lives.* Urbana, IL: University of Illinois Press.

Li, Q., et al. 2011. "Acute Effects of Walking in Forest Environments on Cardiovascular and Metabolic Parameters," *European Journal of Applied Physiology* 111(11): 2845–2853

Louv, R. 2005. *Last Child in the Woods: Saving Our Children from Nature-Deficit Disorder.* Chapel Hill, NC: Algonquin Books.

Majuri, C. E. 2002. "Proposing Horticultural Therapy Programs in Residential of Day Treatment Facilities," *Journal of Therapeutic Horticulture* 13: 82–89.

Marcus, C. C., and M. Barnes. 1999. *Healing Gardens: Therapeutic Benefits and Design Recommendations.* New York: John Wiley and Sons.

Marcus, C. C., and N. A. Sachs. 2014. *Therapeutic Landscapes: An Evidence-Based Approach to Designing Healing Gardens and Restorative Outdoor Spaces*, Hoboken, NJ: John Wiley & Sons.

McKee, B. 2004. "Why Do We Need Research?" (An edited transcript of the paper given at LIRG, in July 2003, as part of the CILIP Umbrella 2003 Conference), *Library and Information Research* 28(88): 3–12.

"Melwood Prevocational Evaluation Form," In *The Melwood Manual: A Planning and Operations Manual for Horticultural Training and Work Co-op Programs*, Upper Marlboro, MD, 1980.

Neuberger, K. 2012. "The Status of Horticultural Therapy around the World: Practice, Research, Education - Austria, Germany, Switzerland," *Acta Horticulturae* 954: 187–189.

Olszowy, D. 1978. *Horticulture for the Disabled and Disadvantaged.* Springfield, IL: Charles C. Thomas.

Orr, B., R. Mattson, N. Chambers, and M. Wichrowski. 2004. "Factors Affecting Choice of Horticultural Therapy at the Rusk Institute of Rehabilitation Medicine," *Journal of Therapeutic Horticulture* 15: 6–14.

Park, S. A., K. C. Son, and W. K Cho. 2012. "Practice of Horticultural Therapy in South Korea," *Acta Horticulturae* 954: 179–185.

Relf, D. 2012. "Advancing Horticultural Therapy through Research and Publishing," *Acta Horticulturae* 954: 13–20.

Relf, P. D. 2016. "Moving into the Future: Understanding Horticultural Therapy and Evidence-Based Practice," Presentation at the XIII International People Plant Symposium, Montevideo, Uruguay, November 10–12.

Sempik, J., J. Aldridge, and S. Becker. 2003. *Social and Therapeutic Horticulture: Evidence and Messages from Research*. Leicestershire, Thrive and CCFR.

Shoemaker, C. 2016. "The Role of Volunteers in Horticultural Therapy." Presentation at the XIII International People Plant Symposium, Montevideo, Uruguay, November 12.

Shoemaker, C. A., and E. R. M. Diehl. 2012. "The Practice and Profession of Horticultural Therapy in the United States," *Acta Horticulturae* 954: 161.

Soga, M., J. Gaston, and Y. Yamaura. 2017. "Gardening Is Beneficial for Health: A Meta-analysis," *Preventive Medicine Reports* 5: 92. doi:10.1016/j.pmedr.201.11.007.

Starling, L. A. 2016. "Horticultural Therapy with Special Populations" In *Urban Horticulture*, ed. T. M. Waliczek, and J. M. Zajicek. Boca Raton, FL: CRC Press, Taylor & Francis Group, p. 7.

Truong, M., Y. Paradies, and N. Priest. 2014. "Interventions to Improve Cultural Competency in Healthcare: A Systematic Review of Reviews," *BMC Health Services Research* 14: 99. doi:10.1186/1472-6963-14-99P.

United States Census Bureau Newsroom Archive. 2011. "2010 Census Shows America's Diversity." Last modified March 24, 2011. https://www.census.gov/newsroom/releases/archives/2010_census/cb11-cn125.html.

US Department of Agriculture, Forest Service. 2018. Urban Nature for Human Health and Well-being: A Research Summary for Communicating the Health Benefits of Urban Trees and Green Space. FS-1096. Washington, DC: US Department of Agriculture, Forest Service, p. 24.

Waliczek, T. M., and J. M. Zajicek, eds. 2016. *Urban Horticulture*. Boca Raton, FL: CRC Press, Taylor & Francis Group.

Williams, F. 2017. *The Nature Fix: Why Nature Makes Us Happier, Healthier and More Creative*. New York: W.W. Norton & Company.

Winterbottom, D., and A. Wagenfeld. 2015. *Therapeutic Gardens: Design for Healing Spaces*. Portland, OR: Timber Press.

section two

Theories supporting horticultural therapy efficacy and practice

Section two explores key theories that are relevant to horticultural therapy. The chapters in this section describe an array of principles, concepts, and ideology to form a basis for effective practice, as well as cornerstones of thought about why and how horticultural therapy has a positive impact on program participants. As another foundational section of the text, it is designed to provide perspective to new and experienced professionals in horticultural therapy and related fields.

chapter five

People–plant response
Theoretical support for horticultural therapy

Beverly J. Brown

Contents

Many gardeners note that they feel better after planting, weeding, or generally tending their green charges. Charles Darwin, the renowned scientist and developer of the theory of evolution, made thoughtful walks in green areas near his home a twice-daily ritual and spoke highly of the benefits of walks in nature (Darwin 1887; Young 2017). J.R.R. Tolkien—whose literary works introduce us to hobbits, ents, and orcs—valued connecting with nature so highly that he intentionally created rich narratives of plant life in his books to help readers connect to nature (Judd and Judd 2017). Temple Grandin, an expert on livestock handling and an advocate for those on the autism spectrum, attributes playing outside as a key factor in her development (Grandin 2017). These individuals have discovered for themselves what researchers in biology, psychology, and other disciplines have found: There is a compelling case for the physical, cognitive, emotional, social, intellectual, and even spiritual benefits of being in and around nature or natural elements (Stevens 2010; Lin 2013).

This chapter introduces a series of events in human evolution and development, which helps to explain why humans may benefit from being in natural settings and working with natural materials, especially plants. It examines several theories and therapeutic approaches, as well as the scientific evidence for their efficacy. Specifically, with regard to evolution, there are three primary aspects of early human evolution that horticultural therapy draws upon: evolving in a life-giving and life-threatening natural environment; maintaining a pre-industrial, pre-technology pace of life; and needing to be part of a group to survive. The first two are discussed in this chapter. For a discussion of the third aspect, see Exhibit 7.3 in Chapter 7. Finally, this chapter presents ways that horticultural therapists can increase the efficacy of their practice by understanding supporting theories and research and by applying these insights to their work.

Being in nature yields diverse benefits

Horticultural therapy research often finds that more than one of the categories mentioned above is positively affected during an activity. A study of four groups with individuals diagnosed with dementia found that both engagement and affect improved in horticultural therapy activities compared to traditional adult day service activities (Gigliotti and Jarrott 2005). Park et al. (2017) investigated the impact of working with and without plants on physiology and psychological relaxation in college-age males. They found that after three minutes of interactions with plants, physiological and psychological relaxation improved compared to the same type of interaction, but without plants. Park et al. (2016) investigated the impact of gardening on fifty women over the age of seventy during a fifteen-session program and found that women who participated in the program significantly reduced their waist size, improved muscle mass, increased aerobic endurance, and increased hand dexterity compared to the control group.

Evolution and human affinity for nature

In the past, hunter-gatherer societies have been considered primitive, with some level of derision inherent in the term (Kaplan 2000; Lancaster 2017). Interpretations of hunter-gatherer life entered a new realm in the 1960s when researchers made claims of hunter-gatherers spending few hours on foraging and hunting and therefore having increased time to socialize and build community. The first to take this view, Sahlins (1998) called hunter-gatherers "The Affluent Society." However, our concept of hunter-gatherers has been almost completely revised (Panter-Brick et al. 2001; Mayor 2012). Closer evaluation of research reveals a wide range in the time hunter-gatherers invest in survival (Panter-Brick et al. 2001). Either way that hunter-gatherers are regarded, there is one central view that most researchers agree on: The earliest hunter-gatherers lived in close association with nature, in relatively small groups, and at the pace of nature (Panter-Brick et al. 2001).

Human origins began on the savannahs in Africa (Orians and Heerwagen 1995). To survive, humans needed to be keen observers of their environments and work in groups rather than as individuals (Orians and Heerwagen 1995; Mayor 2012). Careful observation could lead to better food and medicine quality. Being part of a group would allow the strengths of the group to benefit group members, while the weaknesses of individuals need not affect anyone. Thousands of years living in these conditions would have selected for characteristics that helped individuals survive (Orians and Heerwagen 1995; Appleton 1996) and that subsequently would be carried in their DNA. Once preserved in DNA, these characteristics would be passed onto future generations (Piffer 2013).

Today humans live in many habitats, most of which are quite removed from nature compared to our ancestors' lifestyle. In 2018, 55% of the world's population lived in urban areas. By 2050, experts predict that 68% of the world's population will live in urban areas (United Nations Department of Economic and Social Affairs [UNDESA] 2018). As a species, human brains are programmed for a life in nature. Separation from nature could affect responses to current life events, even though individual humans cannot consciously remember details of their hunter-gatherer heritage. Present-day humans, across multiple cultures, prefer nature scenes over urban scenes (Wypijewski 1997; Corazon et al. 2010; Moreno 2017) and associate nature with relaxation and calm (Ulrich 1986). This makes evolutionary sense. Individuals who survived in the early conditions that hunter-gatherers experienced would have passed down the genes that helped the hunter-gatherers survive to descendants in the modern era (Figure 5.1).

Figure 5.1 Humans prefer nature scenes. (Courtesy of Rebecca Haller.)

Evolution defined

Evolution is a term that has taken on many meanings depending on the context. In this chapter, the term is used in its original, Darwinian sense (Darwin 1950). Darwin proposed that populations—a population being a group of organisms belonging to one species within a defined geographical area—reproduce at exponential rates, potentially creating incredibly large populations. However, resources needed for survival are limited, leading to competition. Individuals who have genetically determined capabilities allowing them preferential access to resources survive, while other individuals die. Eventually, the gene pool of the population changes to reflect the best adaptations for survival within the population. This process is referred to as natural selection and is well supported by research (Milot et al. 2011; Kirchengast 2014).

Let's follow this line of thinking: Since humans evolved in nature, nature being the environment of evolutionary adaptedness (Bowlby 1982) or the adaptively relevant environment (Irons 1998) rather than a conglomeration of buildings, concrete, and asphalt, humans should possess a strong affinity for natural settings. Being away from these natural elements would create a mismatch between the setting in which the species evolved and the setting the species now finds itself (Gluckman and Hanson 2006). This could potentially create a steady state of anxiety. Persisting levels of anxiety might not be perceptible to individuals, but their physical body would live in a sense of high alert (Kirchengast 2014). While humans as a species are well adapted to respond with quick bursts of energy, which would be needed when running from a predator, they are not well adapted to long-term levels of stress. Living in a state of stress, even though imperceptible to the individual, could reduce immune response, affect cognition, and reduce overall health (Khanfer et al. 2012; Toussaint et al. 2016; Schutte et al. 2017).

Spending time in nature calms

The calming effect of nature has been noted for thousands of years. Ancient Egyptian healers advised Pharaohs to walk in gardens for their health (Davis 1998). In recent history, sanitariums around the world provided large green areas where individuals might regain health (Gerlach-Spriggs et al. 1998). Research supports reduced anxiety when:

- Humans are in nature (Bowler et al. 2010; Tyrvainen et al. 2014; Martyn and Brymer 2016).
- A natural view is available (Ulrich 1984; Ulrich et al. 1991; Shin 2007; Raanaas et al. 2012).
- Plants are present in the indoor environment (Lohr et al. 1996; Larsen et al. 1998; Dravigne et al. 2008).
- Images of nature are present (Nanda et al. 2011; Brown et al. 2013).

In hospitals and medical settings, patients with natural views heal more quickly, have fewer complications, and require less medication (Ulrich 1984; Ulrich et al. 1991). Those taking a walk through green areas, compared to a walk through urban areas with little greenery, had lower physiological and/or psychological stress (Tyrvainen et al. 2014; Lanki et al. 2017). Not all studies are conclusive (Bowler et al. 2010; Dosen and Ostwald 2016), and more studies are needed. However, the bulk of research points to the important role that being in nature plays in promoting psychological and physical health (Figure 5.2).

Figure 5.2 Walking in green areas may reduce stress. (Courtesy of Rebecca Haller.)

Prospect–refuge theory

Appleton (1996) proposed the Prospect–Refuge Theory based on the understanding of human evolution already described in this chapter. His primary interest was in developing design guidelines for urban development, but the work clearly relates to horticultural therapy. Humans express higher levels of comfort when they have trees or high bushes around them, especially to their back and sides (Kaplan 1985; Appleton 1996; Gerlach-Spriggs et al. 1998; Marcus and Sachs 2014). Being surrounded by nature equates to a "refuge" or place of safety. From this refuge, a long view to an open area (prospect), with a few trees, a few reminders of friendly humans, some peaceful animals, and clear water (Ulrich et al. 1991; Appleton 1996) is regarded as peaceful and calming. Ulrich et al. (1991) tested recovery from a stressful event using subject self-ratings of effect and multiple physiological measures. They found that recovery was quicker and more complete after subjects viewed natural scenes compared to urban scenes (Stress Recovery Theory). Low cortisol levels, which occur when an individual is calm and relaxed (Miller et al. 2007; Gladwell et al. 2013), help induce a positive mental state and openness to change. Focusing on any reminder of this green refuge establishes a setting where growth and change are more likely to occur. Research supports a human preference for natural scenes or views over urban scenes, and (slight) variability in temperature and wind, rather than a static environmental state (summarized in Ryan et al. 2014). Individuals may be more comfortable at different levels along the prospect–refuge continuum. Those who are depressed, sad, or feeling at risk may prefer to be in a high-refuge area. Those who are feeling confident and exuberant, or who have little experience of harm may prefer an area with a higher level of prospect (Kaplan and Kaplan 1989) (Figure 5.3).

Figure 5.3 The design of seating in this garden offers enclosure and a sense of refuge. (Courtesy of Rebecca Haller.)

Attention restoration theory

While individuals may feel mentally tired following a day of work or at the end of a project, determining which aspect of the day contributed to tiredness can be baffling. Kaplan and Kaplan (1989) drew on the work of James (1892) on attention, identifying two types of attention. The first, involuntary attention, is relatively effortless, occurring when a person watches something interesting, does an activity that she or he loves, or anticipates an interesting event. Directed attention is, in many ways, the opposite. James (1892) proposed that directed attention is established when other stimuli are blocked. Willfully blocking input takes energy and is mentally tiring. This type of fatigue can result in irritability, poor decision making, and the inability to complete complex tasks or tasks that must be done in distracting environments (e.g., responding to an emergency or directing air traffic) (Faber et al. 2012).

Components of attention restoration

Kaplan and Kaplan (1989) incorporated evolutionary theory into their work on the restorative capabilities of natural settings, particularly with regard to restoring attention (Kaplan 1995, 2001). Directed attention would have allowed hunter-gatherers to focus on their environment, spot predators more quickly, and respond appropriately. Those without directed attention abilities would likely not have survived to pass on their genes.

While many activities can be an escape from extended, directed attention, activities that restore energy have four primary components: (1) *being away*, being removed from locations or reminders of activities that require substantial directed attention; (2) *extent*, the feeling of being in a totally novel world (mentally or physically);

(3) *fascination*, which triggers involuntary attention; and (4) *action and compatibility*, being in an environment where one feels comfortable and able to complete any necessary actions. While there are many locations where one or two of these factors could exist, in nature all four readily occur. Thus, time spent in nature using involuntary attention is a restorative experience for most individuals (Kaplan and Kaplan 1989). The publication of Kaplan and Kaplan's work provided a framework for research on why humans find nature appealing.

Insights for horticultural therapy practice

Horticultural therapists are advised to provide a setting that maximizes exposure to nature, elements of nature, views of nature, or at least images of nature. All of these facilitate stress reduction and allow for improved focus. It is also important to provide seating that allows individuals to choose where they would like to sit along the prospect–refuge continuum. This allows individuals to be as comfortable as possible. As sessions progress, it is also effective to encourage individuals to choose seats in different locations to increase their sense of fascination. This can create an ambiance of excitement about new activities or skill development. The more that individuals can use involuntary attention, the greater the feeling of restoration will be and the greater the amount of energy they will have for working toward their treatment goals (Figure 5.4). Exhibit 5.1 describes the restorative benefits of gardens and gardening with people who are incarcerated.

Figure 5.4   Gardens at Thrive in the United Kingdom are designed to accommodate a variety of individuals. (Courtesy of Rebecca Haller.)

EXHIBIT 5.1

Program Example: Corrections, New York, the United States

People who are in the criminal justice system benefit from the restorative environment created by horticultural therapy. At the GreenHouse program at the Rikers Island Correctional Facility, near New York City, we want to strengthen our participants in their entire being and support them as they prepare for their next steps, whether directly back into the community or, for our detainee groups, often into upstate prisons to serve sentences varying in length from one to thirty-five-plus years. Started as vocational training in 1986, the focus of GreenHouse shifted over time to horticultural therapy and each participant's psychosocial, physical, and cognitive well-being. All gardens, 0.5 to 3 acres, have sections for sensory stimulation; raised beds with edibles, which are prepared and consumed in the program; rose gardens; sitting areas; ponds; and woodlands.

Who are people who are incarcerated? They are people from many walks of life: children, mothers, fathers, grandparents, brothers, sisters. As we look for reasons that individuals are in jail or prison, the most common underlying factors include mental illness, prolonged and often early exposure to violence and abuse, childhood neglect and trauma, substance abuse, lack of education and opportunities, poverty, homelessness, and family history of incarceration. Recidivism is as high as 43%–51% for women and 55%–70% for men.

Horticultural therapy is only one piece of a puzzle. The GreenHouse program must partner with community-based organizations and agencies for support with substance abuse issues, housing, mental health services, and more. But horticultural therapy provides very specific opportunities in a jail setting. Participants often say, "In the garden, I forget that I'm in jail. I feel free," and "Working outside helps take my mind off my case. I feel less depressed," and, "I am happy to go home, but sad to leave the garden," and, "I'm treated like a person here. This is my sanctuary." These statements are not measurable outcomes of specific activities, nor are they predictors of a person's successful reentry to the world outside the jail. However, they express the impact that horticultural therapy has on the person.

Being in the garden provides a chance to begin healing, reconsider choices, come to terms with things lost, and gain strength and courage. Participants experience plants thriving under their care and discover that they can create beauty in a harsh environment. Plants do not "judge" the caretaker. Often participants have lost contact with their families or failed to properly take care of someone in their charge. Through horticultural therapy, they learn that they are able to nourish and sustain something, without the pressure of failure, that the responsibility for human beings might bring. Each person's contributions are relevant, profound experiences in an environment that deliberately disempowers individuals. Gardens give second chances. A mistake is not the end of the world and most can be fixed, like many mistakes in life. When participants move on, they leave behind something of themselves as testimony of their presence and of their time spent in the garden. The memory of this space, and the positive impact they made on it, can bring back the feeling of comfort and peace that participants experience while in the program. They know that their presence mattered.

Hilda Mechthild Krus, HTR, MSW
Director GreenHouse and HT Programs, The Horticultural Society of New York

If being in the presence of nature is healing, what happens when individuals spend little or no time in nature? In 2005, Louv wrote at length about children who grow up without spending time in nature. He termed this condition nature deficit disorder. Louv contends that children raised with their attention primarily on electronics (televisions, cell phones, computers) have higher rates of illness, more mental health issues, and incomplete development (summarized succinctly in Natural Learning Institute 2012).

What children growing up away from nature are missing

In 2015, Chawla reported twenty-seven benefits to children when they have access to nature (Figure 5.5). The following list includes benefits when children play in or simply view nature:

- Greater positive health outcomes including increased exercise, improved nutrition, improved eyesight, reduced diabetes, fewer asthma cases, and increased resistance to infection (McCurdy et al. 2010; Wolch et al. 2011; Gladwell et al. 2013; French et al. 2013).
- Better development of social skills, increased attention spans, and improved self-discipline (Wells and Lekies 2006; Carrus et al. 2015).
- Children who have safe, green areas nearby that they can visit, or even marginal views of green areas, show increased self-discipline, especially for girls. Reduced stress leads to better decision making (Taylor et al. 2001).
- Improved academic performance. For instance, those diagnosed with attention deficit disorder show longer periods of attention after being in nature (Taylor et al. 2001; Carrus et al. 2015).
- Improved problem solving (Taylor et al. 2001) and greater creativity (Louv 2011).

Figure 5.5 Natural playgrounds offer children a nature connection. (Courtesy of Rebecca Haller.)

- Increased appreciation and concern for the environment, including increased entry into careers related to the environment (Wells and Lekies 2006; Catchpole and Catchpole 2012).
- Improved motor fitness, balance, and coordination with play in nature (Fjørtft 2004).

What adults living away from nature are missing

As Louv (2011) points out, nature deficit disorder does not just occur in children. Many adults are also spending less time in natural settings and could benefit from viewing or being in nature. These benefits parallel the list for children, with the addition of the following:

- Improved outcomes after surgeries (Ulrich 1984, 2002; Ulrich et al. 1991; Raanaas et al. 2012).
- Reduced stress, increased productivity, or both, in work environments where plants are present (Bringslimark et al. 2007, 2009; Raanaas et al. 2011), natural views are readily available (Kaplan 2001; Raanaas et al. 2011), or images of nature are present (may only be relevant for males) (Kweon et al. 2008).
- Increased vitality (feeling positive and energized as discussed by Ryan et al. 2010) after engaging in physical activity in nature (Korpela et al. 2008).
- Improved and maintained cognition, mood, and physical mobility when natural elements are included in the daily environment. These benefits are especially notable when programming, for example, in a nursing home involves individuals in direct interactions with plants (Chen and Ji 2015; Yao and Chen 2017).

Biophilia

In the early 1960s, concern for the environment was rising, in large part due to the publication of Rachel Carson's (1962) *Silent Spring*. The impact of Carson's work brought environmental issues to public consciousness. Erich Fromm (1973) wrote a strongly worded indictment of humanity's disregard for the earth, which he attributes essentially to the rise of industrialization and urbanization. Fromm states that the biologically normal impulse for humans is to create and grow. He defined biophilia as "...the passionate love of life and of all that is alive; it is the wish to further growth, whether in a person, a plant, an idea, or a social group" (Fromm 1973). Drawing on Darwin's work on evolution (Darwin 1950), Fromm thought that humans needed to return to a nondestructive relationship with the earth (Gunderson 2014). Once humanity saw itself as part of nature rather than separate from it, respect and proper regard for the earth would be reestablished.

E. O. Wilson, noted scientist and activist for environmental sanity, resurrected and embellished the concept of biophilia (Kellert and Wilson 1993; Wilson 2009). Wilson's ultimate hope was similar to Fromm's: that individuals who understood their evolutionary roots and essential connection to the environment would cease destruction of the planet. With a keen, internalized understanding of our dependence on nature, we would no longer destroy species, communities, and ecosystems.

Ecopsychology

Ecopsychological approaches arose in the past forty years in a variety of contexts (Fisher and Abram 2012). Ecopsychology is still in the process of developing a synthesized definition (Sussman 2014). Concepts of ecopsychology appear to fall along a continuum ranging from recognizing that a human's health is deeply, even spiritually, connected to nature (Roszak et al. 1995; Roszak 2001; Buzzell and Chalquist 2009; Stevens 2010; Sussman 2014) and to nature's stresses—largely caused by humans—manifesting as our human illnesses (Conn 1995). Seeking to find essential, uniting tenets, Sussman (2014) has proposed the following five ideas in ecopsychological work:

- Ecopsychology proposes that contact with healthy nature contributes to human physical and psychological well-being (Chalquist 2009; Kahn and Hasbach 2013).
- Ecopsychology proposes that humans have a deep, reciprocal bond with nature and that this bond constitutes a primary psychological relationship (Clinebell 1996; Fisher and Abram 2012).
- Ecopsychology recognizes that the quality of humans' contact with nature and relationship with nature is rapidly declining worldwide (Roberts 1998; Kahn and Hasbach 2013).
- Ecopsychology seeks to address this decline by critically examining social systems, paradigms, and institutions that contribute to it and proposing alternative solutions (Roszak 1992; Fisher 2002).
- Ecopsychology maintains a commitment to social justice and ecological sustainability, with particular attention to the most vulnerable populations and ecosystems (Anthony and Soule 1998; Chalquist 2009; Smith 2013).

Ecotherapy

With the development of ecopsychology as a discipline, professional counselors can apply this approach to their own work, leading to the development of ecotherapy. Buzzell and Chalquist (2009) define ecotherapy as applied ecopsychology, with a simplicity that belies the gamut of approaches currently utilized in the field. The ultimate goal is to bring humans and nature into a reciprocally communicating relationship. Whether this journey to wholeness begins from the individual and moves to the environment, or vice versa, is of little consequence (Buzzell and Chalquist 2009) as long as it uses a person-centered approach (Tudor 2013). In addition to talk therapy, the practice of ecotherapy can include environmental activism, wilderness experiences, shamanistic practices, and rituals.

The diversity of ecotherapeutic approaches has led to wide-ranging studies of its efficacy. Chalquist (2009) lists research that documents economic benefits (homes built with location and sustainability in mind save money); increased productivity; problem solving and creativity with the addition of plants to a workplace; an increased sense of community (Van Herzele and de Vries 2012; Groenewegen et al. 2012); improved mental health, including within health care settings (Van Herzele and de Vries 2012; Groenewegen et al. 2012); improved quality of life for nursing home residents (Chalquist 2009); and positive effects on child development, including reduced life stress (Chalquist 2009).

Evolution and the pace of life

The pace of modern life is far different from that of ancient humans. Being able to cope with quick bouts of stress during a dangerous situation ("fight or flight") is physiologically quite different from being in a state of constant alert. At some point, modern humans—especially those in Western economies—began to feel that they needed to work harder and harder to get ahead. While it may appear that some hunter-gatherers took time for resting, researchers find that relative inactivity is pragmatic. Frequently, these periods are due to inclement weather or a disincentive of mandatory sharing. In these groups, the more one acquires, the more one is led to share. There comes a point where acquiring more food has little benefit to the individual or kin group (Kaplan 2000).

For modern humans, relaxation seems equally elusive

Promises of time-saving technology, allowing more time for relaxation, have not had the desired results. Instead, individuals seem driven to accomplish even more. Children's lives are programmed to take advantage of every possible class and sport. This short-circuits children's opportunities for unscripted play in nature, which is essential for their physical, mental, and social development (Fjørtft 2004; Wells and Lekies 2006; Louv 2008, 2011; Carrus et al. 2015; Chawla 2015). Introducing high levels of stress into children's lives is particularly troubling as stress/adversity in childhood can be a predictor of premature aging and early onset of disease (Puterman et al. 2016).

Adults program their own lives in the same way and still feel they are missing out because they can't accomplish everything on their to-do lists (Keller et al. 2012). Living at high speed challenges human physiological systems. Stress triggers a chemical cascade, with far-reaching effects. Elevated levels of stress chemicals for sustained periods of time can result in increased incidence of anxiety, depression, and heart disease; reduced immune system function; and increased chance of stroke (Esposito and Bianchi 2012) (Figure 5.6).

Figure 5.6 Strolling in a garden offers opportunity to build relationships and enjoy a slower pace. (Courtesy of Moss Cremer.)

Horticultural therapy as an antidote to stress

A hectic lifestyle typically continues until an individual or group consciously decides to step out of it. Horticultural therapy can be a tool to help individuals return to life at a human pace. Plants grow at the speed of life. Horticultural therapists and clients can work at the speed of life in return. Whether tending a garden or a potted plant, proceeding at lightning speed is likely to result in damaged stems and broken flowers. While speed may be necessary for commercial greenhouses or vocational training, it is not an advantage in most horticultural therapy sessions. Simply being around plants can reduce stress and improve or restore cognitive abilities (Raanaas et al. 2011; Berman et al. 2012). Working with plants and nature brings the body into harmony with the hunter-gatherer part of the brain. Even when plants languish with lack of care or succumb to the vagaries of weather or neglect, a clearer mind may help individuals face memories triggered by these smaller losses. A calming connection with plants can allow safe expression of emotions such as anger, grief, and sadness. These feelings can be processed with the help of a horticultural therapist (Figure 5.7).

Working with plants requires movement

Whether coaxing a recalcitrant arm to reach a bit further or tossing a layer of compost over a garden, tending plants and gardens means using muscles. Movement increases blood flow and other chemical cascades, which increase a sense of vitality (Korpela et al. 2017). Physical activity also mitigates stress, potentially protecting individuals from the negative effects of high cortisol levels (Puterman et al. 2011).

Figure 5.7 Connections with plants allow a space for safe expression of emotions. (Photo provided by Pacific Quest.)

A slower pace means more time to make friends

In group settings, horticultural therapists can help clients/patients take advantage of a slower pace by facilitating social interactions. Increasing social interactions, and the social network that can result, reduces the perceived threat of stressors, reducing stress and improving quality of life (Gerich 2014).

The people–plant connection and maximizing horticultural therapy efficacy

The first step is to bring awareness of the people–plant connection to mind while designing horticultural therapy programs or planning and carrying out horticultural therapy interventions. Research indicates that humans have a deep longing for respite from the vagaries of modern life (Clinebell 1996; Puhakka 2014). Horticultural therapy is uniquely positioned to address these needs because, at their center, the need is for greater contact with the natural world. The following are some suggestions for incorporating the theories discussed in this chapter into a horticultural therapy practice. Therapists can:

1. Arrange a simple environment with only the tools and materials needed for the day. Avoid clutter. Include nature in any way possible, including views of nature; natural items such as cones, flowers, rocks, and soil; and even images of nature.
2. Remember that natural sounds and mildly varying environmental attributes are calming. If inside, try to replicate the sounds, breezes, scents, and general atmosphere of a natural environment. Having one of these characteristics in place is good; more than one is better.
3. Establish spaces along the prospect–refuge continuum, whether inside or out, so that each group member can be comfortable. If appropriate for the group, explain Prospect–Refuge Theory to group members and encourage them to change their seat at each meeting according to how they are feeling at that time.
4. Work with clients/patients to find their previous connections to plants and nature. Bring these connections to the fore as activities are planned and carried out. Providing activities that move individuals from directed attention to involuntary attention will be less tiring and more invigorating for those attending the session.
5. Help others slow down and savor their natural environment. Working at the pace of nature reduces the mismatch between the conditions in which humans evolved and the conditions in which most (especially those in Western societies) live in.
6. Focus on where clients/patients are and their therapy goals. If an individual or group does not see the value in working with plants, consider taking a nature-based approach. Talk with the individual(s) about their prior outdoor experiences. Begin by taking very small steps in or around natural settings. Even if they say they don't like nature, their physiology will still respond.
7. Celebrate completed tasks, even small ones, and each new skill acquired. This is not the time to focus on high production, unless that is one of their therapy goals (for example, in vocational training). Meaningful compliments help move to or help keep clients/patients in involuntary attention and therefore provide a richer, more relaxing experience for them.
8. Work in groups and help individuals make connections with others in the group. This increases the healing potential of a horticultural therapy activity.

9. Remember that seeing a plant or other natural item, viewing nature, or being in nature is healing; even seeing a picture of nature has a relaxing effect. Just getting someone to look at a plant, a painting of a natural scene, or out the window at greenery is a step toward assuaging the human need to connect to nature.
10. Remember to apply all of the above when proposing a program. Take small plants and draw open the blinds in meeting rooms. Integrate natural images into the proposal. Be businesslike, and move at a confident, calm pace.
11. Develop an elevator speech that includes the reasons why nurturing our people–plant connection is important. Just about everyone would like to improve their quality of life. Let them know that horticultural therapy can help!

Summary

Many horticultural therapists have observed and personally experienced the profound improvements in quality of life that participating in horticultural therapy activities can bring. When co-treating with a professional from another healing profession, those professionals often comment on how much more quickly improvements occur. There are sound, research-supported reasons for these observations. Consciously using theory-based practice to establish or reestablish the people–plant connection deepens the experience for the horticultural therapist and client/patient alike. More research will provide a greater understanding of the theories and mechanisms that support the efficacy of horticultural therapy. As this new information is incorporated into horticultural therapy practice, horticultural therapists will be able to facilitate even stronger people–plant connections.

Key terms

Co-treat To work with one or more health professionals to treat a living being for relief from physical or mental illness, or to increase general well-being
Nature An area that persists without or in spite of human intervention and recalls at some level the human connection with ecological systems establishing in our hunter-gatherer past
People–plant connection A feeling of belonging with elements of nature, particularly plants, established in our hunter-gatherer ancestors and genetically transmitted to all humans

For more information

Attachment theory

Schwartz, J. 2015. "The Unacknowledged History of John Bowlby's Attachment Theory." *British Journal of Psychotherapy* 31(2): 251–266.

Creating places for nature-play for children

Children and Nature Network, www.childrenandnature.org
Natural Learning Initiative, www.naturalearning.org

Environmental psychology

Steg, L., A. E. van den Berg, and J. I. M. de Groot. 2013. *Environmental Psychology, An Introduction,* Chichester, UK: John Wiley & Sons.

References

Anthony, C., and R. Soule. 1998. "A Multicultural Approach to Ecopsychology." *The Humanistic Psychologist* 26(1–3): 155–161. doi:10.1080/08873267.1998.9976970.

Appleton, J. 1996. *The Experience of Landscape*, Revised ed. Chichester, UK: John Wiley & Sons.

Berman, M. G., E. Kross, K. M. Krpan, et al. 2012. "Interacting with Nature Improves Cognition and Affect for Individuals with Depression." *Journal of Affective Disorders* 140(3): 300–305. doi:10.1016/j.jad.2012.03.012.

Bowlby, J. 1982. *Attachment*, 2nd ed. New York: Basic Books.

Bowler, D. E., L. Buyung-Ali, T. M. Knight, and A. S. Pullin. 2010. "A Systematic Review of Evidence for the Added Benefits to Health of Exposure to Natural Environments." (Research Article) (Clinical Report). *BMC Public Health* 10: 456.

Bringslimark, T., T. Hartig, and G. G. Patil. 2007. "Psychological Benefits of Indoor Plants in Workplaces: Putting Experimental Results into Context." *HortScience: A Publication of the American Society for Horticultural Science* 42(3): 581–587.

Bringslimark, T., T. Hartig, and G. G. Patil. 2009. "The Psychological Benefits of Indoor Plants: A Critical Review of the Experimental Literature." *Journal of Environmental Psychology* 29(4): 422–433. doi:10.1016/j.jenvp.2009.05.001.

Brown, D. K., J. L. Barton, and V. F. Gladwell. 2013. "Viewing Nature Scenes Positively Affects Recovery of Autonomic Function Following Acute-Mental Stress." *Environmental Science & Technology* 47(11): 5562–5569. doi:10.1021/es305019p.

Buzzell, L., and C. Chalquist. 2009. *Ecotherapy: Healing with Nature in Mind*. San Francisco, CA: Sierra Club Books.

Carrus, G., Y. Passiatore, S. Pirchio, and M. Scopelliti. 2015. "Contact with Nature in Educational Settings might Help Cognitive Functioning and Promote Positive Social Behaviour/El Contacto Con La Naturaleza En Los Contextos Educativos Podra Mejorar El Funcionamiento Cognitivo Y Fomentar El Comportamiento Social Positivo." *Psyecology* 6(2): 1–22. doi:10.1080/2 1711976.2015.1026079.

Carson, R. 1962. *Silent Spring*. Boston, MA: Houghton Mifflin.

Catchpole, T., and K. Catchpole. 2012. "Nature Play." *Parks & Recreation* 47(4): 10–13.

Chalquist, C. 2009. "GARDEN as Therapist and Community Organizer." *Communities Winter* 145: 22–23.

Chawla, L. 2015. "Benefits of Nature Contact for Children." *Journal of Planning Literature* 30(4): 433–452. doi:10.1177/0885412215595441.

Chen, Y., and J. Ji. 2015. "Effects of Horticultural Therapy on Psychosocial Health in Older Nursing Home Residents: A Preliminary Study." *Journal of Nursing Research* 23(3): 167–171. doi:10.1097/jnr.0000000000000063.

Clinebell, H. 1996. *Ecotherapy: Healing Ourselves, Healing the Earth*. Minneapolis, MN: Fortress Press.

Conn, S. A. 1995. "When the Earth Hurts, Who Responds?" In *Ecopsychology, Restoring the Earth, Healing the Mind*, ed. T. Roszak, M. E. Gomes, and A. D. Kanner, 156–171. San Francisco, CA: Sierra Club Books.

Corazon, S. S., U. K. Stigsdotter, A. G. C. Jensen, and K. Nilsson. 2010. "Development of the Nature-Based Therapy Concept for Patients with Stress-Related Illness at the Danish Healing Forest Garden Nazadia." *Journal of Therapeutic Horticulture* 20: 34–51.

Darwin, C. 1887. *The Life and Letters of Charles Darwin*, Vol. 1. London, UK: John Murray.

Darwin, C. 1950. *The Origin of Species and the Descent of Man*. New York: The Modern Library.

Davis, S. 1998. "Development of the Profession of Horticultural Therapy." In *Horticulture as Therapy: Principles and Practice*, ed. S. Simson, and M. Straus, 4. Binghamton, NY: Haworth Press.

Dosen, A., and M. Ostwald. 2016. "Evidence for Prospect–Refuge Theory: A Meta-Analysis of the Findings of Environmental Preference Research." *City, Territory and Architecture* 3(1): 1–14. doi:10.1186/s40410-016-0033-1. https://search.proquest.com/docview/1786640263.

Dravigne, A., T. M. Waliczek, R. D. Lineberger, and J. M. Zajicek. 2008. "Effect of Live Plants and Window Views of Green Spaces on Employee Perceptions of Job Satisfaction." *HortScience: A Publication of the American Society for Horticultural Science* 43(1): 183–187.

Esposito, A., and V. Bianchi. 2012. *Cortisol Physiology, Regulation and Health Implications*. Hauppauge, NY: Nova Science Publishers.

Faber, L. G., N. M. Maurits, and M. M. Lorist. 2012. "Mental Fatigue Affects Visual Selective Attention." *PLoS One* 7(10): e48073. doi:10.1371/journal.pone.0048073.

Fisher, A. 2002. *Radical Ecopsychology: Psychology in the Service of Life*. Albany, NY: State University of New York Press.

Fisher, A., and D. Abram. 2012. *Radical Ecopsychology Psychology in the Service of Life*, 2nd ed. Albany, NY: State University of New York Press.

Fjørtft, I. 2004. "Landscape as Playscape: The Effects of Natural Environments on Children's Play and Motor Development." *Children Youth and Environments* 14(2): 21–44. doi:10.7721/chilyoutenvi.14.2.0021.

French, A. N., R. S. Ashby, I. G. Morgan, and K. A. Rose. 2013. "Time Outdoors and the Prevention of Myopia." *Experimental Eye Research* 114: 58–68. doi:10.1016/j.exer.2013.04.018.

Fromm, E. 1973. *The Anatomy of Human Destructiveness*. New York: Holt, Rinehart and Winston.

Gerich, J. 2014. "Effects of Social Networks on Health from a Stress Theoretical Perspective." *Social Indicators Research* 118(1): 349–364. doi:10.1007/s11205-013-0423-7.

Gerlach-Spriggs, N., R. E. Kaufman, and S. B. Warner. 1998. *Restorative Gardens: The Healing Landscape*. New Haven, CT: Yale University Press.

Gigliotti, C. M., and S. E. Jarrott. 2005. "Effects of Horticulture Therapy on Engagement and Affect." *Canadian Journal on Aging/Revue Canadienne Du Vieillissement* 24(4): 367–377. doi:10.1353/cja.2006.0008.

Gladwell, V. F., D. K. Brown, C. Wood, G. R. Sandercock, and J. L. Barton. 2013. "The Great Outdoors: How a Green Exercise Environment Can Benefit All" (Review). *Extreme Physiology & Medicine* 2: 3.

Gluckman, P. D., and M. A. Hanson. 2006. *Mismatch: Why Our World No Longer Fits Our Bodies*. Oxford, UK: Oxford University Press.

Grandin, T. 2017. "LISTEN: Temple Grandin, 2017 Women's Hall of Fame Inductee." last modified September 6, accessed September 27, 2017, http://interactive.wxxi.org/node/387433.

Groenewegen, P. P., D. B. Van, J. Maas, R. A. Verheij, and S. De Vries. 2012. "Is a Green Residential Environment Better for Health? If So, Why?" *Annals of the Association of American Geographers* 102(5): 996–1003. doi:10.1080/00045608.2012.674899.

Gunderson, R. 2014. "Erich Fromm's Ecological Messianism." *Humanity & Society* 38(2): 182–204. doi:10.1177/0160597614529112.

Irons, W. 1998. "Adaptively Relevant Environments Versus the Environment of Evolutionary Adaptedness." *Evolutionary Anthropology: Issues, News, and Reviews* 6(6): 194–204. doi:AID-EVAN2>3.0.CO;2-B.

James, W. 1892. *Psychology: Briefer Course*. New York: Henry Holt and Company. https://archive.org/stream/psychologybriefe00willuoft#page/n7/mode/2up.

Judd, W. S., and G. A. Judd. 2017. *Flora of Middle-Earth: Plants of J.R.R. Tolkien's Legendarium*. New York: Oxford University Press.

Kahn, P. H., and P. H. Hasbach. 2013. *The Rediscovery of the Wild*. Cambridge, MA: MIT Press.

Kaplan, D. 2000. "The Darker Side of the 'Original Affluent Society.'" *Journal of Anthropological Research* 56(3): 301–324. doi:10.1086/jar.56.3.3631086.

Kaplan, R. 1985. "Nature at the Doorstep: Residential Satisfaction and the Nearby Environment." *Journal of Architectural & Planning Research* 2(2): 115–127.

Kaplan, R. 2001. "The Nature of the View from Home: Psychological Benefits." *Environment and Behavior* 33(4): 507–542. doi:10.1177/00139160121973115.

Kaplan, R., and S. Kaplan. 1989. *The Experience of Nature: A Psychological Perspective*. New York: Cambridge University Press.

Kaplan, S. 1995. "The Restorative Benefits of Nature: Toward an Integrative Framework." *Journal of Environmental Psychology* 15(3): 169–182. doi:10.1016/0272-4944(95)90001-2.

Keller, A., K. Litzelman, L. E. Wisk, et al. 2012. "Does the Perception that Stress Affects Health Matter? The Association with Health and Mortality." *Health Psychology* 31(5): 677–684. doi:10.1037/a0026743.

Kellert, S. R., and E. O. Wilson. 1993. *The Biophilia Hypothesis*. Washington, DC: Island Press.

Khanfer, R., D. Carroll, J. M. Lord, and A. C. Phillips. 2012. "Reduced Neutrophil Superoxide Production among Healthy Older Adults in Response to Acute Psychological Stress." *International Journal of Psychophysiology* 86(3): 238–244. doi:10.1016/j.ijpsycho.2012.09.013.

Kirchengast, S. 2014. *Physical Inactivity from the Viewpoint of Evolutionary Medicine*, Vol. 2. Basel, Switzerland: MDPI AG. doi:10.3390/sports2020034.

Korpela, K., J. De Bloom, M. Sianoja, T. Pasanen, and U. Kinnunen. 2017. "Nature at Home and at Work: Naturally Good? Links between Window Views, Indoor Plants, Outdoor Activities and Employee Well-Being over One Year." *Landscape and Urban Planning* 160: 38–47.

Korpela, K. M., M. Yln, L. Tyrvinen, and H. Silvennoinen. 2008. "Determinants of Restorative Experiences in Everyday Favorite Places." *Health and Place* 14(4): 636–652. doi:10.1016/j.healthplace.2007.10.008.

Kweon, B., R. Ulrich, V. Walker, and L. Tassinary. 2008. "Anger and Stress. The Role of Landscape Posters in an Office Setting." *Environment & Behavior* 40(3): 355–381. doi:10.1177/0013916506298797.

Lancaster, J. 2017. "The Case against Civilization." *The New Yorker*, September 18, 2017. Accessed February 25, 2018, from https://www.newyorker.com/magazine/2017/09/18/the-case-against-civilization.

Lanki, T., T. Siponen, A. Ojala, et al. 2017. "Acute Effects of Visits to Urban Green Environments on Cardiovascular Physiology in Women: A Field Experiment." *Environmental Research* 159: 176–185. doi:10.1016/j.envres.2017.07.039.

Larsen, L., J. Adams, B. Deal, and E. Tyler. 1998. "Plants in the Workplace: The Effects of Plant Density on Productivity, Attitudes, and Perceptions." *Environment and Behavior* 30(3): 261–281. doi:10.1177/001391659803000301.

Lin, C. 2013. "A Review of Horticultural Therapy and Caregiver's Burden." *International Journal of Organizational Innovation* 5(4): 138. https://search.proquest.com/docview/1355887388.

Lohr, V. I., C. Pearson-Mims, and G. K. Goodwin. 1996. "Interior Plants May Improve Worker Productivity and Reduce Stress in a Windowless Environment." *Journal of Environmental Horticulture* 14(2): 97–100.

Louv, R. 2008. *Last Child in the Woods: Saving Our Children from Nature-Deficit Disorder*. New York: Algonquin Books.

Louv, R. 2011. *The Nature Principle: Human Restoration and the End of Nature-Deficit Disorder*, 1st ed. Chapel Hill, NC: Algonquin Books.

Marcus, C. C., and N. Sachs. 2014. *Therapeutic Landscapes: An Evidence-Based Approach to Designing Healing Gardens and Restorative Outdoor Spaces*. Hoboken, NJ: John Wiley & Sons.

Martyn, P., and E. Brymer. 2016. "The Relationship between Nature Relatedness and Anxiety." *Journal of Health Psychology* 21(7): 1436–1445. doi:10.1177/1359105314555169.

Mayor, T. 2012. "Hunter-Gatherers: The Original Libertarians." *Independent Review* 16(4): 485–500.

McCurdy, L. E., K. E. Winterbottom, S. S. Mehta, and J. R. Roberts. 2010. "Using Nature and Outdoor Activity to Improve Children's Health." *Current Problems in Pediatric and Adolescent Health Care* 40(5): 102–117. doi:10.1016/j.cppeds.2010.02.003.

Miller, G. E., E. Chen, and E. S. Zhou. 2007. "If It Goes Up, Must It Come Down? Chronic Stress and the Hypothalamic-Pituitary-Adrenocortical Axis in Humans." *Psychological Bulletin* 133(1): 25–45. doi:10.1037/0033-2909.133.1.25.

Milot, E., F. M. Mayer, D. H. Nussey, M. Boisvert, F. Pelletier, and D. Réale. 2011. "Evidence for Evolution in Response to Natural Selection in a Contemporary Human Population." *Proceedings of the National Academy of Sciences of the United States of America* 108(41): 17040. doi:10.1073/pnas.1104210108.

Moreno, J.M., B. Moura, W.S. Ferreira, Jr., T.C. da Silva, U.P. Albuquerque. 2017. "Landscapes preferences in the human species: insights for ethnobiology from evolutionary psychology." *Ethnobiology and Conservation* 6(10). doi:10.15451/ec2017-07-6.10-1-7.

Nanda, U., S. Eisen, R. S. Zadeh, and D. Owen. 2011. "Effect of Visual Art on Patient Anxiety and Agitation in a Mental Health Facility and Implications for the Business Case." *Journal of Psychiatric and Mental Health Nursing* 18(5): 386. doi:10.1111/j.1365-2850.2010.01682.x.

Natural Learning Institute. 2012. "Benefits of Connecting Children with Nature: Why Naturalize Outdoor Learning Environments." College of Design, North Carolina State University, last modified January, accessed October 8, 2017.

Orians, G. H., and J. H. Heerwagen. 1995. "Evolved Responses to Landscapes." In *The Adapted Mind*, ed. J. H. Barkow, L. Cosmides, and J. Tooby, pp. 555–579. Oxford, UK: Oxford University Press.

Panter-Brick, C., R. Layton, and P. Rowley-Conwy. 2001. *Hunter-Gatherers: An Interdisciplinary Perspective*. Cambridge, UK: Cambridge University Press.

Park, S.-A., A.-Y. Lee, K.-C. Son, W.-L. Lee, and D. Kim. 2016. "Gardening Intervention for Physical and Psychological Health Benefits in Elderly Women at Community Centers." *HortTechnology* 26(4): 474–483.

Park, S.-A., C. Song, Y.-A. Oh, Y. Miyazaki, and K.-C. Son. 2017. "Comparison of Physiological and Psychological Relaxation Using Measurements of Heart Rate Variability, Prefrontal Cortex Activity and Subjective Indexes after Completing Tasks with and without Foliage Plants." *International Journal of Environmental Research and Public Health* 14: 1082. doi:10.3390/ijerph 14091087.

Piffer, D. 2013. "Correlation of the COMT Val158Met Polymorphism with Latitude and a Hunter-Gather Lifestyle Suggests Culture-Gene Coevolution and Selective Pressure on Cognition Genes due to Climate." *Anthropological Science* 12(3): 161–171. doi:10.1537/ase.130731.

Puhakka, K. 2014. "Intimacy, Otherness, and Alienation: The Intertwining of Nature and Consciousness." In *Ecopsychology, Phenomenology and the Environment*, ed. D. A. Vakoch, and F. Castrillón. New York: Springer.

Puterman, E., A. Gemmill, D. Karasek, et al. 2016. "Lifespan Adversity and Later Adulthood Telomere Length in the Nationally Representative US Health and Retirement Study. (Psychological and Cognitive Sciences: Cell Biology)." *Proceedings of the National Academy of Sciences of the United States* 113(42): E6335. doi:10.1073/pnas.1525602113.

Puterman, E., A. O'Donovan, N. E. Adler, et al. 2011. "Physical Activity Moderates Effects of Stressor-Induced Rumination on Cortisol Reactivity." *Psychosomatic Medicine* 73(7): 604–611. doi:10.1097/PSY.0b013e318229e1e0.

Raanaas, R. K., K. H. Evensen, D. Rich, G. Sjstrm, and G. Patil. 2011. "Benefits of Indoor Plants on Attention Capacity in an Office Setting." *Journal of Environmental Psychology* 31(1): 99–105. doi:10.1016/j.jenvp.2010.11.005.

Raanaas, R. K., G. G. Patil, and T. Hartig. 2012. "Health Benefits of a View of Nature through the Window: A Quasi-Experimental Study of Patients in a Residential Rehabilitation Center." *Clinical Rehabilitation* 26(1): 21–32. doi:10.1177/0269215511412800.

Roberts, E. J. 1998. "Place and the Human Spirit." *The Humanistic Psychologist* 26(1–3): 5–34. doi:10.10 80/08873267.1998.9976964.

Roszak, T. 1992. *The Voice of the Earth: An Exploration of Ecopsychology*. New York: Simon & Schuster.

Roszak, T. 2001. *The Voice of the Earth: An Exploration of Ecopsychology*. 2nd ed. Grand Rapids, MI: Phanes.

Roszak, T., M. E. Gomes, and A. D. Kanner, eds. 1995. *Ecopsychology: Restoring the Earth, Healing the Mind*. San Francisco, CA: Sierra Club Books.

Ryan, C., W. Browning, J. Clancy, S. Andrews, and N. Kallianpurkar. 2014. "Biophilic Design Patterns: Emerging Nature-Based Parameters for Health and Well-Being in the Built Environment." *ArchNet-IJAR: International Journal of Architectural Research* 8(2): 62–75.

Ryan, R. M., N. Weinstein, J. Bernstein, K. W. Brown, L. Mistretta, and M. Gagn. 2010. "Vitalizing Effects of Being Outdoors and in Nature." *Journal of Environmental Psychology* 30(2): 159–168. doi:10.1016/j.jenvp.2009.10.009.

Sahlins, M. 1998. "The Original Affluent Society." In *Limited Wants, Unlimited Means: A Reader on Hunter-Gatherer Economics and the Environment*, ed. J. Gowdy. Washington, DC: Island Press, pp. 5–41.

Schutte, A. R., J. C. Torquati, and H. L. Beattie. 2017. "Impact of Urban Nature on Executive Functioning in Early and Middle Childhood." *Environment and Behavior* 49(1): 3–30. doi:10.1177/0013916515603095.

Shin, W. S. 2007. "The Influence of Forest View through a Window on Job Satisfaction and Job Stress." *Scandinavian Journal of Forest Research* 22(3): 248–253. doi:10.1080/02827580701262733.

Smith, J. P. 2013. "Ecopsychology: Toward a New Story of Cultural and Racial Diversity." *Ecopsychology* 5(4): 231–232. doi:10.1089/eco.2013.0093.

Stevens, P. 2010. "Embedment in the Environment: A New Paradigm for Well-Being?" *Perspectives in Public Health* 130(6): 265–269. doi:10.1177/1757913910384047.

Sussman, R. 2014. "Deeper Than Our Differences: The Five Common Factors of Ecopsychology." *Ecopsychology* 6(1): 48–49. doi:10.1089/eco.2014.0004.

Taylor, A. F., F. E. Kuo, and W. C. Sullivan. 2001. "Coping with ADD." *Environment and Behavior* 33(1): 54–77. doi:10.1177/00139160121972864.

Toussaint, L., G. S. Shields, G. Dorn, and G. M. Slavich. 2016. "Effects of Lifetime Stress Exposure on Mental and Physical Health in Young Adulthood: How Stress Degrades and Forgiveness Protects Health." *Journal of Health Psychology* 21(6): 1004–1014. doi:10.1177/1359105314544132.

Tudor, K. 2013. "Person-Centered Psychology and Therapy, Ecopsychology and Ecotherapy." *Person-Centered & Experiential Psychotherapies* 12(4): 315–329. doi:10.1080/14779757.2013.855137.

Tyrvainen, L., A. Ojala, K. Korpela, T. Lanki, Y. Tsunetsugu, and T. Kagawa. 2014. "The Influence of Urban Green Environments on Stress Relief Measures: A Field Experiment." *Journal of Environmental Psychology* 38: 1.

Ulrich, R. S. 1984. "View through a Window May Influence Recovery from Surgery." *Science* 224: 420.

Ulrich, R. S. 1986. "Human Responses to Vegetation and Landscapes." *Landscape and Urban Planning* 13: 29–44. doi:10.1016/0169-2046(86)90005-8.

Ulrich, R. S. 2002. "Health Benefits of Gardens in Hospitals." In *Paper for conference, Plants for People International Exhibition Floriade* 17(5): 2010.

Ulrich, R. S., R. F. Simons, B. D. Losito, E. Fiorito, M. A. Miles, and M. Zelson. 1991. "Stress Recovery during Exposure to Natural and Urban Environments." *Journal of Environmental Psychology* 11(3): 201–230. doi:10.1016/S0272-4944(05)80184-7.

United Nations Department of Economic and Social Affairs (UNDESA). 2018. "Revision of World Urbanization Prospects." Accessed June 29, 2018, from https://www.un.org/development/desa/publications/2018-revision-of-world-urbanization-prospects.html.

Van Herzele, A., and S. de Vries. 2012. "Linking Green Space to Health: A Comparative Study of Two Urban Neighbourhoods in Ghent, Belgium." *Population and Environment* 34(2): 171–193. doi:10.1007/s11111-011-0153-1.

Wells, N. M., and K. S. Lekies. 2006. "Nature and the Life Course: Pathways from Childhood Nature Experiences to Adult Environmentalism." *Children Youth and Environments* 16(1): 1–24. doi:10.7721/chilyoutenvi.16.1.0001.

Wilson, E. O. 2009. *Biophilia.* Cambridge, MA: Harvard University Press.

Wolch, J., M. Jerrett, K. Reynolds, et al. 2011. "Childhood Obesity and Proximity to Urban Parks and Recreational Resources: A Longitudinal Cohort Study." *Health and Place* 17(1): 207–214. doi:10.1016/j.healthplace.2010.10.001.

Wypijewski, J., ed. 1997. *Painting by Numbers.* New York: Farrar, Straus, Giroux.

Yao, Y., and K. Chen. 2017. "Effects of Horticulture Therapy on Nursing Home Older Adults in Southern Taiwan." *Quality of Life Research* 26(4): 1007–1014. doi:10.1007/s11136-016-1425-0.

Young, D. 2017. *How to Think about Exercise.* New York: Pikasta.

chapter six

Brain, mind, and relationship

Implications for horticultural therapy

Jay Stone Rice

Contents

This chapter discusses how horticultural therapy supports brain functioning and human well-being. The human brain has undergone successive evolutionary reorganizations, which have accentuated the importance of integration in optimal human development. The emergence of the field of interpersonal neurobiology has emphasized the key role of relationships in human brain structure and operation. The interweaving of brain, mind, and relationship with plants will be considered to identify the unique contribution of horticultural therapy.

Neuroscience: Understanding the brain

The past thirty years have been an era of unprecedented growth in the understanding of the human brain. A multitude of structural and functional imaging techniques such as computed tomography (CT), magnetic resonance imaging (MRI), and positron-emission tomography (PET) scans are continually uncovering new information that challenges or modifies existing models of the brain. The brain's capacity to formulate new synaptic connections suggests adaptability and change are fundamental characteristics. The models utilized in this chapter have been chosen for their explanatory relevance for horticultural therapy.

The evolutionary brain

Paul MacLean (1973) was an evolutionary neuroanatomist who developed a brain model that illuminated the challenges that the brain faces in achieving integration. He hypothesized that the brain is vertically comprised of three distinct brains that developed during the course of evolution to address different demands. This sequence is repeated in human neonatal brain development. MacLean's evolutionary model has been superseded by the recent neuroscientific view that the brain has undergone periods of vertical and horizontal reorganization to meet each emerging evolutionary necessity (Butler 2009). While it is more accurate to discuss an integrated brain, recognizing the purpose and functions of each of the "brains" posited by MacLean will enhance the understanding and effectiveness of horticultural therapy interventions.

Reptilian

The oldest part of the brain is known as the Reptilian brain because it first developed in reptiles 250 million years ago. This brain encompasses areas such as the spinal cord, the brainstem, and the cerebellum. It fulfills several critical roles in regulating essential life functions, including respiration, heart rate, swallowing, and states of arousal. This part of the brain also partially influences the degree of expression of biological temperament traits such as activity/intensity, focus/distractibility, sensory sensitivity/insensitivity, and rigidity/flexibility. Evolutionary theory suggests that each temperament contributes to the survival of the human species. It is important to differentiate between temperament and diagnostic categories. A hummingbird is not hyperactive; it is expressing its nature. The same can be said for people with different temperaments. The horticultural therapist, with experience, can match each participant's temperaments with appropriate gardening tasks (Figure 6.1).

The Reptilian brain enhances survival by triggering neural circuits to respond to perceived threat and danger by fight, flight, or freeze behaviors. The Reptilian brain also responds to states of safety by activating somatic receptivity. Stephen Porges (2004) named this process *neuroception* in his polyvagal theory. Porges's theory suggests that the Reptilian brain receives input from the parasympathetic and sympathetic nervous systems via the vagus nerve, which determines the type of response behavior that is activated. While elements of this model have not been proven conclusively, it has contributed significantly to the understanding and development of treatments for trauma.

Limbic brain

MacLean noted that while reptiles have elaborate behaviors for courting, mating, and territorial defense, they do not care for their young. Paul Broca, a French surgeon and neuroanatomist, discovered in 1870 that there was a lobe of the brain that all mammals share in common, which he called the *limbis*, Latin for "edge," "border," or "ring" (Goleman 1995). This name derives from its location surrounding the reptilian brain and also represents the separation between reptiles and mammals over 100 million years ago. The development of this limbic brain arose to address the extended postbirth care mammals require.

Figure 6.1 Client–therapist relationship: the therapist as mentor and co-participant. (Courtesy of Kathryn Grimes.)

The limbic brain is an interconnected system of brain structures that manage emotions, behavior, long-term memory, motivation, and olfaction. It includes the hippocampus, where memories are processed; the amygdala, where primary emotions are regulated; and the hypothalamus, which adjusts hormonal responses (Siegel 2008). Daniel Goleman (1995) suggests these limbic structures contribute significantly to the experience of feelings within oneself and within others. In order for mammals to raise their young successfully, they must have a capacity to perceive the internal experience of another.

Neocortex

The third brain, the neocortex, evolved 100,000 years ago. *Neo* (Greek, meaning "new," Morris 1981b, p. 880) and *cortex* (Latin, for "bark," Morris 1981a, p. 300) contains four lobes and two hemispheres. It is the largest part of the brain in the more recently evolved mammals. The cortex is largely responsible for higher brain functions, including sensation, voluntary muscle movement, thought, planning, and abstract reasoning. The back of the cortex processes complex sensory information, including seeing and hearing. The frontal lobe is responsible for motor actions, planning, thinking, and attention (Siegel 2008).

Siegel (2007, 2010) locates the mind's capacity for self-awareness in the neocortex. He provides the following as an example: If someone is asked to become aware of their right foot, the somatosensory strip on the left side of the prefrontal cortex activates. When this area is active, sensation is registered from the right foot. This is how the mind's awareness evokes brain activity as well as receptivity to new information. Similarly, if the horticultural therapist draws attention to elements of plant care that promote healthy growth and their correspondence with human needs, this stimulates receptivity and reflection within the participant's neocortex.

Nonhierarchical integration and nonjudgment

Failure to appreciate the role each brain plays in human functioning frequently contributes to the experience of miscommunication. If the development of the triune brain is viewed as a hierarchy, then it would be easy to give primacy to the neocortex. However, it is possible to see that each of these brains provides information critical to human well-being. For example, the Reptilian brain focuses on immediate survival. Within it, evolution's bias toward acting first and thinking afterward is observed. If early humans took too long to determine if the animal running toward them was a saber-toothed cat, further evolution would have been doubtful (Serangeli et al. 2015).

As a result of this survival mechanism, humans are prone to react too quickly to perceived threats that are either physical or psychological. When this reaction occurs, the limbic brain's emotional response of fear or anger may block the reasoning function of the neocortex. However, the mammalian limbic brain also provides humans with the capacity to feel the effects of their actions on others. Once the emotional state that accompanied the initial response has subsided and a sense of safety has been restored, a person can apologize for their Reptilian brain erroneously responding to an imagined threat with an ill-placed survival reaction. The neocortex introduces the human potential for self-reflection regarding this sequence, which in turn enhances the integration of brain functioning and fosters the experience of well-being.

Triune brain and horticultural therapy

The triune brain model may be utilized by horticultural therapists to better understand how to interact with their horticultural therapy clients. For example, if a client is feeling fearful, the therapist might initially want to address the client's concerns. Unfortunately, fear can block the functioning of the reasoning neocortex, so this intervention may be unsuccessful. However, if the therapist recognizes fear has been activated, he or she can modulate vocal tone to provide emotional reassurance and support through the relational limbic brain. This facilitates the client's Reptilian brain to shift over time from the fight, flight, or freeze response to one of receptivity. When this occurs, discussion and reasoning become possible.

The triune brain model can also inform the horticultural therapist's use of plants. A client may be in an anxious, fearful state due to a life circumstance that has activated the Reptilian brain. The therapist can invite the client to focus attention on an element of the garden that provides a calm, pleasurable sensory experience. The nervous system brings this information to the Reptilian brain where neuroception triggers the neural circuits associated with safety. The horticultural therapist simultaneously provides a calming influence through limbic attunement by speaking in a firm and soothing voice. The combination of these somatic and relational inputs supports the functioning of the client's neocortex, which brings forward new thought and awareness. The use of horticultural therapy to support neuroception in homeless children is discussed in Exhibit 6.1 (Figure 6.2).

EXHIBIT 6.1

Program and Case Example: Utilizing a Sensory Garden to Cultivate the Experience of Safety in Homeless Children

Emotions frequently run high at Vogel Alcove, a childcare center in Dallas, Texas, for children affected by homelessness. For these infants, young children, and their parents, chronic stress and daily chaos are part of life, and typical cognitive functioning is affected by the more immediate needs to feel safe, secure, loved, and fed. Children experiencing homelessness are three times more likely to have emotional and behavioral problems than their nonhomeless peers (National Child Traumatic Stress Network 2005) and four times more likely to have delayed development (National Center on Family Homelessness 2009).

Hypervigilance and altered sensory integration are two manifestations of the toxic stress these children endure, and both affect their ability to benefit from the learning environment. Interaction with the sensory garden in "The Backyard," the center's outdoor learning environment, has been helpful in addressing these concerns. With aromatic herbs such as energizing rosemary and calming lavender, varied textures provided by fuzzy Lamb's Ear and curly kale, the sound of rustling grasses from Inland Sea Oats, and the towering visual delight of sunflowers, the sensory garden offers opportunity for the children's focus to turn from internal distress to external input.

As a typical example, when "Steven" first arrived at Vogel Alcove at age 3, connecting with his teacher and the children in his class was a challenge. His eyes darted around him, he moved from center to center without engaging in play. When his anxiety escalated, he would use his arm to sweep all the toys off the shelves or hurt his classmates. When a child accidentally touched him during circle time, he would lash out, "Don't touch me!" and either run away or respond with a punch. His teacher learned that the first step toward teaching Steven was connecting with him, and this was a gradual process aided by the sensory garden.

In "The Backyard" one of Steven's favorite places was the mud kitchen. Here, he and his teacher would pretend to cook up aromatic delights, and Steven would let cool water and bright lemon balm trickle through his fingers as he lifted his face to receive the warmth of the sun. His teacher modeled taking deep breaths of the scent while relaxing and releasing to the experience. At times, she soothed his arm with a soft leaf of Lamb's Ear as they discussed a story they had read, and other times they hunted insects on the pathway between the herbs to let the rosemary and oregano tickle their legs and disarm Steven's reactive impulse to the sense of touch.

Recovery is not guaranteed for these children affected by the trauma of homelessness. In many cases the effects will continue to be present in the child's adult life; however, progress is evident, usually within just a few months. Aided by the sensory garden and his teacher's care, Steven became more able to relax and engage in learning activities with his peers.

By Kathryn E. Grimes, MAT, HTR

Figure 6.2 Relieving stress and anxiety through sensory-rich aromatic activities. (Courtesy of Kathryn Grimes.)

It is important to recall that, for the purpose of description, the three brains have been discussed as if they functioned separately. However, the whole brain is operating organically. In the next section, the field of interpersonal neurobiology is reviewed to provide horticultural therapists with a more nuanced and comprehensive view of brain, mind, and relationships.

Interpersonal neurobiology

Daniel Siegel and Alan Shore coined the term *interpersonal neurobiology* to describe the dynamic interconnection of brain, mind, and relationships (Siegel 1999). Siegel (2010) observes that the experience of well-being arises when there is optimal integration of brain and mind functioning. The quality of significant relationships during the brain's development enhances or undermines its capacity to accomplish the increasingly complex tasks of integration that human life requires.

Siegel (2008) describes the brain as a complex system of energy and information that flows, via electrical and chemical neuron-to-neuron transmission, throughout the body. DNA from each parent combines to formulate the brain's basic architecture. However, the brain's neuronal structure is then further developed significantly by the quality of attachment relationships experienced. The brain, generally understood to be contained within the skull, actually extends throughout the body's nervous system. From the field of neurocardiology, we learn that there is a concentration of neurons surrounding the heart that processes information independently from the part of the brain located in the skull (Armour 2003). There is also a weblike neuronal center surrounding the gut, known as the enteric nervous system, which performs similar functions (Gershon 1998). Contradicting subjective experiences generated from the different neuronal processing areas can lead to the experience of internal conflict and self-doubt. For example, Susan is invited to dinner by her colleagues after a long day at work. She experiences reluctance to saying yes deriving from a sensation located in her gut region. However, she also has an emotional, heart-centered concern that saying no will affect her future relationships with her co-workers. She determines that she should go even though she does not feel up to it. She returns home quite late and awakens the following morning with a fever and chills. She questions why she talked herself out of her first adverse reaction to the invitation. Often these inner conflicts from the different processing centers create the experience of self-doubt. Understanding and integrating intelligence from these different processing centers can reduce the second-guessing that frequently occurs when a choice has been made from one center only. Perhaps a mind-body connection refers to these different areas of neuronal activity that process the moment-to-moment experience of aliveness. An ecological understanding entails seeing how the interweaving of these different information centers supports the experience of health and balance. Stevens (2010) suggests that mind-body integration is inseparable from human-nature integration in the experience of well-being.

Integrating brain, mind, and relationships

Neuroplasticity

The brain has the capacity for continued development throughout the lifespan. This capacity to grow new neural connections, which enables the brain to reshape itself, is known as neuroplasticity. Supportive relationships and compassionate self-reflection stimulate this regenerative capacity in part through the release of oxytocin (Graham 2013). Oxytocin is a neuropeptide that is produced in response to warm affectionate touch, music, and scents. Oxytocin levels have been shown to increase during breastfeeding, orgasm, petting animals, and practicing self-compassion. Further research is vital for determining if horticultural therapy also stimulates the release of oxytocin through the care of plants (Rice 2012). When well-being, trust, and connection are experienced repeatedly, the accompanying release of oxytocin enables the brain to rewire itself (Ecker 2010). Exhibit 6.2 describes how horticultural therapy is utilized to aid neuroplasticity in people who have experienced trauma.

EXHIBIT 6.2

Program Example: The Interface between Horticultural Therapy Trauma Treatment and Somatic-Oriented Mental Health Therapy

At the Mental Health Center of Denver (MHCD), we work to provide effective and compassionate mental health treatment so that people can live full lives and achieve their goals. The horticultural therapy program addresses emotional, behavioral, relationship, and psychiatric goals. We provide individual, family, and group psychotherapy for children, teens, families, and parents. Many of the children and adults who seek psychotherapeutic support have experienced minor traumatic experiences, and we frequently treat individuals who have experienced multiple, life-threatening, ongoing, or developmental (occurring during early life stages) traumatic events. We integrate best practices in trauma-informed care, recognizing and responding to the effects of trauma on individuals, and empowering the people we serve to take the lead in identifying their treatment goals.

Our horticultural therapy program takes place in therapy offices, classrooms, and group meeting rooms, and makes use of our horticultural therapy garden, urban farm, pollinator garden, container plants, and raised beds. Horticultural therapy services are provided directly by a licensed mental health practitioner who is also a registered horticultural therapist. Therapeutic horticulture is also provided by other mental health professionals on staff, who receive consultation with the horticultural therapist about how to integrate horticultural therapy activities into their therapeutic work.

There is a growing recognition within the field of trauma treatment that talk therapy alone is limited in its efficacy. Effective treatment engages the whole body because memories include thoughts; sense perceptions including visual images, sounds, smells, tastes, and tactile sensations; and somatic fight, flight, or freeze responses. Physical restlessness, hypervigilance, numbing or dissociation, and inability to recognize or tolerate one's own somatic states frequently cause trauma survivors to feel they cannot trust their own body, and they often react to nontraumatic situations as if they were in danger again.

In the garden, the smell of an herb or the feel of a flower can be starting points to ask, "Where in your body do you notice your response to that?" The process of helping people pay attention to their somatic responses to external stimuli is a crucial step toward their being able to recognize and tolerate internal emotional states. In addition, calming, rhythmic activities such as mixing potting soil with amendments, stirring compost tea, or digging a hole create a soothing counterpart to the fight-or-flight response. The neural systems in the brain that are active during experiences of curiosity, sensory seeking, and physical movement have little overlap with the neural systems that are active during fight, flight, or freeze responses. Repeated activation of the brain's engagement systems builds stronger neural networks to tolerate inner and outer sensations, form healthier relationships, and learn to trust self and other again.

Carol LaRocque, LPC, HTR
Horticultural Therapist, Mental Health Center of Denver

Mirror neurons and empathy

The field of social neuroscience studies the social circuitry networks of the brain (Freedberg and Gallese 2007; Cacioppo and Berntson 1992). Mirror neurons are one component of this network. As is often the case, the initial excitement over the discovery of these neurons led to overstating their role in a complex network of brain activity. Nevertheless, they seem to be a key component of our capacity to learn from others. The mirror neurons were discovered by a group of Italian researchers while studying Macaque monkeys (Rizzolatti and Fabbri-Destro 2010). They observed that when a monkey acted intentionally to pick up a peanut, neurons in an area of their frontal lobe called the motor strip would activate. The same neurons in the motor strip of an observing companion monkey would become active in preparation for the same action. Mirror neurons have been observed and recorded in multiple human systems, including the cortex, hippocampus, and motor areas (Mukamel et al. 2010). For example, when an observing participant watches another participant move their hand to pick up a glass of water, neurons in the observer's motor strip activates and the experience of thirst is reported.

Mirror neurons have been shown to mimic feelings (Enticott et al. 2008). Siegel (2008) maps the neural pathway for this process. A person observes another person about to cry. This information moves from the frontal lobe into the insula, a part of the middle layers of the cortex that connects the cortex with the limbic system and the body. The information then emerges as a feeling within the body of the observer. This feeling rises through another part of the insula back into the cortex. When the mind becomes conscious of the feeling that the body is communicating, it uses this information to make a guess about what the other person is feeling. Thus, insight into one's own feelings creates a map of what may be in the mind of the person being observed. In essence, this is a significant part of the neural mechanism of empathy. The following observation of a county jail inmate working in the greenhouse of the Garden Project at the San Francisco Sheriff's Department, may highlight the interplay of mirror neurons and empathy in horticultural therapy.

> All of these trays are seeds that are germinating. Mainly what seeds need in order to germinate is water and light, and someone to pay attention to them and make sure that they don't have any problems…. This program is like that—it starts a germination in your mind, to try and instill the tools you need to grow and to make changes in your life (Richards 1991, 1).

Social neuroscience provides us with the evidence that we are structurally wired for connection to others. Horticultural therapy is predicated on the recognition that this includes our relationships with plants. This challenges the human exceptionalism narrative that humans are separate and isolated. Siegel (2010) suggests that interpersonal neurobiology can be used to understand our relatedness, beyond the interpersonal, to other living creatures and to our whole planet.

Trauma's influence on brain functioning: Implications for horticultural therapy

Current best practices in the treatment for psychological trauma emphasize a combination of somatic (body-oriented), attachment, and cognitive or insight-based interventions. This reflects a recognition that trauma may have an adverse impact on integrated brain

Figure 6.3 Horticultural therapy provides a safe environment. (Photo provided by Pacific Quest.)

functioning. Traumatic interactions most frequently occur between humans. Early traumatic experiences can impair the capacity for healthy attachment. In developing their sensorimotor psychotherapy for the treatment of attachment and trauma, Ogden and Fisher (2015) have identified two responses to threat and danger in addition to fight, flight, or freeze. These attachment-oriented responses are "submit" and "attachment cry for help" (Figure 6.3).

Horticultural therapy may affect a participant's thoughts, physical experience, and capacity for attachment. Cultivating plants introduces a relationship that may feel less threatening and therefore may evoke the experience of safety. Safety opens the participant to new information and self-reflections. It is important to note that nature at times may also introduce trauma into a person's life through major destructive weather events such as fires and floods. Exhibit 6.3 describes how horticultural therapy fosters the restoration of healthy emotional attachment essential for the healing of trauma.

EXHIBIT 6.3

Perspective: Using Horticultural Therapy to Recover from Trauma

The vital role of attachment in complex trauma is increasingly recognized (Cozolino 2010; De Zulueta 2009; van der Kolk 2002, 2007, 2014). Therapists understand that using our authentic self to connect with the client is essential to repairing attachment bonds that have been broken or damaged through traumatic experiences of abuse or neglect. But forging a good enough connection with clients who may be extremely wary, distrustful, avoidant, or even paranoid is a particularly challenging aspect of trauma treatment (Mearns and Cooper 2005). A therapeutic model of horticultural therapy introduces a third element into this

(Continued)

EXHIBIT 6.3 (Continued)

Perspective: Using Horticultural Therapy to Recover from Trauma

relationship—nature—that offers the possibility of buffering, or augmenting, the client–therapist therapeutic alliance (Jordan 2015; Jordan and Hinds 2016).

As this chapter explains, we are deeply hardwired to connect with nature, and our nervous systems continually scan our environment for potential signs of danger, or for benign landscapes rich in shelter and sources of food and water (Kaplan and Kaplan 1989; Ulrich 1993). Evidence-based therapeutic landscape design knowingly taps into this hardwiring to provide surroundings we can experience as both safe and stimulating (Cooper Marcus and Sachs 2014). Once we sense we are in safe territory, we can relax and allow our attention to be engaged by fascinating natural elements, such as the color and texture of plants; the movement of bees, butterflies, and birds; and the sounds of flowing water or windblown foliage. Repeated, simple, and familiar actions such as walking, digging, pruning, and potting up, when carried out mindfully, engage all the senses in the present moment, calm and regulate our nervous system, and may help us to process traumatic memories (Redwood 2011; Thompson et al. 2014). From this more relaxed state, we are also likely to feel more open toward engaging with other friendly animals, such as dogs or horses, and with other friendly humans, thus helping to reduce the isolation and loneliness that frequently accompany trauma.

Nature offers a wide range of activities that, as research evidence has also shown, help promote individual mental and physical health, and where the experience of caring human contact and trusted connections can gradually be introduced and enjoyed (Bratman et al. 2012; Corbett and Milton 2011; Pretty et al. 2005). But more than this, working with nature therapeutically offers a three-way connection and, through it, the possibility of transitioning away from an individual, pathology-based model of trauma recovery and cure toward more secure and healthy attachments involving restoration, renewal, and resourcing (Jordan 2015; Milton 2010). Using a therapeutic model of horticultural therapy in this way, a more integrated and broader experience of recovery awaits, involving reciprocal healing relationships not simply between individual client and therapist but also potentially with our wider community, society, and ultimately our world.

Joanna Wise, DCounsPsych, MBPsS, HCPC ACAT EMDRIA, DipSTH

Paul Gilbert (2009) hypothesized that humans have three emotional regulation systems. The *threat and protection* system creates feelings of anxiety, fear, or disgust in response to potentially threatening stimuli. The system entails the regulation of serotonin. The *drive and excitement* system motivates seeking that which is needed, such as food, sex, territory, and status. Pleasure is experienced when this drive is satisfied. This system is influenced by dopamine. The *contentment, soothing, and social safeness* system is a peaceful state that is generally associated with having internalized the experience of secure attachment as a child. Caring and soothing behavior are regulated through the opiate and oxytocin systems.

Gilbert suggests that early adverse experiences can lead to an imbalance in these systems, with a heightened reactivity toward perceptions of internal or external threat. In addition,

Gilbert observes that modern society overly stimulates the threat and protection system. This imbalance generates increased thoughts of self-criticism and shame. Gilbert developed a mindfulness-based Compassion Focused Therapy to reinstate balance in these emotional regulation systems. Compassion Focused Therapy proposes that fostering healthier drive and excitement, along with contentment, soothing, and social safeness systems functions, will act as antagonists to lessen the reliance on the threat and protection system.

Horticultural therapy introduces a connection with nature that may stimulate the drive and excitement system with curiosity and interest. At the same time, the caring for and nurturing of plants stimulates the Contentment, soothing, and social safeness system. Joanna Wise, author of *Digging for Victory*, (Wise 2015) suggests that these two horticultural therapy influences may reduce the threat and protection system, thus support the balancing of these emotional regulation systems (personal communication with the author, March 10, 2018). Cathrine Sneed, founder of the San Francisco Sheriff's Department's Garden Project, taught plant care to chemically dependent county jail inmates. A study of the Garden Project found that participants had experienced significant trauma in their early childhood (Rice and Remy 1994). Sneed (1992) observed that some of her students would re-offend upon release so that they could come back and work in the garden. This likely represented their recognition of the need for further assistance in balancing their emotional regulation systems. Similarly, actor Charles Dutton recalls how his fascination for nature propelled him from the inner city to reform school.

> I remember the older guys would tell you how great reform school was. It was also a place where inner-city kids could have fun in a rural area. To me, reform school was a gas. Sometimes we would plan our juvenile activities so we could go to court in March or April and get sentenced to six months and spend spring and part of the summer in reform school. It was fun to get away from the city; you had your buddies there. (Rothstein 1990, B3)

Exhibit 6.4 presents a neurosequential model for utilizing horticultural therapy to support optimal brain functioning in an intensive outdoor behavioral health care program for adolescents and young adults with behavioral and emotional difficulties. Table 6.1 identifies specific horticultural activities that support each brain's functional domain (Figure 6.4).

EXHIBIT 6.4

Program Example: Neurosequential Approach to Horticultural Therapy

Pacific Quest (PQ) is an intensive outdoor behavioral health care program specializing in the treatment of adolescents and young adults struggling with emotional and behavioral difficulties. Immersed in a sensory-rich garden setting on the island of Hawaii, therapists and paraprofessionals at PQ utilize a neurodevelopmental approach to horticultural therapy (Freedle and Slagle 2018). Drawing on the rich natural history and cultural values of the Hawaiian Islands, PQ's approach to horticultural therapy is

(Continued)

EXHIBIT 6.4 (Continued)

Program Example: Neurosequential Approach to Horticultural Therapy

aimed at implementing therapeutic interventions that activate the senses and capture the imagination.

At PQ, the Neurosequential Model of Therapeutics (Perry 2006, 2016) provides an empirical basis for therapeutic garden design, clinical interventions, and program structure (Freedle and Slagle 2018). Guided by the principles and best practices of experiential learning, rites of passage, whole-person wellness, and evidence-based treatment, PQ provides an integrative approach to horticultural therapy and healthy community living. Specific horticultural activities are chosen based on an understanding of brain development and regulation of affect. This sequential method of therapeutic engagement begins by targeting earlier-developing brain regions, such as the brainstem, diencephalon, and limbic system. Subcortical regions of the brain are activated through rhythmic and repetitive somatosensory experiences (Perry 2006).

Improving the functioning of subcortical brain systems optimizes the functioning of higher-order brain systems responsible for decision making, problem solving, and abstract reasoning. According to Perry (2006), higher-order brain systems in the cortex depend on the integrity of lower, subcortical systems. The emphasis on the sequential process of neurodevelopment, and the interdependent relationship between human health and environmental health highlights the integral role that nature has in shaping the nervous system.

Perry (2006, 2009), identifies four functional domains of the Neurosequential Model of Therapeutics that correlate with specific regions of the brain. At PQ, the Neurosequential Model of Therapeutics functional domains coincide with a sequential approach to horticultural interventions, with an emphasis on engaging the senses and cultivating a reciprocal relationship with the natural environment. For an overview of the integration of horticultural activities and the neurosequential model, see Table 6.1.

Horticultural activities, such as weeding, planting, and collecting flowers and herbs, can be further enhanced through bilateral movement, diaphragmatic breathing, incorporating music, and modulating gross-motor and fine-motor activity. With a foundation in somatosensory processing and self-regulation, the focus of horticultural intervention can shift toward more skill building and problem solving. Horticultural activities targeting the limbic system and neocortex increase in complexity, incorporating more mindfulness skills, problem solving, symbolic meaning, and cognitive behavioral techniques. By utilizing a neurosequential approach, future horticultural therapists may increase the therapeutic impact and sophistication of nature-assisted intervention and provide a clinical rationale for horticultural therapy aimed at optimizing neuronal activity and increasing relational health.

Travis Slagle, NCC, LMHC
Horticultural Therapy Director, Pacific Quest

Table 6.1 Techniques: Neurosequential model of therapeutics and horticultural activities

NMT functional domain	Brain region	Horticultural activities
Sensory integration	Brainstem	Multisensory garden walks Weeding and watering Shoveling, digging, and raking Aromatherapy, flower arrangements Picking and grinding herbs Garden meditation and bilateral movement
Self-regulation	Diencephalon	Handicrafts, tool building, weaving Playing with soil, barefoot garden projects Breaking apart and putting together garden Hauling, lifting, sifting soil and rocks Pushing a wheelbarrow, building walls Building and deconstructing garden beds Music in the garden, rhythmic activity
Relational functioning	Limbic system	Nursery care, tree and animal husbandry Earth art, symbolic garden projects Planting, harvesting, fertilizing Garden ceremony, ritual, and festivities Cultural sharing, storytelling, ethnobotany Paired/small-group gardening projects Garden games, competition, celebration
Cognitive problem solving	Neocortex	Natural science, botany, irrigation Composting, grafting, pest management Companion planting, square-foot gardening Following a planting/harvesting schedule Project management, experimentation Vertical gardens, hydroponics, aquaculture farmers markets, and garden recipes

Source: Freedle, L. R., and Slagle, T., Application of the Neurosequential Model of Therapeutics® (NMT) in an integrative outdoor behavioral health care program for adolescents and young adults, in *Proceedings of the Second International Neurosequential Model Symposium*, CF Learning Press, Lennox, SD, 2018.

Figure 6.4 Pacific Quest participants harvesting beans. (Photo provided by Pacific Quest.)

Horticultural therapy: Cultivating well-being and integration

According to Siegel (2010), the mind represents the embodied regulation and processing of information from throughout the nervous system. In *The Mindful Brain*, he notes that well-being depends on a coherent mind, empathic relationships, and neural integration. Siegel (2010) delineates nine domains of neural integration that affect human well-being over the course of life. Horticultural therapy supports many of these domains of integration, including vertical, horizontal, narrative, temporal, and transpirational integration. While the conventions of language necessitate listing them individually, in actual experience, these domains function as an integrated whole (Parker 2008).

> *Integration of consciousness:* sensation, observation, conceptualization, knowing
> *Vertical integration:* mind-body connection, centers of knowing
> *Horizontal integration:* right/left brain, perceptions/feeling with linear/logical thinking
> *Memory integration:* implicit sensory memory with explicit episodic/narrative memory
> *Narrative integration:* coherent emotionally connected narrative over lifespan
> *State integration:* coherent awareness of our states of mind and intentions
> *Temporal integration:* living with awareness of our mortality and impermanence
> *Interpersonal integration:* attuned alignment with others on an emotional level
> *Transpiration integration:* feeling that a person is part of a much larger whole

Plants and the brain

E. O. Wilson (1984, 2002) hypothesized that the human brain evolved as part of the entire biosphere. Biophilia would suggest that humans have a complex interrelationship with plants, which provides each with benefits that support well-being. In order to appreciate how plants may affect the brain, it is important to look beyond the human exceptionalism narrative, a paradigm that emphasizes the unique differences between humans and the rest of life. As noted in Chapter 3, a narrative is sustained by recognizing and linking together information that supports the narrative. In this process, information that does not fit the narrative goes unnoticed. For example, in *The Botany of Desire*, Michael Pollan states that plants are masters of co-evolutionary development (2002). Since plants are unable to move on their own, they use their remarkable capacity for creating compounds to develop various means of attracting insects to them so that their pollen can be disseminated, thereby ensuring their reproduction. However, the conceit inherent in the human exceptionalism narrative excludes the recognition that plants may actually engage humans as well.

Psychologist David Abrams (1977) wrote *The Spell of the Sensuous* to explore how humans cognitively separate themselves from nature. While studying folk magic and healing in Bali, Abrams apprenticed with local shamans. During the course of his training, he became deeply immersed in an ongoing communication with the natural world. Six months after he returned to the United States he found that this conversation completely ceased.

Drawing from research in linguistics, he noted that written languages initially utilized alphabets whose letters were representations of figures in nature. The basis of these alphabets suggests that the formulators were rooted in the natural world. The Greeks were the first to develop an alphabet with vowels, which shifted the emphasis of language to be a representation of the human voice. In the ensuing period, Greek philosophers became the first to conceptualize ideal states that existed supranature.

The modern era may best be defined by the development of tools and technologies that seemingly free people from the limits of the natural world. Electric lights, indoor heating and air conditioning, industrial farming, antibiotics, airplanes, etc., emerge from this narrative. Human technological ingenuity is driven by the neocortex. Albert Einstein (1995) cautioned that it is important not to deify the intellect. Einstein suggests the intellect is a powerful muscle that can serve but should not lead. Indeed, the scientific era has given rise to the development of numerous technologies that have disturbed the balance within human families and nature.

Plant intelligence

Does the human narrative of exceptionalism hinder and/or obscure recognition of the integration with nonhuman intelligent life? Recent research identifies signs of plant intelligence and awareness. A study by scientists at the University of Birmingham published in the *Proceedings of the National Academy of Sciences* (Topham et al. 2017) found evidence of a group of decision-making cells in plant embryos that assess environmental conditions and dictate when the seed germinates. Researchers have observed an aphid-infested plant chemically warns other plants so that they can produce protective chemicals to ward off the aphids traveling toward them (Tsuji et al. 2010).

Plants engage the world by sensing stimuli such as light waves and olfactory signals. Plants express different genes depending on the amount of light they receive. This process determines the timing of flowering. Plants also discern color. Humans have two photoreceptors that perceive colors, but plants have eleven (Chamovitz 2012). A parasitic plant has been shown to utilize olfactory cues in choosing its host plant (Runyon 2006).

Plants and relationship

Agriculture and animal husbandry arose with the evolution of the limbic brain. The human infant evokes parental attachment through behaviors that call forth the parent's attention, such as cooing, reaching out with hands, following with eyes, and smiling. Pollan notes that plants call for attention through their color and fragrance. As noted earlier, olfaction is located within the relational limbic brain. He wonders if agriculture was a human idea or the plant's strategy for procreating and thriving. Yet this either-or sensibility may obscure recognition of a field of mutual attraction and ecological interconnectedness that has been described in oral earth-based traditions as a process of co-creation (Star and Weaver 2009).

Plants and the mind

Stephen Kaplan (1978, 1983) suggests that humans need restorative natural environments that are conducive to reflection and provide a sense of safety, coherence, fascination, and a feeling of being away from one's surroundings and routines. For Kaplan, this fascination is an example of James's (1985) concept of indirect attention, which does not require effort or an act of will. He writes, "People are fascinated with issues that pertain to the self on the one hand and the cosmos on the other" (Kaplan 1978). Having opportunities to reflect has been essential for human adaptation and survival. (See Chapter 5 for more detail on this topic.)

Mindfulness, gardening, and interbeing

The recent integration of mindfulness and gardening has brought new appreciation to the quality of flow that many gardeners experience (Johnson 2008; Murray 2012; Redwood 2011; Wilfong 2010). Mindfulness is a meditative practice that cultivates awareness of mind and body in the present moment without judgment (Stahl and Goldstein 2010). In 1979, Jon Kabat-Zinn founded the Mindfulness Based Stress Reduction Clinic at the University of Massachusetts Medical Center. His model for utilizing meditation, yoga, and somatic awareness has generated positive research findings for the reduction of stress, increased relaxation, and improvement of quality of life (Kabat-Zinn 2013). Mobile electroencephalography research has shown that entering green spaces, such as gardens, stimulates similar brain waves found during meditation (Aspinall et al. 2015). Dan Siegel (2010) observes that mindfulness supports the development of mindsight, a term he created to describe the essential quality of self-awareness that fosters integration and well-being.

It is easy to become overwhelmed by the frantic pace of life. (Refer to Chapter 5 for more detail and an evolutionary basis.) The vast interconnectivity made possible via the internet can significantly increase the experience of information overload. The nonhuman world is obscured by the din. When a person begins to meditate, he or she often profoundly realizes how seldom the mind rests. Working with plants focuses attention on the interconnectivity with other life forms, which provides balance and a different sense of time. It is easy to understand the integration of mindfulness practices and gardening. In the truest sense, horticultural therapy represents an understanding that cultivating a garden is intrinsically linked with cultivation of self (Johnson 2008). This idea is beautifully expressed in the poem "Interrelationship" by Thich Nhat Hanh, the Vietnamese Zen teacher and poet (1999).

Interrelationship

> You are me and I am you.
> It is obvious that we are inter-are.
> You cultivate the flower in
> yourself so that I will be beautiful.
> I transform the garbage in myself so
> that you do not have to suffer.
> I support you you support me.
> I am here to bring you peace
> you are here to bring me joy.

Key terms

Human exceptionalism A paradigm that considers humans as essentially different from other life forms.

Interpersonal neurobiology An interdisciplinary field bringing together many different scientific disciplines to broadly understand human experience.

Neuroception Stephen Porges defines this term as the process where neural circuits determine if situations or people are safe, dangerous, or life threatening.

Neuroplasticity The ability of the brain to form or reorganize synoptic connections in response to learning and experience or following an injury.

Social neuroscience An interdisciplinary study of social and biological approaches to understanding human experience and behavior.

References

Abrams, D. 1997. *The Spell of the Sensuous: Perception and Language in a More-Than-Human World*. New York: Vintage Books.

Armour, A. 2003. *Neurocardiology: Anatomical and Functional Principles*. Boulder Creek, CA: HeartMath Institute, Publication No. 03-011.

Aspinall, P., P. Mavros, R. Coyne, and J. Roe. 2015. "The Urban Brain: Analysing Outdoor Physical Activity with Mobile EEG." *British Journal of Sports Medicine* 49(4): 272–276.

Bratman, G. N., J. P. Hamilton, and G. C. Daily. 2012. "The Impacts of Nature Experience on Human Cognitive Function and Mental Health." *Annals of the New York Academy of Sciences* 1249: 118–136.

Butler, A. B. 2009. "Triune Brain Concept: A Comparative Evolutionary Perspective." *Reference Module in Neuroscience and Biobehavioral Psychology, Encyclopedia of Neuroscience*, 1185–1193. http://www.sciencedirect.com/science/article/pii/B9780080450469009840.

Cacioppo, J. T., and G. G. Berntson. 1992. "Social Psychological Contributions to the Decade of the Brain: Doctrine of Multilevel Analysis." *American Psychologist* 47: 1019–1028.

Chamovitz, D. 2012. *What a Plant Knows: A Field Guide to the Senses*. New York: Scientific American/ Farrar, Straus and Giroux.

Cooper Marcus, C., and N. A. Sachs. 2014. *Therapeutic Landscapes: An Evidence-Based Approach to Designing Healing Gardens and Restorative Outdoor Spaces*. Hoboken, NJ: Wiley.

Corbett, L., and M. Milton. 2011. "Ecopsychology: A Perspective on Trauma." *European Journal of Ecopsychology* 2: 28–47.

Cozolino, L. 2010. *The Neuroscience of Psychotherapy: Healing the Social Brain*, 2nd ed. New York: Norton.

De Zulueta, F. 2009. "Post-Traumatic Stress Disorder and Attachment: Possible Links with Borderline Personality Disorder." *Advances in Psychiatric Treatment* 15: 172–180.

Ecker, B. 2010. "The Brains Rules for Change." *Psychotherapy Networker*. 34(1): 43–45, 60.

Einstein, A. 1995. *The Goal of Human Existence: Out of My Later Years*. New York: Carol Publishing Group.

Enticott, P. G., P. J. Johnston, S. E. Herring, K. E. Hoy, and P. B. Fitzgerald. 2008. "Mirror Neuron Activation is Associated with Facial Emotional Processing." *Journal of Neuropsychologia* 46(11): 2851–2854.http://www.sciencedirect.com/science/journal/00283932?sdc=1.

Freedberg, D., and V. Gallese. 2007. "Motion, Emotion and Empathy in Aesthetic Experience." *Trends in Cognitive Science* 11(5): 197–203.

Freedle, L. R., and T. Slagle. 2018. "Application of the Neurosequential Model of Therapeutics (NMT) in an Integrative Outdoor Behavioral Healthcare Program for Adolescents and Young Adults." In *Proceedings of the Second International Neurosequential Model Symposium*. Lennox, SD: CF Learning Press.

Gershon, M. 1998. *The Second Brain: The Scientific Basis of Gut Instinct and a Groundbreaking New Understanding of Nervous Disorders of the Stomach and Intestines*. New York: Harper.

Gilbert, P. 2009. "Introducing Compassion-Focused Therapy." *Advances in Psychiatric Treatment* 74: 116–143.

Goleman, D. 1995. *Emotional Intelligence: Why It Can Matter More Than IQ*. New York: Bantam Books.

Graham, L. 2013. *Bouncing Back: Rewiring Your Brain for Maximum Resilience and Well-Being*. Novato, CA: New World Library.

Hanh, T. N. 1999. *Call Me by My True Names: The Collected Poems of Thich Nhat Hanh*. Berkeley, CA: Parallax Press.

James, W. 1985. *Psychology: The Briefer Course*, ed. Gordon Allport. Notre Dame, IN: University of Notre Dame Press. (Original work published in 1892).

Johnson, W. 2008. *Gardening at the Dragon's Gate: At Work in the Wild and Cultivated World*. New York: Bantam Books.

Jordan, M. 2015. *Nature and Therapy: Understanding Counselling and Psychotherapy in Outdoor Spaces*. London, UK: Routledge.

Jordan, M., and J. Hinds, eds. 2016. *Ecotherapy: Theory, Research and Practice*. London, UK: Palgrave.

Kabat-Zinn, J. 2013. *Full Catastrophe Living: Using the Wisdom of Your Body and Mind to Face Stress, Pain, and Illness* (Revised ed.). New York: Bantam Books.

Kaplan, R., and S. Kaplan. 1989. *The Experience of Nature: A Psychological Perspective.* New York: Cambridge University Press.

Kaplan, S. 1978. "Attention and Fascination: The Search for Cognitive Clarity." In *Humanscapes: Environments for People*, ed. S. Kaplan, and R. Kaplan. North Scituate, MA: Duxbury Press, pp. 84–90.

Kaplan, S. 1983. "A Model of Person-Environment Compatibility." *Environment and Behavior* 15: 311–332.

MacLean, P. D. 1973. *A Triune Concept of Brain and Behavior.* Toronto, CA: University of Toronto Press.

Mearns, D., and M. Cooper. 2005. *Working at Relational Depth in Counselling and Psychotherapy.* Thousand Oaks, CA: Sage.

Milton, M. 2010. "Coming Home to Roost: Counselling Psychology and the Natural World." In *Therapy and Beyond: Counselling Psychology Contributions to Therapeutic and Social Issues,* ed. M. Milton. Chichester, UK: Wiley Blackwell.

Morris, W., ed. 1981a. Cortex. *The American Heritage Dictionary of the English Language.* 300. Boston, MA: Houghton Mifflin, p. 300.

Morris, W., ed. 1981b. Neo. *The American Heritage Dictionary of the English Language.* 880. Boston, MA: Houghton Mifflin, p. 880.

Mukamel, R., A. D. Ekstrom, J. Kaplan, M. Iacoboni, and I. Fried. 2010. "Single-Neuron Responses in Humans during Execution and Observation of Actions." *Current Biology* 20: 750–756. http://www.cell.com/current-biology/fulltext/S0960-9822(10)00233-2.

Murray, Z. 2012. *Mindfulness in the Garden: Zen Tools for Digging in the Dirt.* Berkeley, CA: Parallax Press.

Ogden, P., and J. Fisher. 2015. *The Norton Series on Interpersonal Neurobiology. Sensorimotor Psychotherapy: Interventions for Trauma and Attachment* (Deborah Del Hierro & Athony Del Hierro, Illustrators). New York: W. W. Norton & Co.

Parker, J. 2008, February. "The Emotional Brain and Well-Being." *Therapy in LA.* http://therapyinla.com/book-review/.

Perry, B. D. 2006. "Applying Principles of Neurodevelopment to Clinical Work with Maltreated and Traumatized Children: The Neurosequential Model of Therapeutics." In *Working with Traumatized Youth in Child Welfare. Social Work Practice with Children and Families,* ed. N. B. Webb. New York: Guilford Press, pp. 27–52.

Perry, B. D. 2009. "Examining Child Maltreatment through a Neurodevelopmental Lens: Clinical Application of the Neurosequential Model of Therapeutics." *Journal of Loss Trauma* 14: 240–255.

Perry, B. D. 2015. *Train-the-Trainer Advanced Training Series: NMT 2015 Update.* The Child Trauma Academy. Retrieved from http://nmt.childtrauma.org/SupportResource.

Perry, B. D. 2016. *The Neurosequential Model of Therapeutics as Evidence-Based Practice.* The Child Trauma Academy. Retrieved from childtrauma.org/wp-content/uploads/2015/05/NMT_EvidenceBasedPract_5_2_15.pdf.

Pollan, M. 2002. *The Botany of Desire: A Plant's Eye View of the World.* New York: Random House Books.

Porges, S. W. 2004. "Neuroception: A Subconscious System for Detecting Threats and Safety." *Zero to Three* 24(5): 19–24.

Pretty, J., J. Peacock, M. Sellens, and M. Griffin. 2005. "The Mental and Physical Health Outcomes of Green Exercise." *International Journal of Environmental Health Research* 15(5): 319–337.

Redwood, A. 2011. *The Art of Mindful Gardening: Sowing the Seeds of Meditation.* East Sussex, UK: Leaping Hare Press.

Rice, J. S. 2012. "The Neurobiology of People-Plant Relationships: An Evolutionary Brain Inquiry." *Acta Horticulturae* 954: 21–28.

Rice, J. S., and L. L. Remy. 1994. "Evaluating Horticultural Therapy: The Ecological Context of Urban Jail Inmates." *Journal of Home and Consumer Horticulture* 1(2,3): 203–224.

Richards, T. 1991. "Bright Wisdom and New Life at the County Jail." *Toward A Human Future* 4(2): 1–5.

Rizzolatti, G., and M. Fabbri-Destro. 2010. "Mirror Neurons: From Discovery to Autism." *Brain Research* 200(3–4): 223–237.

Rothstein, M. 1990. "Dutton: A Star of 'Piano Lesson' Began Life Anew on Stage." *New York Times* 139, B3, April 19.

Runyon, J. B., M. C. Mescher, and C. M. De Moraes. 2006. "Volatile Chemical Cues Guide Host Location and Host Selection by Parasitic Plants." *Science* 313(5795): 1964–1967.

Serangeli, J., T. Kolfschoten, B. M. Starkovich, and I. Verheijen. 2015. "The European Saber-Toothed Cat (Homotherium Latidens) Found in the 'Spear Horizon' at Schöningen (Germany)." *Journal of Human Evolution* 8: 005.

Siegel, D. J. 1999. *The Developing Mind: Toward a Neurobiology of Interpersonal Experience*. New York: Guilford.

Siegel, D. J. 2007. *The Mindful Brain: Reflection and Attunement in the Cultivation of Well-being*. New York: Norton.

Siegel, D. J. 2008. *The Neurobiology of "We": How Relationships, the Mind, and the Brain Interact to Shape Who We Are* (unabridged edition). Louisville, CO: Sounds True Incorporated.

Siegel, D. J. 2010. *The Mindful Therapist: A Clinician's Guide to Mindsight and Neural Integration*. New York: Norton.

Sneed, C. 1992. *Growing Season: The Movie*. Directed by Nicholas Wellington. Olney, PA: Bullfrog Films.

Stahl, B., and E. Goldstein. 2010. *A Mindfulness-Based Stress Reduction Workbook*. Oakland, CA: New Harbinger Publications.

Star, B., and D. Weaver. 2009. *Coyote Goes Global: A Modern Journey of Forgotten Ways*. Hampshire, UK: O Books.

Stevens, P. 2010. "Embedment in the Environment: A New Paradigm for Well-Being?" *Perspectives in Public Health* 130(6): 265–269.

Topham, A. T., R. E. Taylor, D. Yan, E. Nambara, I. G. Johnston, and G. W. Bassel. 2017. "Temperature Variability is Integrated by a Spatially Embedded Decision-Making Center to Break Dormancy in Arabidopsis Seeds." *PNAS* 114(25): 6629–6634.

Tsuji H., K.-I. Taoka, and K. Shimamoto. 2010. "Regulation of Flowering in Rice: Two Florigen Genes, a Complex Gene Network, and Natural Variation." *Current Opinion in Plant Biology* 14(1): 45–52.

Ulrich, R. S. 1993. "Biophilia, Biophobia and Natural Landscapes. In *Biophilia Hypothesis* ed. S. R. Gellert, and E. O. Wilson. Washington, DC: Island Press.

Van der Kolk, B. A. 2007. "Clinical Implications of Neuroscience Research in PTSD." *Annals of the New York Academy of Sciences* 1071(1): 277–293.

Van der Kolk, B. A. 2014. *The Body Keeps the Score: Brain, Mind and Body in the Healing of Trauma*. New York: Viking.

Ward Thompson, C., P. Aspinall, and S. Bell, eds. 2014. *Innovative Approaches to Researching Landscape and Health*. New York: Routledge.

Wilfong, C. 2010. *The Meditative Gardener: Cultivating Mindfulness of Body, Feelings, and Mind*. Putney, VT: Heart Path Press.

Wilson, E. O. 1984. *Biophilia: The Human Bond with Other Species*. Cambridge, MA: Harvard University Press.

Wilson, E. O. 2002. *The Future of Life*. New York: Knopf.

Wise, J. 2015. *Digging for Victory: Horticultural Therapy with Veterans for Post-Traumatic Growth*. London, UK: Karnac Books.

chapter seven

Theories that inform horticultural therapy practice

Matthew J. Wichrowski

Contents

Horticultural therapy is used to benefit participants in a wide array of settings who have an extensive variety of needs and interests. Historically, its practice has been informed by fields of allied health, including occupational therapy, psychology, social work, as well as vocational rehabilitation and horticulture (Davis 1998). Many different disciplines provide knowledge and information that can be helpful in horticultural therapy practice. The way it is ultimately practiced depends on the setting, practitioner, patient, and program goals. Practitioner style and technique also tend to vary widely based on previous educational, professional, and personal experience. The variety of settings in which horticultural therapy is practiced, from in-room visits with a plant cart to classrooms and sunrooms, to greenhouses and larger outdoor spaces, also necessitates diversity in the skill sets needed for success. Practitioners can benefit from surveying peripheral areas for helpful information to broaden their understanding of the treatment milieu in which they work. This chapter outlines various theories applicable in horticultural therapy work. Practice recommendations are also provided. Overall, information is offered from a broad array of disciplines aimed at informing horticultural therapy practice and helping therapists achieve full potential in their work.

People use nature-based activities and therapies in many countries around the world. While each country's approach may be distinct due to formative influences and practice environments, there is a great overlap in the ways these practices affect the human condition. The focus of practices in different parts of the world tends to be holistic due to the great potential for nature-based activities to affect multiple dimensions of human health and well-being. (See Chapters 2 and 5 for more information on nature-based therapies, and Chapter 1 for more on international organizations.) Also, programs are effective in a variety of health care environments, ranging from palliative care to rehabilitation, to wellness. In order to conceptualize horticultural therapy for this chapter, practice can be modeled as being the interaction of beneficial features of the environment, therapeutic use of self, and utilization of gardening and horticultural activities as prescribed exercise that meet the needs of the participant (see Figure 7.1).

By maximizing potential in these areas, therapists ensure optimal function and competence in their work. A main goal of practice is to maximize benefits for clients, and a therapist's skill level plays a large part in achieving success in therapy. Surveying various theories and best practices allows therapists to sharpen their skills and adapt clinically relevant techniques utilized in other disciplines. (See Chapter 14 for more information on the use of research to inform practice.) This also helps better integrate horticultural therapy goals with overall team goals for patient and program. Exploration of specific subject matter to enrich one's practice abilities is guided by the needs of the population treated, and the focus of the overall program in which the horticultural therapist works guides exploration of specific subject matter. This empowers therapists to achieve success in work with patients, success for the horticultural therapy program, and success for the organization.

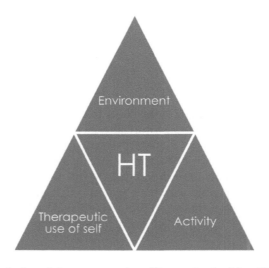

Figure 7.1 Model of horticultural therapy practice. (Illustration by Matt Wichrowski.)

Therapeutic environments

The environment has great potential to have a positive impact on human well-being. The *biophilia hypothesis* (Kellert and Wilson (1993)) proposes that humans have a connection with the natural world, responding favorably to elements that support survival and responding with fear to elements deleterious to survival. Further inquiry supports that biophilic elements of nature enhance psychological well-being (Gullone 2000). In an epidemiological study from Harvard University, James et al. (2016) found that higher levels of greenness in the community were associated with lower rates of mortality. Frumpkin et al. (2017), in a thorough review of the benefits of nature, conclude that contact with nature may offer a range of human health benefits in many areas across the lifespan.

Horticultural therapists specialize in using the environment as part of their work, and a garden environment is typically the clinical space used for programming. The great diversity in environments in which horticultural therapy is provided creates a unique set of challenges for therapists to fashion the optimal setting to support their efforts. This is most apparent when creating a program in an existing setting with potential constraints. While newer health care designs tend to consider views of nature and the importance of creating a more homelike feel, there is typically potential to maximize therapeutic components in an environment that enhances support of the needs and well-being of the people utilizing it. McSweeney et al. (2014) posit that exposure to indoor nature, which people perceive as attractive, pleasing, and pleasant to the senses, has positive effects on psychological and physiological health. Schweitzer (2005), in a survey of environmental features that can have an impact on human health, postulates an array of salutary factors ranging from the design of units and patient rooms to providing views of and access to nature. Benefits from passive experiences of views of nature include positive distraction, which can improve mood and promote restoration from stress (Ulrich 1999). Ulrich (2003) further recommends providing outdoor and indoor gardens, views of nature through windows, and artwork of nature scenes to relieve stress. By utilizing biophilic design principles, one can maximize the therapeutic potential of her or his environment and let nature assist in providing a health-supporting environment. (For a more thorough consideration of these theories, see Chapter 5.)

Ideally, horticultural therapists should strive to utilize nature found in gardens to create a sense of awe in participants. Therapists gain advantage from maximizing the potential benefit of employing nature as a co-therapist in their work. The natural world has the potential to aid horticultural therapists in many ways. Stress reduction, mood enhancement, cognitive restoration, building rapport for traumatized or especially sensitive individuals, and utilizing nature metaphors in the therapeutic process to help achieve goals are all outcomes enhanced by the natural world (Berger 2006; Bratman et al., 2015; Frumpkin 2017). The wondrous sensory potential of the natural world is ripe for exploration. The wise therapist cultivates this potential and is open to the assistance the environment can provide in therapy (Figure 7.2).

People process sensory stimuli in different ways. People may be highly sensitive with some senses and not with others. Individual preference and tolerances of sensory stimulation vary greatly, and medical conditions and treatments sometimes influence these experiences. It is important for the therapist to be aware of and consider these variations in tolerance. Exhibit 7.1 provides more information on sensory processing.

Figure 7.2 Gardens offer multisensory and awe-inspiring experiences. (Courtesy of Stephane Lanel.)

EXHIBIT 7.1

Perspective and Technique: Sensory Processing

Horticultural therapists incorporate knowledge of the five senses—hearing, sight, taste, smell, and touch—in programming, but the sensory systems of the human body encompass much more. Sensory processing is a research subject in several fields, including neuroscience and occupational therapy (Chang et al. 2016; Dunn 2001; Miller et al. 2009).

Winnie Dunn, PhD, OTR, FAOTA, author of Dunn's (1997) model of sensory processing explains behavior from a sensory processing perspective. The model is built on two constructs: neurological thresholds and self-regulation. A threshold is the point at which there is enough sensory input to cause an active response. Dunn theorizes that people have a threshold continuum for every type of sensory input. A low sensory threshold means a person notices and responds to stimuli often because the system is easily set in motion with little input. A high threshold means a person misses stimuli that others notice because the system needs more intense input before it is activated. This construct explains why a person may be sensitive to touch (low threshold for touch) but not notice another sensation such as odor as easily (high threshold for smell). It also explains why two people can have a different response to the same sensory event. A person with a low threshold for sound may cover his ears at a concert, while a person with a high threshold may want to be closer to the stage.

Self-regulation is a behavioral construct that is also explained as being on a continuum. One end is described as passive, the other as active. Applied to sensory processing, a person using a passive strategy would do nothing to control or change

(Continued)

EXHIBIT 7.1 (Continued)

Perspective and Technique: Sensory Processing

what is being experienced. A person using an active strategy would take action to control the amount and type of input causing discomfort. Dunn (1997) illustrates the intersection of these two continuum constructs in her model diagram and the four basic patterns of processing that they create. She places the neurological thresholds continuum on a vertical axis and the self-regulation strategies continuum on a horizontal axis. She labels the resulting quadrants based on the intersection of the neurological thresholds with the self-regulating strategies. Here are the descriptions of her four patterns of sensory processing:

- *Bystander*—a high neurological threshold with a passive self-regulation strategy.
- *Seeker*—a high neurological threshold with an active self-regulation strategy.
- *Avoider*—a low neurological threshold with an active self-regulation strategy.
- *Sensor*—a low neurological threshold with a passive self-regulation.

Horticultural therapists can utilize information from this model to tailor environmental factors and activity process when sensory processing behaviors are observed. For a bystander client, brightly colored and boldly fragrant plant material assists in activating responses. Plan sessions with multiple, fast-paced tasks and opportunities for movement for a seeker. An avoider wants control over sensory stimuli, so provide a workspace away from the group that allows for participation with separation. More details about sensory patterns and how they affect everyday activities are explained in Winnie Dunn's (2008) *Living Sensationally.*

Maria Gabaldo, MEd, OTR/L, HTR
Occupational Therapist for North Suburban Special Education District

The environment should be assessed for its potential to be utilized by therapists to grow or store materials to support their work. Developing high-quality plant and garden displays in and around the setting appeals to the senses and invites exploration for visitors and patients. Patients tend to respond favorably to a diverse, sensory-rich environment, with increased engagement in the group process, aiding in goal achievement. Many images can be found on the internet to provide visual aids to display in the horticultural therapy space, such as a variety of different species from a particular genus or flower form if the specimen is not in bloom. In addition, fresh materials and plants can be brought in as needed. These enhance the informational component of the session and can make it more meaningful for participants. Additional information, such as cultural recommendations and sociocultural uses of the plant, builds richness into the activity. Dried specimens, essential oils, and other props can be stored and utilized when required for a particular session. Plants can be grown in many places around the setting to provide additional examples of a certain specimen or to be used periodically a source for cuttings. Displays around the work setting can provide therapeutic activities for trainees, may reduce stress for staff members and visitors, and bring positive attention to the horticultural therapy program.

Horticulture as prescribed activity

Although many benefits may be realized through passive experience of the environment, horticultural therapy is primarily an activity-based therapy, utilizing horticultural tasks to meet therapy goals. Exercise is a foundation of rehabilitation medicine, general health and wellness, and overall successful aging. Bauman (2016), in a review of factors associated with healthy aging, states that physical activity is the most important part of active aging and has a major role in improving quality of life and reducing disability in later life. Important dimensions of physical activity include aerobic exercise and strength and balance training, which play a major role in health promotion and disease prevention (Selternrich 2017). Gardening and other horticultural pursuits provide a plentiful source of many healthy and enjoyable lifestyle activities (Park et al. 2011).

Horticulture activities also have the potential for a wide range of therapeutic work in the major domains of human function—physical, emotional, cognitive, social, and spiritual. All activities, including gardening, can be conceptualized as a series of steps sequenced together, each having a certain level of difficulty. By being familiar with the array of horticultural activities applicable in a particular setting, horticultural therapists can prescribe a task with the correct type of exercise and level of difficulty. The activity should be challenging, but not frustrating, for the patient and achieve clinical goals (Wichrowski 2005).

While gardening often has a focus on physical activity, a horticultural therapy session has the potential to exercise cognitive functions such as sequencing ability, memory, and problem solving that generalize to many areas of human function. The myriad issues of growing plants in the diverse growing conditions found in horticultural therapy sites present many problems to solve during the course of the season. These create the potential for therapeutic work and provide goals and objectives for those with certain needs. The emotional benefits of stress reduction, mood enhancement, anxiety reduction, and improved self-image and confidence through successful completion of the task are all in the realm of possibility. Therapist-mediated group interactions provide a wide array of benefits in the social realm. It's helpful always to have something to put out on the table to wow participants and encourage exploration in some fashion. It's also important to have a menu of potential activities with different levels of skill requirements so that there is always something to do to meet patient needs. Many garden calendars are available to provide ideas for useful activities. Below is a list of recommended attributes for horticultural therapy activities:

- Ability to meet the therapeutic needs of the patient.
- Multistep, gradeable, and adaptable for different levels.
- A mix of the familiar and comfortable and the unique and exotic.
- Interesting subject with relevant information.
- Sensory involvement.
- Challenging yet attainable for age and ability.
- Opportunities for socialization.

Therapeutic use of self

Therapeutic use of self is a foundation of human service work and a central tenet in horticultural therapy practice (Haller 2017). It is central to the practice of therapy that the relationship between client and therapist is the primary tool used for change (Edwards

Figure 7.3 Young adults with developmental disabilities benefit from the stimulation offered by a garden spider on a bouquet of flowers. (Courtesy of Life Enrichment Center.)

and Bess 1998). Therapeutic use of self takes into account the personal history, personality, and skill sets practitioners utilize in their interactions with clients. Being skilled in this area is very important in the human services field. The first and most important objective of any client-practitioner interaction is the establishment of rapport (Leach 2005). This alliance between therapist and client establishes the foundation for change, growth, and achievement of determined goals (Figure 7.3). (Chapter 3 discusses many aspects of the relationship between the client and therapist.)

Therapist effectiveness

Certain qualities contribute to therapist effectiveness. Therapists identified as master therapists by their peers exhibit a number of similar qualities. They are voracious learners, drawing heavily on accumulated experience. They value cognitive complexity and ambiguity; are emotionally receptive, mentally healthy, and mature' and tend to their own emotional well-being. Aware of how their emotional health affects their work, master therapists possess strong relationship skills and believe in working alliances. Finally, they are experts at using their exceptional skills in therapy (Jennings and Skovhalt 1999).

In a review of studies examining therapist qualities, Ackerman and Hilsenroth (2003) identified the following characteristics as having a positive impact on the therapeutic alliance: flexibility, experience, honesty, respectfulness, trustworthiness, confidence, interest, alertness, and friendliness with a warm and open aspect. Good listening skills are very important and contribute significantly when building a connection between therapist and patient.

Rogerian theory

Carl Rogers's work is very influential in many areas of allied health practice and provides great insight in the therapeutic use of self. In his seminal article, "The Necessary and Sufficient Conditions of Therapeutic Personality Change," he writes that if the following conditions are met and continue over time, constructive personality change can occur:

- Two people are in psychological contact.
- The client is in a state of incongruence, being vulnerable or anxious.
- The therapist is congruent or integrated into the relationship.
- The therapist experiences unconditional positive regard for the patient.
- The therapist experiences empathic understanding of the client's frame of reference and communicates this to the client.
- Communication to the client of the empathic understanding and unconditional positive regard is achieved to a minimal degree (Rogers 1957).

Cultivating a healing presence

McDonough-Means et al. (2004) expand on the concept and definition of healing and on the skills useful in work toward this end. They explore the value of the further reaches of human nature and the spiritual aspects of healing, describing characteristics and development of the healing presence. Healing presence is defined as an interpersonal-, intrapersonal-, and transpersonal-to-transcendent phenomenon that leads to a beneficial, therapeutic, and/or positive spiritual change within another individual and also within the healer. Nature has been shown to promote and enhance spiritual experiences (Williams and Harvey 2001) and provides a good foundation and wide potential for a variety of spiritually oriented pursuits for patient and therapist alike (Fredrickson and Anderson 1999).

Effectiveness of the common factors

There has long been a question in the field of psychology regarding what type of therapy works best. Hubal et al. (1999) examine the many factors related to change in the context of psychotherapy in the book *Heart and Soul of Change.* Through a thorough analysis of a vast amount of research in psychology, they state that 30% of improvement is related to the common factors inherent in most therapeutic relationships, including empathy as the most predictive indicator of being an ineffective or effective therapist (Lafferty et al. 1989), followed by understanding, involvement, warmth, and friendliness. These factors have been linked to positive outcome and client satisfaction (Hubal et al. 1999). Their analysis demonstrates the importance of the above-mentioned factors in providing a positive, clinically relevant, therapeutic experience. Proficiency in therapeutic use of self should be a strong part of every therapist's skill repertoire. Exhibit 7.2 provides an additional perspective on the therapeutic relationship and the role of the horticultural therapist.

EXHIBIT 7.2

Perspective: Patient–Provider Common Ground in Horticultural Therapy Work

Horticultural-based individual therapy and group interaction are critical factors along the rehabilitation route for medically involved patients. Horticultural activities provide stimulating memory associations and meanings, and they instill motivation and in turn further cooperation in the activity. Positive interaction with nature significantly improves pain level (Verra et al. 2012), mood, well-being, and quality of life (Clatworthy 2013). In my experience, individuals form friendly associations toward treatment options that foster their progress and highlight their abilities. Patient–provider empathy is vital to developing trust, and horticultural therapy offers a meaningful diversion that allows individuals to express themselves more freely.

In order to fully assist a patient, discussions about treatment goals and rehabilitation progress are necessary, and should be affiliated with positive feelings of recovery. Patient–provider interaction should supply adequate stimulation in order to reintegrate, reinforce, and enhance skills. Plants are also intricately and directly associated with the seasons and therefore evoke memory and communication. Using plants as a rehabilitation tool not only provides the physiological and emotional connection to healing but also offers a tangible form of progress in the growth of the plants that allows an individual to feel in control, along with embodying behaviors like commitment and responsibility. An important aspect of horticultural therapy is focusing on how an individual has responded to therapeutic presence instead of on the fruits of one's labor. Positive horticultural therapy interactions are beneficial to self-esteem and strengthen self-expression and creativity. Working through plant-based activities is also an opportunity for reflection and building confidence in the process of healing and growth.

Gardening in groups or communally allows the freedom and independence for individuals to reinforce their own positive, healing presence. Entering a garden space, especially in a hospital, unconsciously puts people more at ease. For example, an herb garden provides a nature view and sensory stimulation, which immediately lowers blood pressure. Patients are also gathered for group interactions with other patients who have similar rehabilitation and treatment goals in order to communicate; share experiences; and, most important, take a break from stress and relax. The group dynamic of therapeutic horticulture provides an exciting recreational activity as well as an opportunity for socialization. This type of therapy is attractive to many people, from beginners to avid gardeners. It offers a great source of practice to reinforce adaptive skills learned in other therapies, provides a variety of vocational possibilities postdischarge, and offers the opportunity to become involved in a popular leisure activity supporting a healthy lifestyle.

For horticultural therapists, it is essential to enhance the positive, therapeutic atmosphere that horticulture already provides. One of the goals of horticultural therapy is to plant the seed of positive mental attitude either within the activity instruction or via the therapist's encouragement. Instilling the values of positivity aids individuals in becoming their strongest, most optimistic, motivating force in the face of their trauma and throughout their rehabilitation. Once recovery has been achieved, individuals are not only able to maintain assurance in seemingly

(Continued)

EXHIBIT 7.2 (Continued)

Perspective: Patient–Provider Common Ground in Horticultural Therapy Work

fruitless circumstances but also are more inclined to help and empathize with others. Group gardening within a medical environment induces a positive atmosphere and provides a plant-rich space in which to share values, establish common ground, and to sanction personal community contributions. In the patient-provider dynamic, horticultural therapy is a healing process that is emotionally and physiologically marked by positive associations with personal growth.

Amanda R. Rodriguez-Mammas
Certificate in Horticultural Therapy, Horticultural Therapist,
Visiting Nurse Association of Somerset Hills

Intentional relationship

The Intentional Relationship Model (Taylor 2008) was designed to conceptualize and model therapist-client interactions in the context of an occupational therapy practice. This model takes into account the client's personal history, developmental level, coping style, and current situation. The model helps the therapist to understand the person in treatment and identify factors that may affect treatment. The model also takes the interpersonal style of the therapist into account with an assortment of therapeutic modes utilized in the treatment process, including advocating, collaborating, empathizing, encouraging, instructing, and problem solving. The goals of this model include self-reflection by the therapist for client-therapist intentionality, increasing therapist self-awareness, and understanding the various therapeutic modes utilized in therapy so that the therapist becomes proficient and flexible in their use in the context of treatment (Bonsaken et al. 2013).

This model helps therapist conceptualize and think about their practice. Knowing one's impact within a therapeutic relationship and having self-awareness of their strengths and weaknesses helps sharpen therapists' skill sets. Being comfortable with a range of modalities allows therapists to choose what modality will work best with a certain patient at specific stages in the therapeutic process. There is an emphasis on continued growth and learning to assist therapists in achieving their potential as a helping professional.

Group work

Many horticultural therapy programs utilize group work as the therapeutic format. Working with a group allows for an expanded range of goals and benefits for participants. The therapist can encourage these benefits through the interplay of interpersonal skills and group dynamics. During horticultural therapy sessions, when appropriate opportunities present themselves, therapists can introduce issues, concepts, and observations for discussion. Yalom (1995) speaks of therapeutic qualities of group work. Among those cited, universality, instillation of hope, development of socialization techniques, altruism, and imitation learning are readily applicable in horticultural therapy groups (Figure 7.4).

- *Universality*—We all face challenges in life. This is an inevitable part of the human condition. It is part of human nature to become self-focused while coping with stressful situations and to get bogged down in thinking about one's situation and challenges, losing perspective and feeling sorry for oneself. This can affect mental

Figure 7.4 Working in a group enhances interpersonal skills for work or leisure. (Courtesy of John Murphy.)

status and motivation for improvement. Being with others who are facing challenges and seeing them work and improve their situations demonstrate that challenge is part of life; the client is not alone in facing a particular issue. A wonderful quality of the human condition is that we can work to improve our situation and heal.

- *Instillation of hope*—When someone is facing a challenge, it helps to see the light at the end of the tunnel. Without hope for positive change and outcome individuals might lose motivation to work toward their goals. Someone in the early phases of recovery can observe and interact with others facing both similar and unique challenges who are farther along in the healing process. They might gain motivation from seeing that there is potential for improvement by observing skills and strategies utilized by other group members.
- *Development of socialization techniques*—Some individuals have not experienced supportive conditions in which to learn appropriate functional social skills. Others have lost skills due to injury or disease. The group setting is very conducive to social skills work. Therapists can model specific skills and discuss specific goals for individuals, while patients can watch and learn from the interactions and practice their skills in a supportive, nonthreatening environment. Basic skills such as providing eye contact, taking turns, listening considerately, and sharing can be part of the rules of the group.
- *Altruism*—Altruism can be a powerful factor because many clients, as a result of their disabilities, challenges, and life circumstances, feel that they have little of value to offer. Creating situations that allow someone to help another person fosters self-esteem, self-value, and motivation. Often, the best way to help someone is to let them help you (Yalom 1995). In horticultural therapy, the participant has opportunities to care for plants, as well as to create beauty, which is provided to others in the form of gardens to enjoy, flower arrangements to share, and products of their work to share.
- *Imitation learning*—We learn a lot from watching others. The group context has great potential for teaching interpersonal skills. Group members look to the therapist as the group leader, which allows the therapist to model different skills and techniques for

group members. Group members who are further along on the healing continuum or are more proficient at a particular skill also serve as role models. It helps for the therapist to call attention to and reinforce adaptive behaviors.

Exhibit 7.3 provides some insight into the evolutionary background of the importance of social life in human experience.

EXHIBIT 7.3

Perspectives: Evolution and the Need to Belong

Evolution has left humans with an innate preference for belonging to a small group, or at the very least, being in a significant relationship. Current thinking recognizes that there is a wide range of social structures that fall under the hunter-gatherer label (Panter-Brick et al. 2001). Most agree that the hunter-gatherer life involved working in or around kin-based groups. Daily tasks were largely based on gender. Men hunted in groups, while women foraged for food and medicines. Groups could survive as a community by relying on each other for food, medicine, and childcare. Individuals would find it a tough life on their own. The need to be part of a group, driven by the instinct to survive, appears to be a basis in modern human DNA.

Bowlby (1984) first presented the idea that many psychological issues could be traced back to when young children were separated from their primary caregiver. He proposed Attachment Theory based on Darwinian evolution. He asserted that human infants are born with a series of genetically based behaviors that keep them near a primary caregiver. Staying near a primary caregiver would have kept an infant safe from the many dangers in preagricultural and preindustrial life, such as predation. The primary caregiver would serve as a secure base from which the infant, as it develops into childhood, could safely explore. The child would explore farther and farther from the secure base, running back whenever danger presented itself or the unknown became too scary. Bowlby asserts that this need for a secure base stays with us throughout life.

Johnson (2011) builds on Bowlby's work and applies Darwinian evolution to couples therapy (Couples Emotionally Focused Therapy). When in a relationship, an individual becomes attached to her or his partner. If the partner leaves, either by breaking off the relationship, dying, or intimating that the relationship is in peril, individuals frequently say that they are in pain. Johnson traces this feeling of pain and loss back to early human evolution, which required human connections to survive. Although we don't recall this drive to be attached as being necessary for maintaining life, we certainly react as if our very lives were threatened. Taking this evolutionary approach to couples counseling, Johnson focuses on helping couples listen to each other's emotional needs, which are also driven by fear of losing attachment. By understanding the sore spots—areas where their partner's fear of being unattached are particularly strong—the attachment between the couple deepens, and their relationship is strengthened.

If therapy is to be successful, an attachment of some strength must also be established between the therapist and the patient/client. Fostering and maintaining an appropriate attachment is an aspect of effective therapy that all therapists need to keep in mind as they work. Likewise, when ending therapy, care needs to be taken so that client and therapist alike have closure (Hjorngaard and Taylor 2010).

Beverly J. Brown, PhD, HTR

Social-ecological model

Within the context of clinical practice, it helps the therapist to be aware of factors that can have an impact on the therapy process. The person sitting in front of you and the many specifics of the therapist-patient interaction are the culmination of multiple systems of influence. It's important to acknowledge that each person is an individual with a unique genetic makeup, personality, and history. It's also important to understand that each individual is part of a number of different systems that currently influence her or his life. Understanding the influence of these factors helps the therapist design programs specifically tailored to the identities and needs of participants. Understanding how the systems in their lives influence people also helps the therapist establish realistic parameters for her or his work and realistic expectations of response to therapy. (Information on this model is also provided in Chapter 3.) To further illustrate how this might apply to horticultural therapy practice, see Exhibit 7.4, which describes a project with refugees utilizing a social-ecological (ecological) model.

EXHIBIT 7.4

Program Example: Using Ecological Theory with a Refugee Population

When exploring the many theories related to emotional well-being, one theory that has been found to be quite effective is *ecological theory*. This theory proposes that each person's environment, the context in which the person was raised, and the extent to which there is either discord or concordance within his or her life are all considered when working with the individual. If there is discordance, this theory finds ways to meet one's needs more effectively and help the individual achieve balance in his or her life. One holistic counseling model that can be employed for many populations because it is rooted in ecological theory is horticultural therapy.

An excellent example of horticultural therapy in action is a program in Cincinnati, Ohio, at the Lighthouse Community School, a high school specifically designed for at-risk youth. The curriculum of study includes all the academics, but it also includes time each day for work in the garden where the students and counselors can be together. During this time, the students have an opportunity to speak freely, expend physical energy, and see tangible results of their labor. It has become a therapeutic time of the day, allowing a low-stress environment and open dialogue for the students.

Between 2011 and 2015, many refugees from the African nation of Burundi and the Central American nation of Guatemala began settling in Cincinnati. Many refugees came to St. Leo the Great, a local parish that also has a food pantry. This writer conducted his doctoral research at the University of Cincinnati with a study identifying their on-site garden project as a sustainable intervention. Using ecological theory as the umbrella for horticultural therapy, the refugees were given the opportunity to experience wellness and counseling as well as increase income, food security, and adjustment to their new community. The refugees from both Burundi and Guatemala face many similar challenges, including low income, language barriers, cultural barriers, average family size of eight, and low levels of education for

(Continued)

EXHIBIT 7.4 (Continued)

Program Example: Using Ecological Theory with a Refugee Population

heads of household. Their needs include childcare, education, income, and resources for their families. The overall program goals of the garden project include:

- Decreasing the overall need for access to desired foods from the home country of the refugees.
- Increasing the ability to interact with volunteers and staff members from St. Leo's parish in order to become more proficient in English.
- Increasing their self-sufficiency in terms of work to be done throughout the day and their ability to provide food for their family.

The garden uses the tenets of trauma-informed care with sensory-based activities: working in the soil, stretching to weed and pick fruit, and aromatherapy for relaxation. In addition to the garden project on the grounds of St. Leo, many of the refugees also take the excess starts of tomato and pepper to their four-by-five-foot garden in the backyard of their apartments. This allows them to supplement their diet by having twenty-four-hour access to the foods they enjoy.

The garden plots that were originally farmed for the first two years were donated by Finley Market Garden, but transportation was a challenge due to distance and accessibility at the top of a large hill. The current, more sustainable garden is on the hillside behind the church. The fresh produce grown can also be taken to the food pantry located across the street from the church. In this church garden, the refugees are closer to more resources and support from the St. Leo staff members, including counselors and caseworkers. Access is very easy, and since it is owned by St. Leo, rent is free and there is much greater security for the tools, produce, and their well-being. By using the seeds donated by the Civic Garden Center of Cincinnati, the refugees have many options for the crops they can grow.

The garden has indeed provided not only financial supplement but also been the place for finding emotional support and sense of purpose. This gardening project serves as a model for anyone working with a refugee population.

Jonathan Trauth, EdD, LISW-S
Visiting Assistant Professor, Miami University

Environmental press

The importance of environment is also influential in the theory of Environmental Press. This theory can inform horticultural therapy practice. Environmental Press examines the many-layered demands placed on us by the surrounding environment during the course of life. We must constantly navigate and negotiate the elements of our world in meeting our needs, from shelter, safety, and food to social acceptance, esteem, and on to self-actualization. Skill and mastery in these interactions foster accomplishment and success in life's activities. See Exhibit 7.5 for an explanation and example of Environmental Press in practice (Figure 7.5).

EXHIBIT 7.5

Techniques: Environmental Press

The Environmental Press (EP) model is one we use quite regularly in the therapeutic horticulture program at the University of Florida. The model works well in therapeutic horticulture and horticultural therapy settings because of the inherent opportunities that exist to vary the demands it places on clients. Just as there are many ways to adapt a horticulture activity to successfully challenge a client, there are numerous ways to manipulate the horticulture environment to match its demands to the client's needs and goals.

Although the EP model was developed with older adults in mind, it can be used effectively with many populations. The model describes the outcome of the interaction between a person and his or her environment. It looks at two major components: (1) an individual's *competencies,* such as physical health, cognitive, and sensory-perceptual abilities, ego strength, and the resulting behaviors, and (2) the impacts of *press of the environment* on that individual. (*Press* includes demands or limiting aspects as well as stimulation levels of the environment.)

The demands of a garden or greenhouse setting can be varied in many ways to increase or decrease its EP. Once the therapist understands the client's competencies, the optimal therapeutic environment can be created through the manipulation of the setting's layout, material access, activity demands, and even environmental distractions.

For example, movement is an important goal for many of our clients with Parkinson's disease or those who are recovering from a stroke. By spreading the tools and materials throughout the greenhouse, we can increase the setting's EP and the clients' level of movement during the session. We can also increase the EP by placing things so that they must reach for them and engage their range of movement. We can decrease the EP by ensuring that there are balance supports throughout the greenhouse as the clients move through the space. At a worktable setting, we can increase or decrease the EP by placing items within or out of reach to address range of motion, bilateral coordination, perceptual neglect, and social interaction.

It is essential, of course, to understand the competencies of the client in order to hit that optimal zone of EP: an environment that promotes reasonable challenge but doesn't frustrate. If a client is exposed to a low level of EP, it can cause boredom, lethargy, and a decrease in competence. If the level of EP is too high, the client may respond with adverse behavior and negative emotional states. In the optimal zone of EP, the client is simultaneously supported and stimulated, which leads to positive affect and adaptive behavior.

Elizabeth R. M. Diehl, PLA, HTM
Director of Therapeutic Horticulture, Wilmot Gardens at the University of Florida

Flow

Flow Theory, developed by Csikszentmihalyi (1990), also has wide applicability in horticultural therapy practice. All activity has a specific difficulty level, and all performers of activity have a level of skill applicable to complete that specific activity. When the level of difficulty matches the level of skill in the right proportion, a condition called flow can exist. Csikszentmihalyi (1990) defines flow as a sense that one's skills are adequate to cope

Figure 7.5 The environmental press model encourages a sense of mastery. (Courtesy of Beth Bruno.)

with the challenges at hand, in a goal-directed, rule-bound action system that provides clear clues about how well someone is performing. Concentration is so intense that there is no attention left over to think about anything irrelevant or to worry about problems. Self-consciousness disappears, and the sense of time becomes distorted. An activity that produces such experiences is so gratifying that people are willing to do it for its own sake. See Figure 7.6 on flow.

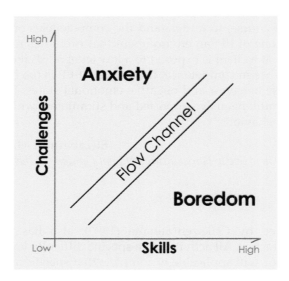

Figure 7.6 Flow diagram.

EXHIBIT 7.6

Techniques: Flow in a Horticultural Therapy Session

On numerous occasions while running a horticultural therapy group, it has become very quiet. As the group leader, I worry, am I doing something wrong? Should I talk about the plant more? Should I stir up some conversation? But when I look around the room and see the participants fully concentrating, intently focused on their task, with smiles and looks of contentment on their faces, I realized that this is a good thing. They were in the flow channel. At the end of these sessions, patients often comment that they enjoyed the session even if they didn't garden at home or state that they enjoyed plants much at all. Now I try to set up opportunities for flow and promote it whenever I can.

Matthew Wichrowski, HTR

Horticultural therapy and gardening activities have great potential for flow experiences. Garden tasks have wide levels of difficulty, with activities varying from tending plants on the windowsill to earnestly double-digging an outdoor vegetable bed in clay soil. You can grow a single plant or a thousand, depending on your time, energy, skill level, and therapeutic need. From a cognitive standpoint, there is always something new to learn and challenges to overcome. When the therapist organizes an activity where challenge meets ability in the right proportions, participants can experience the benefits of flow, relaxation, restoration, and satisfaction during their therapeutic exercises. See Exhibit 7.6 for an example in practice.

Resiliency

Horticultural therapists work with individuals experiencing challenges. Each person's coping abilities tend to vary, and some individuals are more proficient or resilient at bouncing back effectively and overcoming obstacles in life. Theoretical understanding of resilience came from the study of children at risk (Green 2002). Theories of resilience have also been informed by research and therapy with survivors of highly traumatic experiences and the promotion of positive psychosocial adaptation (Green 2002). White et al. (2008) define resilience as qualities that allow individuals to flourish in the face of adversity. Thompkins and Schwartz (2009) add that it is the process of human growth and development emerging from successful adaptability and healthy coping skills. It is one's ability to overcome stressors, misfortunes, and unforeseen circumstances. Richardson (2002) states resilience is the internal force that drives each individual to seek wisdom, self-actualization, altruism, and inner peace.

Resiliency includes a person's individual competencies, such as one's strengths, positive character attributes, determination, and realistic perspective. It also includes one's social competencies such as social skills and adaptability. Spirituality has also been cited as a large contributor to resilient coping skills (Green et al. 2002). Note that resiliency skills are not superhuman in nature and can be learned and developed by anyone. The group setting allows for therapist feedback, modeling, and imitation learning from watching other group members, which aids in skill development.

Horticultural therapists can promote and enhance the use and learning of resiliency skills in the course of their group work. Calling attention to character strengths exhibited by group members is one way. Pointing out the beneficial nature of these strengths in surviving a tough situation, coping well, and getting back to a good life reinforces the effort by the individual. This also points out effective coping skills to others in the group. It is important to remember that a person is much more than their diagnosis; all people have assets and abilities that aid in personal growth. Developing a sense of purpose also has been shown to help people develop great strength when experiencing challenging times. Having something to live for provides the motivation to push through tough times. See Exhibit 7.7 for an account of the use of horticultural therapy to promote resilience.

EXHIBIT 7.7

Case Example—Promoting Resilience in Children

In 2016, I started a social and therapeutic program at Mensajeros de la Paz, in Lima, Peru.

Mensajeros de La Paz is a Spanish nonprofit organization that gives integral cognitive and emotional support to two hundred children from various shanty towns in Lima. The main focus is to prevent and reduce domestic violence. This organization has been working in Peru since 1993. Children attend the center every afternoon after school. They receive lunch and are assigned a teacher who gives them support with their homework.

The horticultural therapy program was offered three days a week. We worked with thirty children every afternoon, in groups of ten, for thirty minutes. The program goals were: (1) to enhance social interaction and group work among students because they came from different schools, (2) reduce violent outbursts (many of the children came from very dysfunctional homes), (3) connect children with nature, (4) enhance nurturing of another living organism, and (5) improve knowledge in science. Every week we planned different activities, which included seed germination, identification of plant parts, transplanting vegetable seedlings in a small vegetable garden, flower pressing, succulent propagation, pot decoration, sprout production, and terrarium planting, among others.

The program had 100% attendance and acceptance. Teachers and other staff members and psychologist were impressed by the children's willingness to participate in the activities. Our program gave children the opportunity to be creative in a nonthreatening setting in which they were able to channel their aggressive behavior. But above all, it gave them a sense of worth, which was really meaningful for these children who were growing in very poor conditions, in very hostile environments. They felt loved, cared for.

This pilot program demonstrated the importance of gardening for this population. These children need to know that there are other aspects of life that are not violent or aggressive. Gardening offers them an outlet to feel good and valued.

Daniela Silva-Rodriguez Bonazzi
Biologist and Horticultural Therapist,
Director, Asociacion de Horticultura Terapéutica y Social del Peru (AHTSP)

Gardening activities provide a wide array of potential to acquire and utilize resiliency skills. It's motivational to engage in meaningful activities that are enjoyed by both the client and therapist. It is useful to explore the personal and cultural history and use this information to develop meaningful activities. The therapist can encourage curiosity and exploration of the environment. Horticultural therapists, through their training and work focus, become docents of the natural world. Sharing this knowledge through gardening experiences illuminates the wonders of nature for participants. A microscope or portable magnifying device can be helpful in these explorations of miniature worlds of wonder. It's a rewarding experience when the therapist sees a participant light up with curiosity, excitement, and awe when exploring nature. Keeping up his or her fund of information helps the therapist create a rich experience for participants.

Mindfulness

For the therapist

Kabat-Zinn (1990) describes mindfulness as a "consciousness discipline" that can be explained "as the intentional cultivation of non-judgmental moment to moment awareness." This practice has the potential to help cultivate skills and attitudes for therapists to enhance their therapeutic use of self, as well as provide exercises and activities for therapists to utilize with patients. (See Chapter 6 for background theory.)

For therapists, mindfulness can be developed through mindfulness meditation, a meditative exercise aimed at focusing on present experience in the role of a witness. The experiences might be sensual, physical, emotional, or mental experiences. Over time, this exercise helps the practitioner build skills in nonjudgmental acceptance, nonattachment, beginner's mind, nonstriving, gentleness, and kindness. Mindfulness helps the therapist become "better able to experience feelings of compassion, calm, and competence along with a deeper understanding of other people" (Schmidt 2004). Meditation can be a very useful tool to build empathy, increase presence of focus, and develop other important skills useful in the counseling setting. See Exhibit 7.8 for a further explanation of mindfulness.

EXHIBIT 7.8

Techniques: Mindfulness—Stopping to Smell the Roses

Mindfulness refers to a state of awareness of self in the present moment or to a supporting therapy seeking behavioral change. Mindfulness techniques, or attention regulation exercises, train the mind to present moment awareness by resting attention on a sensory anchor such as the breath, sounds, sights, or sensations. Evidence suggests that developing nonjudgmental attention deliberately and with intention to here-and-now experience positively reinforces awareness of self and is a pathway toward decreasing stress, better compliance with medically induced lifestyle changes, or improving the management of pain.

Mindfulness-based stress reduction (MBSR) is a well-known protocol, popularized by Jon Kabat-Zinn, researcher at the University of Massachusetts Medical School, who first developed the eight-week protocol with medical doctors experiencing burnout in the 1980s (1990).

(Continued)

EXHIBIT 7.8 (Continued)

Techniques: Mindfulness—Stopping to Smell the Roses

Mindfulness-based cognitive therapy (MBCT) traces its pedigree to cognitive behavioral psychology. The protocol, developed by Segal et al. (2002), targets the regulation of emotions and, more specifically, addresses relapse into depression. Unlike classic cognitive behavioral therapy, which seeks to help patients reframe their ways of thinking and exercise better control of thoughts, sensations or emotions, MBCT teaches patients to recognize triggers and anticipate the first signs of emotional reactivity.

Mindfulness-based art therapy (MBAT), promoted by Rappaport (2014), is particularly salient for the horticultural therapist. The sensory qualities of art media are the basis on which mindfulness instructions are built. Acute attention to materials strengthens opportunities to attain flow and thus to reduce negative emotions and disruptive ruminative thinking, of past or future. The here and now of art making fosters individual attunement to personal values, which in turn may support goals of increased psychosocial well-being.

A horticultural therapist may incorporate mindfulness into professional practice in order to maximize clients' scope of immersion (extent) while being engaged effortlessly (soft fascination) (Kaplan and Kaplan 1989). *Noticing games* are a primary tool. For example, at the beginning of a horticultural therapy session participants may be invited to do their own internal *weather report*, using the moment to arrive, settle, and notice how they feel physically or emotionally. This may take the form of a short meditation or silent self-contemplation. During the session, the therapist provides cues for *slow down or stop*. Supplies are distributed in a controlled fashion. Time is taken for examining elements separately. In my practice with stressed urban professionals who have little or no experience of gardens or plants, I anchor group activities heavily in the sensory. During a planting exercise, for instance I may invite clients to explore materials thoroughly, with as many senses as appropriate. Bringing here-and-now attention to object weight, color, density, asperities, and textures, they also turn to themselves, checking in, attending to the palm of their hands, their skin, their nails, and the space between their fingers. Clients may be asked to describe succinctly what they observe—a sensation, a thought, a feeling, pleasant or unpleasant. The suggestion to photograph or sketch details may be a creative corollary. Clients invariably report that the unhurried pace of the program is salutary.

Tamara Singh, MA
Horticultural Therapist, France, Clinical Practitioner, Hopital Privé D'Antony

For clients and patients

Mindfulness exercises can be utilized with program participants as well. The garden setting provides many opportunities for this practice. Try the following:

- Find a comfortable place in the garden, one that isn't associated with familiar tasks or activities. It should be a place that becomes associated with the exercise. It is also possible to do this as a walking meditation through the garden.

- Prepare for the experience by clearing the mind of thoughts. Taking five slow, deep cleansing breaths can assist in this transition. The next steps will be to utilize all of the senses one-by-one to experience the presence of the garden.
- Sit with eyes closed and listen to the symphony of the garden. Listen to the qualities of the different sounds. Try to differentiate the many different parts of the orchestra. This activity also works well at night.
- Smell the fragrances of the garden. Try to experience and differentiate the different notes of each scent in the garden. This can also be done as a walking activity, visiting the various sensory plants in the garden and sitting with them for a few minutes to experience their aroma.
- Experience the tastes of the garden. Contemplate the essence of taste and flavor. What does it actually mean to be sweet or bitter, to taste like mint or basil?
- Touch the various textures in the garden. Get to know the differences in leaves and bark; the soft, the fuzzy, the smooth, the waxy, the crisp, and the sharp. Focus on the essence of the texture. What does it mean to be waxy, fuzzy, soft, or hard?
- Look from far away to close up. Examine the variegation of the leaves. See all the different greens in the garden. How many different shades of green can you see? Use a magnifying glass or microscope to uncover the hidden worlds around you. Sit for a while and watch as the birds and other garden inhabitants come out. Be quiet and still, witnessing to the vast complexity of life. If one sits under a tree for a day, and carefully observes, one could write a thousand pages describing the experience.

The concept of forest bathing fits well with mindfulness exercises. One can walk into the forest with the intention of leaving one's situation behind, to be in the present with an open mind to explore the wonders of the forest and become immersed in its myriad sensory potential. Research on forest bathing has demonstrated many health benefits as well, including lowered blood pressure and increased immune function (Tsunetsugo 2013).

Mindfulness techniques have a huge range of potential uses for the horticultural therapist. They can be practiced in a range of settings, from bedside to the greenhouse to the outside garden and on to larger outdoor settings. Overall mindfulness techniques are helpful in developing skill in the therapeutic use of self as well as providing a range of activity ideas for patient care.

Positive psychology

Psychology has many useful techniques to help with pathological conditions, but it can also be used to assist people in living a contented, fulfilled life. Positive psychology explores factors that make life worth living and the human strengths that enable individuals to confront challenges, appreciate others, and regard daily experiences as meaningful (Dunn and Dougherty 2005). (Chapter 3 also describes positive psychology.) In rehabilitation, subjective appraisal plays a major role in recovering or regaining strength after a disabling injury (Elliot 2002). People with an optimistic attitude about their rehabilitation are usually able to ward off depression and hopelessness that can compromise progress in therapy (Seligman and Csikszentmihalyi 2000). The health benefits of a positive attitude even extend to improved immune function (Aspinwall and Tedeschi 2010).

Positive mood may be elevated through physical, social, and mental activities, which encourage an outward focus and active engagement in one's environment (Watson 2002).

Positive psychology techniques include altruism, finding meaning, and benefit finding (Nolen-Hoeksema and Davis 2002). These concepts and skills can be introduced and reinforced by the therapist in discussion during the activity and often result organically due to the nature of being part of a supportive group and through cultivating plants.

Horticultural therapy goals and techniques mesh well with the goals, philosophies, and techniques of positive psychology. Generally speaking, horticulture, and gardening are considered enjoyable hobbies. Mood tends to shift in a positive direction. People often report positive experiences (Bratman et al. 2015), transcendental experiences (Williams and Harvey 2001), relief from stress (Ulrich et al. 1991), and improvement in self-rated health and happiness (Collins and O'Callaghan 2008). Planting is associated with the hope and promise of good things in the future. Many enjoyable aesthetic and sensory experiences can be found in nature and the well-designed garden. The therapist's role is to create an array of positive activities and therapeutic experiences to share with participants.

Philosophers have long pondered the purpose of life and how to be happy (Denier and Seligman 2004). Dunn (2008) has thoroughly explored this concept and described the good life, which includes enjoying pleasures of the flesh and the mind in balanced proportions, making connections with others, developing positive traits, performing purposeful acts for self and others, recognizing beauty, acquiring knowledge, and enjoying friendship. These activities, associated with a contented, fulfilled existence, all have the potential to be realized in the garden setting with a trained therapist to facilitate group and socially derived benefits (Figure 7.7).

Figure 7.7 Flower arranging can be a positive and therapeutic activity. (Courtesy of Debra Edwards.)

Finding meaning

Finding meaning after traumatic and life-altering experiences is a significant part of coping with life's challenges. Finding meaning in this context includes reattribution and causal understanding, perceptions of growth or positive change, integration of stressful experience into one's identity, and restored or changed meaning in life (Park 2010). Working in the nonjudgmental context of nature with an understanding and supportive therapist fosters this growth. Meaning in life has been associated with a higher quality of life, a greater sense of well-being, and lower levels of depressive symptoms in patients facing a variety of physical and psychological challenges (Dezutter et al. 2013).

Meaning also is a factor in providing interesting and motivating activities for patients. It's important for horticultural therapists to have thorough horticultural knowledge regarding the elements of their activity. It is also important for therapists to know the steps of the activity so that they can customize it for the skill level and interests of the group. In some cases, steps need to be modified to make them easier so that patients with limitations can be successful or to make them a bit harder to present the appropriate amount of challenge in order to achieve goals. A motivated patient greatly assists the work of therapy.

Thorough horticultural knowledge allows therapists to impart information. This makes the activity more interesting and meaningful for the patient. In some cases, the patient doesn't have interest in planting or keeping plants at home. In these situations, the patient may be motivated and engaged with a fascinating or thought-provoking story about the plant or with a thorough sensory exploration and discussion. Relating the interesting history of coffee; tasting the many different varieties of tea; smelling jasmine, lavender, or roses and discussing the nature of scent; or discussing the effects of the spice trade on the Renaissance in Europe or the great tulip crash are just a few of the possibilities. Sometimes it's not so much the product but the process that is more important.

Applying theories in a horticultural therapy session

This part of the chapter discusses the previous theories and techniques in the context of a hypothetical horticultural therapy group of patients in a physical rehabilitation program. Although the patients have specific needs defined by their medical conditions, they also share general human needs. Horticultural therapy has the potential to meet a wide range of needs, including physical, emotional, cognitive, and social (Relf 1981). An activity is set up and patients begin to arrive. The first patient to arrive is new. Matt is a 75-year-old client who had a heart attack and subsequent coronary artery bypass graft (CABG) after testing revealed blockages, which were limiting oxygen to his heart muscle. The therapist greets him warmly and welcomes him to the horticultural therapy program. The therapist inquires if he is a gardener or if he has plants at home and also asks how his rehabilitation stay is going. In the discussion with Matt, the therapist explains the structure of the horticultural therapy program and how it fits into the rehabilitation regimen. The therapist discusses the benefits for Matt's specific condition and what goals can be accomplished with the activity. This helps Matt feel comfortable and become aware of the potential benefits if he is not familiar with horticultural therapy. This also activates expectations of benefits and instills hope, which has been shown to influence progress. The next two patients arrive. The first is Rachel, a 40-year-old woman, in rehabilitation recovering from the effects of Guillain-Barre syndrome, which has left her with residual weakness in her four extremities. The other is Jean, a 70-year-old woman who had a hip fracture after a fall in her bathroom. They have attended a few sessions and are motivated to maximize the benefits of their rehabilitation stay. They attend all sessions they can and work hard. They also have an optimistic attitude toward their prognosis. The last to arrive is John, a 67-year old man who had a stroke. He has been coming to group for two weeks and has made good progress in his recovery. He is getting ready to go home in a few days. The therapist introduces the new member to the rest of the group and engages in some small talk. The therapist asks how their therapies have been going and comments on the progress in Rachel's endurance and in Jean's standing tolerance.

Then the activity is introduced. Today it is dividing and planting a selection of sweet basil, lemon basil, and cinnamon basil seedlings, allowing for further explorations of scent, and purple basil for color. A few of the patients enjoy cooking and look forward to taking the herb seedlings home. Along with cultural requirements for the plants, a little history is provided, and discussion is encouraged to help build group cohesion, enhance the meaningfulness of the activity, and increase motivation. This is helpful for John, whose attention and focus has been affected by the stroke. He also has mild dysarthria, which affected his speech ability. Participating in the discussion in a supportive nonjudgmental group allows Matt to practice more comfortably the recommendations and strategies from his speech therapist. Jean feels relaxed in the garden setting and comfortable with the therapist and group members. She shares how she was stuck in her bathroom for two days when she fell. She had nothing to eat or drink and was worried that she would die there. Fortunately, her family checked on her when she didn't answer her phone, and she made it through her traumatic experience. The therapist praises her strength and resilience. She is a tough survivor. Whenever a participant exhibits adaptive coping techniques, the therapist acknowledges their benefit. It also highlights positive coping strategies and demonstrates the universality of human challenge in its various forms for fellow group members. This is encouraging for Rachel, who is still uncertain of her final prognosis and concerned if she will return to complete independence. The therapist has analyzed the steps of the activity and designed it to present a challenging but not frustrating level of difficulty. Assistance and adaptive techniques are provided when needed to avoid frustration, enabling patient abilities and encouraging flow experiences. The therapist is aware of individual goals and objectives (such as increased endurance and stress management for Matt, bilateral upper extremity exercise for independence in activities of daily living and reducing anxiety for Rachel, psychosocial support for Jean, and bilateral upper extremity exercise and improved sequencing ability for John) and encourages their achievement in the group context.

As the activity moves toward completion, the therapist might engage in more process and encourage interaction. Questions are answered, loose ends tied up. The therapist comments on individual progress. It also gives people who work more slowly the time to finish. The time also allows for exploration of patient interests in order to help plan future interesting and motivational activities that will engage participants. Last, the therapist speaks with the new patient to reinforce the potential benefits from participation and wishes all of them a good day.

The skills and recommendations exemplified by this physical rehabilitation group can apply to many different group types with diverse goals and individual objectives.

Putting it all together

Horticultural therapy is exercised in many creative ways: It is influenced by the skills and therapeutic style of the therapist and by the wide variety of settings in which it is practiced. Developing best practices is informed by an array of supporting theories from many different allied health perspectives. While knowledge of the core skills in horticultural therapy is a necessary foundation for competent practice, keeping current with peripheral theories, practice recommendations, and new developments helps the horticultural therapist stay current and offers opportunities to improve practice. Maximizing skill sets in the areas of environmental enhancement; developing clinically relevant, interesting, meaningful, and motivating horticultural activities; and therapeutic use of self can maximize therapeutic effectiveness in therapists' horticultural therapy practice.

Key terms

Activities of daily living (ADL) A term in physical rehabilitation pertaining to those activities most people perform to accomplish their typical routines in life.

Dysarthria Difficulty in articulating words caused by weakness in the muscles used for speech.

Self-actualization-concept From Abraham Maslow, developing one's skills and abilities and applying this toward achieving one's maximum potential.

Sequencing Linking together of component steps of a specific activity in order to ensure its successful completion.

References

Ackerman, S. J., and M. J. Hilsenroth. 2003. "A Review of Therapist Characteristics and Techniques Positively Impacting the Therapeutic Alliance." *Clinical Psychology Review* 23: 1–33.

Aspinwall, L. G., and R. G. Tedeschi. 2010. "The Value of Positive Psychology for Health Psychology." *Annals of Behavioral Medicine* 39: 4–15.

Bauman, A., D. Merom, D, F. Bull, D. Buchner, and H.M. Fiataron-Singh. 2016. "Updating the Evidence for Physical Activity: Summative Reviews of the Epidemiological Evidence, Prevalence, and Interventions to Promote Active Aging." *The Gerontologist 56* (supp.2): S. 268-S. 280.

Berger, R. 2006. "Using Contact with Nature, Creativity, and Rituals as a Therapeutic Medium with Children with Learning Disabilities: A Case Study." *Emotional and Behavioral Difficulties* 11(2): 135–146.

Bonsaken, T., K. Vollestad, and R. Taylor. 2013. "The Intentional Relationship Model: Use of Therapeutic Relationship in Occupational Therapy Practice." *Ergoterapeuten* 5: 26–31.

Bowlby, J. 1984. "Psychoanalysis as a Natural Science." *Psychoanalytic Psychology* 1(1): 7–21. doi:10.1037/0736-9735.1.1.7.

Bratman, G. N., P. Hamilton, K. S. Hahn, G. C. Daily, and J. J. Gross. 2015. "Nature Experience Reduces Rumination and Subgenual Prefrontal Cortex Activation." *Proceedings of the National Academy of Sciences* 112(28): 8567–8572.

Chang, Y.-S., M. Gratiot, J. P. Owen, A. Brandes-Aitken, S. S. Desal, S. S. Hill, A. B. Arnett, J. Harris, E. J. Marco, and P. Mukherjee. 2016. "White Matter Microstructure Is Associated with Auditory and Tactile Processing in Children with and without Sensory Processing Disorder." *Frontiers in Neuroanatomy* 9: 169.

Clatworthy, J., J. Hinds, and P. M. Camic. 2013. "Gardening as a Mental Health Intervention: A Review." *Mental Health Review Journal* 18(4): 221–225.

Collins, C., and A. O'Callaghan. 2008. "The Impact of Horticultural Responsibility on Health Indicators and Quality of Life in Assisted Living." *HortTechnology* 18(4): 611–618.

Czikszentmihalyi, M. 1990. *Flow: The Psychology of Optimal Experience*. New York: Harper Collins.

Davis, S. 1998. "Development of the Profession of Horticultural Therapy." In *Horticulture as Therapy: Principles and Practice*, ed. S. Simson and M. Straus. Binghamton, NY: Haworth Press, pp. 3–18.

Denier, E., and M. E. P. Seligman. 2004. "Beyond Money: Towards an Economy of Well-being." *Psychological Science in the Public Interest* 5: 1–31.

Dezutter, J., S. Casalin, A. Wacholtz, L. Koen, J. Hekking, and W. Vandewiele. 2013. "Meaning in Life: An Important Factor for the Psychological Well-being of Chronically Ill Patients." *Rehabilitation Psychology* 58: 334–341.

Dunn, D. S., and C. Brody. 2008. "Defining the Good Life after Acquired Disability." *Rehabilitation Psychology* 53: 413–425.

Dunn, D. S., and S. B. Dougherty. 2005. "Prospects for a Positive Psychology of Rehabilitation." *Rehabilitation Psychology* 53: 305–311.

Dunn, W. 1997. "The Impact of Sensory Processing Abilities on the Daily Lives of Young Children and Their Families: A Conceptual Model." *Infants and Young Children* 9: 23–35.

Dunn, W. 2001. "The Sensations in Everyday Life: Theoretical, and Pragmatic Considerations." *American Journal of Occupational Therapy* 55: 608–620.

Dunn, W. 2008. *Living Sensationally: Understanding Your Senses.* London, UK: Jessica Kingsley Publishers.

Edwards, J. K., and J. M. Bess. 1998. "Developing Effectiveness in Therapeutic Use of Self." *Clinical Social Work Journal* 26: 89–105.

Elliot, T. R. 2002. "Defining Our Common Ground to Reach New Horizons." *Rehabilitation Psychology* 47(2): 131–143.

Fredrickson, L., and D. Anderson. 1999. "A Qualitative Exploration of the Wilderness Experience as a Source of Spiritual Inspiration. Developing Effectiveness in Therapeutic Use of Self." *Journal of Environmental Psychology* 19(1): 21–39.

Frumpkin, H., G. N. Bratman, and S. J. Breslow. 2017. *Environmental Health Perspectives.* 125(7): 075001. doi:10.1289/EHP1663

Green, R. R. 2002. "Human Behavior Theory: A Resilience Orientation." In *Resilience: An Integrated Approach to Practice, Policy, and Research,* ed. R. R. Green. Washington, DC: NASW Press, pp. 1–28.

Green, R. R., N. J. Taylor, M. L. Evans, and L. Smith. 2002. "Raising Children in an Oppressive Environment." In *Resilience: An Integrated Approach to Practice, Policy, and Research,* ed. R. R. Green. Washington, DC: NASW Press, pp. 241–275.

Gullone, E. 2000. "The Biophilia Hypothesis in the 21st Century: Increasing Mental Health or Pathology." *Journal of Happiness Studies* 1(3): 293–322.

Haller, R. 2017. "Working with Program Participants." In *Horticultural Therapy Methods: Connecting People and Plants in Health Care, Human Services, and Therapeutic Programs,* ed. R. L. Haller and C. L. Capra. Boca Raton, FL: CRC Press, Taylor & Francis Group, pp. 70–74.

Hjorngaard, T., and B. S. Taylor. 2010. "Parent Perspectives: The Family-Therapist Relationship and Saying Good-Bye." *Physical & Occupational Therapy in Pediatrics* 30(2): 79–82. doi:10.3109/01942630903543625.

Hubal, M. A., B. L. Duncan, and S. D. Miller. 1999. *Heart and Soul of Change: What Works in Therapy.* Washington, DC: Island Press.

James, P., J. E. Hart, R. F. Banay, and F. Laden. 2016. "Exposure to Greenness and Mortality in a Nationwide Prospective Cohort of Women." *Environmental Health Perspectives* 124(9): 1344–1352.

Jennings, L., and T. Skolvolt. 1999. "The Cognitive, Emotional and Relational Characteristics of Master Therapists." *Journal of Counseling Psychology* 46: 3–11.

Johnson, S. 2011. *Hold Me Tight: Seven Conversations for a Lifetime of Love.* Boston, MA: Little Brown & Co.

Kabat-Zinn, J. 1990. *Full Catastrophe Living: Using the Wisdom of Your Body and Mind to Face Stress, Pain and Illness.* New York: Delacorte Press.

Kaplan, R., and S. Kaplan. 1989. *The Experience of Nature: A Psychological Perspective.* Cambridge, UK: Cambridge University Press.

Kellert, S. R., and E. O. Wilson. 1993. *Biophilia Hypothesis.* Washington, DC: Island Press.

Lafferty, P., Beutler, L. E., and Crago, M. 1989. "Differences Between More and Less Effective Therapists: A Study of Select Therapist Variables." *Journal of Consulting and Clinical Psychology* 57(1): 76–88.

Leach, M. J. 2005. "Rapport: A Key to Treatment Success." *Complementary Therapies in Clinical Practice* 11: 262–265.

McDonough-Means, S. I., M. J. Krietzer, and I. R. Bell. 2004. "Fostering a Healing Presence and Investigating Its Mediators." *Journal of Alternative and Complementary Medicine* 10(Suppl 1): S-25.

McSweeney, J., D. Rainham, S. A. Johnson, S. B. Sherry, and J. Singelton. 2014. "Indoor Nature Exposure (INE): A Health Promotion Framework." *Health Promotion International*. 30(1): 126–139.

Miller, L. J., D. M. Nielsen, and S. S. Schoen. 2009. "Perspectives on Sensory Processing Disorder: A Call for Translational Research." *Frontiers in Integrative Neuroscience* 3: 22.

Nolen-Hoeksema, S., and C. G. Davis (2002). "Positive Responses to Loss: Perceiving Benefits and Growth," in *Handbook of Positive Psychology*, ed. C. R. Snyder and S. J. Lopez. New York, NY: Oxford University Press, pp. 598–606.

Panter-Brick, C., R. Layton, and P. Rowley-Conwy. 2001. *Hunter-Gatherers: An Interdisciplinary Perspective*. New York: Cambridge University Press.

Park, C. L. 2010. "Making Sense of the Meaning Literature: An Integrative Review of Meaning Making and Its Effects on Adjustment to Stressful Life Events." *Psychological Bulletin* 136(2): 257–301.

Park, C. L., K.-S. Lee, and K.-C. Son. 2011. "Determining Exercise Intensities of Gardening Tasks as a Physical Activity Using Metabolic Equivalents in Older Adults." *HortScience* 46: 1706–1710.

Rappaport, L., ed. 2014. *Mindfulness and the Arts Therapies: Theory and Practice*. London, UK: Jessica Kingsley Press.

Relf, D. 1981. "Dynamics of Horticultural Therapy." *Rehabilitation Literature* 42: 147–150.

Richardson, G. 2002. "The Metatheory of Resilience and Resiliency." *Journal of Clinical Psychology* 58(3): 307–321.

Rogers, C. 1957. "The Necessary and Sufficient Conditions of Therapeutic Personality Change." *Journal of Consulting Psychology* 21: 95–103.

Schmidt, S. 2004. "Mindfulness and Healing Intention: Concepts, Practice, and Research Evaluation." *Journal of Alternative and Complementary Medicine* 10(Suppl 1): S-7.

Schweitzer, M., L. Gilpin, and S. Frampton. 2005. "Healing Spaces: Elements of Environmental Design that Make an Impact on Health." *Journal of Alternative and Complementary Medicine*, 10 (Supp.1): S-71-S-83.

Segal, Z., J. Teasdale, and M. Williams. 2002. *Mindfulness-Based Cognitive Therapy for Depression*. New York: Guilford Press.

Seligman, M. E., and M. Csikszentmihalyi. 2000. "Positive psychology: An introduction." *American Psychologist* 55: 5–14.

Selternrich, N. 2017. "From Intuitive to Evidence Based: Developing the Science of Nature as a Public Health Resource." *Environmental Health Perspectives* 125(11). doi:10.1289/EHP2613.

Taylor, R. R. 2008. *The Intentional Relationship: Occupational Therapy and the Use of Self*. Philadelphia, PA: F.A. Davis.

Thompkins, S. M. and R. C. Schwartz. 2009. "Enhancing Resilience in Youth at Risk: Implications for Psychotherapists." *Annals of the American Psychological Association*. 12(4): 32–39.

Tsunetsugo, Y. 2013. "Physiological and Psychological Effects of Viewing Urban Forest Landscapes Assessed by Multiple Measurements." *Landscape and Urban Planning* 113: 90–93.

Ulrich, R., R. Simons, and B. Losito. 1991. "Stress Recovery during Exposure to Natural and Urban Environment." *Journal of Environmental Psychology* 11(3): 201–230.

Ulrich, R.S. 1999. *"Effects of gardens on health outcomes: Theory and Research in Healing Gardens, Theoretical Benefits and Design Implications."* New York: John Wiley and Sons.

Ulrich, R.S., R.F. Simmons, and M.A. Miles. 2003. *"Effects of Environmental Stimulation and Television on Blood Donor Stress."* Journal of Architecture and Planning 20(1): 38–47.

Verra, M., F. Angst, T. Beck, S. Lehmann, R. Brioschi, R. Schneiter, and A. Aeschlimann. 2012. "Horticultural Therapy for Patients with Chronic Musculoskeletal Pain: Results of a Pilot Study." *Alternative Therapies* 18(2): 44–50.

Watson, D. 2002. Positive Affectivity: The Disposition to Experience Pleasurable Emotional States. In Snyder, C.R. & Lopez, S.L. (Eds.) *Handbook of Positive Psychology*, p. 106–134. New York : Oxford Press.

White, B., S. Driver, and A. M. Warren. 2008. "Considering Resilience in the Rehabilitation of People with Traumatic Disabilities." *Rehabilitation Psychology* 53(1): 9–17.

Wichrowski, M. 2005. "Skills and Theories to Inform Horticultural Therapy Practice." *Journal of Therapeutic Horticulture* 17: 48–53.

Williams, K., and D. Harvey. 2001. "Transcendent Experience in Forest Environment." *Journal of Environmental Psychology* 21: 249–260.

Yalom, I. 1995. *Theory and Practice of Group Psychotherapy*. New York: Harper Press.

section three

Practice within program models

Section three characterizes the three primary models of horticultural therapy programs. The chapters in this section describe the foci, contexts, and applications of each model, and provide examples, considerations, and guidance for programming. This section is organized by model rather than by population served (as described in the preface) in order to convey essential practices and approaches to treatment that are applicable across an assortment of settings, situations, and diagnoses. Generally, there is more similarity in practice across populations in any of the three models—therapeutic, vocational, and wellness—than difference of approach for each population. It is also important to note that more than one model may be applied in the same setting. To illustrate, the wellness model may be applied to address behavioral and leisure skills in the same organization whose primary focus uses the therapy model to address physical rehabilitation. Extensive exhibits and photos are included to illustrate actual practice goals, methods, and results. This section describes professional practice on the horticultural therapy end of the "Use of Horticulture as Therapy Spectrum" (Figure 2.6) described in Chapter two, yet many concepts, considerations, and processes are equally applicable in therapeutic horticulture and the full spectrum of intervention types.

chapter eight

Therapeutic model

Jonathan Irish and Pamela Young

Contents

This chapter describes a therapeutic model of horticultural therapy, explains some distinctive aspects of this type of practice, and illustrates overlap with other models. The American Horticultural Therapy Association (AHTA, 2013) describes horticultural therapy as "the participation in horticultural activities facilitated by a registered horticultural therapist to achieve specific goals within an established treatment, rehabilitation, or vocational plan. Horticultural therapy is an active process, which occurs in the context of an established treatment plan where the process itself is considered the therapeutic activity rather than the end product." Rehabilitation, recovery, vocational training, or horticultural therapy programs with a therapeutic focus do not just happen from a hospital bed, inside a gym, at a school, or within the confines of the four walls of a facility. Horticultural therapists practice in hospitals, rehabilitation and vocational facilities, skilled care agencies and senior centers, community gardens, botanical gardens, schools, horticultural businesses, prisons, and other settings (AHTA 2013). In an effective therapeutic model of horticultural therapy, it is important to note that the physical setup that allows individuals the opportunity to get back to nature is a major influence that leads to physical and psychological recovery.

In exploring the therapeutic model of horticultural therapy practice, this chapter focuses on those programs that seek physical or psychological outcomes for participants. It is important to note that therapeutic programs also address a full range of issues and seek varied outcomes in addition to these emphases. Horticultural activity interventions have led to physical, psychological/emotional, social, behavioral, cognitive, and educational benefits (Relf 1999), and any of these may be addressed using a therapeutic model of horticultural therapy. Specific applications, in varied settings, with a sampling of diverse populations and considerations for the horticultural therapy practitioner are introduced.

In the therapeutic model, the range of documented specific health outcomes is extensive. Psychological and physical outcomes within the therapeutic model occur in a variety of very diverse settings, with a broad range of people seeking recovery and positive health outcomes. Documented studies of horticultural therapy and time spent in a garden setting have reported improvements in attention, reduction of pain, lessening of stress, modulation of agitation, reduced antipsychotic and pain medications, and a reduction in the number of falls (Detweiler 2012). Patients who benefit include those with cardiovascular disease, infectious diseases, cancer, postsurgery healing, respiratory diseases, obesity, chronic pain and musculoskeletal complaints, migraines, depression and anxiety disorders, diabetes mellitus, attention-deficit/hyperactivity disorder (ADHD), and many others. Approaches of this model are described in this chapter, including the therapeutic process, treatment settings, treatment team makeup, evaluation and goal-setting processes and procedures, and considerations for the horticultural therapy practitioner (Figure 8.1).

Figure 8.1 The therapeutic model targets specific health outcomes. (Courtesy of Mary Newton.)

Overview of the therapeutic model

Horticultural therapy treatment sessions in the therapeutic model focus on recovery from illness or injury and on treatment of mental health issues. Sessions often occur based on a medical model of care, which focuses on the physical, mental, or biological aspects of disease and illness, and emphasizes clinical diagnosis and the medical intervention associated with the diagnosis and its symptoms. Conditions are considered intrinsic to the individual and may reduce the individual's quality of life and cause clear disadvantages. It is important for horticultural therapy practitioners to recognize how these conditions affect levels of function and wellness. Often a task-oriented, holistic approach is utilized to optimize physical and mental health for those participating in a horticultural therapy session in the therapeutic model. Therapists use a whole-person and strength-based approach to address a variety of treatment goals, including, but not limited to, physical and mental health. Participants are not defined by disease, diagnosis, or disability. The focus is on the interrelatedness of the participant, the task, and the recognition of goals and needs of the whole person in order to maximize functioning. A program for children with disabilities is described in Exhibit 8.1 to illustrate a range of goals addressed in this therapeutic approach.

EXHIBIT 8.1

Program Example: Horticultural Therapy and Children with Developmental Disabilities

Program and Population

The horticultural therapy program at the Children's Center primarily serves children from birth to fifth grade who are experiencing physical disabilities, orthopedic challenges, and other health impairments. As part of the school's curriculum, a registered horticultural therapist leads gardening and nature lessons, both in the classroom and outside in the accessible garden space. Activities can be easily adapted to accommodate all ability levels.

Goals

Each student at the Children's Center has a customized, individual care plan that supports her or his goals outlined in the individual educational plan (IEP), individualized family services plan (IFSP), and/or early learning accomplishment profile (E-Lap). The focus of the horticultural therapy program is to support these goals by designing activities that provide a unique opportunity to develop and master the skills highlighted in their plans. The horticultural therapy program brings nature-based activities into the school setting using hands-on lessons to stimulate sensory, motor, cognitive, and communication skills.

(Continued)

EXHIBIT 8.1 (Continued)

**Program Example: Horticultural Therapy and
Children with Developmental Disabilities**

Horticultural Therapy in Action

Each student brings with them a unique set of skills, preferences, and challenges. Horticultural therapy provides a wonderful toolkit to support growth by using nature in ways designed to bring out students' strengths and encourage them to develop new skills. Here are some of the lessons found in nature experiences in the garden:

- Strength—Planting seeds and pulling weeds can develop fine motor and gross motor skills. Standing balance can be safely practiced by using a raised bed for planting. Building upper body strength is incorporated by watering, increasing the volume of water as strength increases.
- Curiosity—Children learn by exploring their environment. Many of the students come to school with sensory defensiveness, which creates a barrier to learning. We are able to provide a variety of novel sensory experiences to tempt students to explore. The garden is full of herbs to smell, vegetables to taste, birds to scan visually and hear in the environment, and soil to dig in and explore.
- Self-expression—Nature brings with it a whole new vocabulary and an opportunity for self-expression. While gardening, we learn new ways to describe what we see. We also practice making choices and expressing our opinions and preferences.
- Courage—It takes a lot of courage to touch something you have never seen before. Wiggling worms and slimy snails call out to be investigated. We celebrate each step of the way, whether students are just watching worms from a safe distance or holding them gently in their hands.
- Math and science skills—There are seeds to count, stalks to measure, and gourds to weigh. Observing life cycles of butterflies and seeds teaches basic observational skills and inspires a sense of wonder.

All children, regardless of ability level, learn by doing. Goals and milestones can be reached while having fun exploring the wonders of gardens and nature. Horticultural therapy complements the work of teachers and other therapists while supporting the needs of our students.

JoAnn Yates, HTR, MBA, CHTP
Horticultural Therapist, The Centers for Exceptional Children

Focus on psychological outcomes

This section addresses overarching common characteristics of mental health and psychological concerns within a horticultural therapy program and provides guidelines and best practices for the treatment of psychological issues from a horticultural therapist's perspective. It also highlights some practices that have particular application or relevance to the people–plant connection during therapy.

Psychological concerns affect all humans at some point in their lives, either directly or indirectly. These issues are present in horticultural therapy group participants to varying degrees, regardless of whether or not the primary goal of the horticultural therapy program is designed to address psychological concerns. Horticultural therapy programs may specifically focus on psychological health, but horticultural therapists working in any setting should be equipped to recognize psychological concerns and must be prepared to consult with or refer a client to an appropriate mental health professional if issues arise beyond their expertise.

When there is an imbalance in a person's life that is not addressed properly, the imbalance tips the person toward the help he or she needs. Who or what tips the scale—a doctor recommending treatment or an individual seeking treatment after a stressful life event—is less important than the fact that the imbalance occurs. This movement creates uncertainty, anxiety, and fear in most everyone who comes to terms with the fact that there is something out of order. People who experience these feelings may seek professional assistance, and a horticultural therapy program is an ideal place to process and heal from mental health challenges. Horticultural therapists should have a working knowledge of the mental health concerns within their practice in order to engage patients successfully in meaningful activities that help restore psychological balance.

Psychological factors may come into play even in horticultural therapy programs focused on other health issues. For example, a participant in a horticultural therapy program for at-risk youth with substance abuse problems may experience physical issues from withdrawals, cognitive impairment, social changes, family conflict, and emotional issues that show up through the participation in the group. Exhibit 8.2 describes how a horticultural therapist in Lima, Peru, utilizes horticultural therapy activities to reach at-risk

EXHIBIT 8.2

Program Example: Addiction Treatment, Lima, Peru

In 2012, Daniela Silva-Rodriguez Bonazzi started a horticultural therapy program for people with varying addictions, mainly adolescents with cocaine addiction, at Humana, Rehabilitation Center, in Lima, Peru. This was the first horticultural therapy program in Peru, where this practice is still unknown. The setting was a nursery and greenhouse adjacent to the center. The participants attended the program for 60 to 90 minutes, two times per week.

Although the use of gardening as part of a rehabilitation program is a new concept in Peru, the acceptance and participation rates were 100%. All participants engaged in the activity planned for the day expressed their desire to stay for longer periods of time in the nursery and to attend the program on a daily basis instead of twice a week. Even the nurses participated in the activities. Some activities offered by the program included plant propagation, cuttings, seed sowing, kokedamas, mandalas with flower petals, terrariums, succulent propagation, bonsai, and topiaries.

Most of the participants hadn't had a previous encounter with gardening but were willing to learn and nurture a plant. For some, who were unsuccessful expressing their issues in traditional therapies, the nursery setting was a place where they could talk about their problems, most of them expressing that they felt a sense of calm as they engaged in each session's activity.

(Continued)

EXHIBIT 8.2 (Continued)

Program Example: Addiction Treatment, Lima, Peru

The program goals were to promote and engage participants in plant care to enhance self-worth and self-esteem, and to promote gardening as a tool to distract the mind from negative and self-defeating thoughts. These goals were met through various activities during each session. For example, during the first session participants were encouraged to choose a plant from the nursery in order to shift roles: From that day on, the participant would become the plant caregiver. This approach was taken with great enthusiasm by most of the participants. Some even chose a plant for a family member. While this approach may seem simple, in actuality, skillful session planning and therapeutic use of self were keys to positive outcomes for participants.

Daniela Silva-Rodriguez Bonazzi

Biologist and Horticultural Therapist Director,

Asociacion de Horticultura Terapeutica y Social del Peru (AHTSP)

youth in a city where daily exposure to plants is minimal, and horticultural therapy is relatively unrecognized. Similarly, a participant who has had a brain injury and is treated in a horticultural therapy program at a rehabilitation hospital will likely have psychological concerns that can be wide ranging, unpredictable, and cyclical as therapy progresses. The main goals and objectives in these programs may be vocational or physical in nature, but mental health must also be recognized and addressed in an appropriate manner.

Comparatively, horticultural therapists in programs in settings such as inpatient psychiatric hospitals or mental health centers must be aware of the compounding physical conditions, life transitions, developmental levels, cognitive abilities, and past and current events in their patients' lives. There must be an overall holistic understanding of patients and therapeutic approaches. The client (patient) should be regarded as a unique and multifaceted individual.

Mental health diagnoses

Outside a mental health program, horticultural therapists are most likely to work with people who experience mild symptoms. In programs that exist specifically to address mental health, psychological concerns may range from mild to severe and must be addressed purposefully and skillfully. Normalizing the feelings, social changes, and life difficulties is vital for treatment. When developing horticultural therapy sessions, therapists must consider the emotional and physical stability of patients; appropriate fit to group directive; safety (suicidal or homicidal thoughts, actions, or intentions); necessary modifications to equipment, tools, and/or space; and how exactly the group will benefit the psychological well-being and overall treatment progress of each individual. The therapist must seek to create a supportive environment conducive to change. A supportive environment is created when a clearly defined issue is being addressed by properly trained professionals. Particularly in mental health settings, horticultural therapists work with clients who have a range of psychiatric diagnoses: from neurodevelopmental disorders to those related to mood, anxiety, or depression; trauma; eating or addictions; or disorders such as personality, obsessive-compulsive, conduct, or schizophrenia spectrum as well as other psychotic disorders.

While understanding the symptoms and issues typically experienced with various diagnoses, therapists should use a strengths-based approach, always remembering that a person is not her or his disorder. Exhibit 8.3 presents a case example that recognizes the client as a whole and unique person and highlights this strength-based approach.

EXHIBIT 8.3

Case Example: Inpatient Psychiatric

A 22-year-old male with significant history of childhood trauma and post-traumatic stress disorder (PTSD), along with obsessive-compulsive disorder, was admitted for suicidal ideation and harm to oneself. The patient lives with his mother and is unemployed. The patient recently graduated from culinary school and is interested in finding work.

The patient has been unsuccessful in finding work, which has resulted in self-harming behaviors, including head-banging and hitting self. He lacks effective coping skills to ease his depression and often resorts to using alcohol or illicit drugs, such as marijuana or methamphetamines. He was not engaging in group offerings and was isolated in his room for much of the day at the hospital.

The provider assigned to the patient asked the horticultural therapist to meet individually with him to see if talking about plants or gardening would help encourage him to look for meaningful activity related to growing vegetables or culinary herbs. The patient was receptive and became interested in working on a culinary herb and vegetable garden plan for the new garden area at the hospital. The therapist offered books and garden catalogs to help him choose the plants for the garden space and took the patient to the garden area to plot out the space.

The interdisciplinary treatment team working with the patient noticed that he was smiling more and wanted to share his planting plans with anyone who came into his room. The provider noted that he seemed to exhibit more self-worth and confidence in himself when asked to help with the garden project.

The patient spent time out of his hospital room and used the general living area to draw up his planting plans. He attended groups other than horticultural therapy and became more engaged in his treatment during the admission. He reduced harm to self by 80% and was engaged in positive conversations with staff members and visitors.

When the therapist interviewed the patient after the plans were drawn up, the patient was thankful and stated, "I never felt like I could use my culinary skills to help with a hospital garden. I want to come back here and volunteer when I am all better." The patient was grateful for finding a new purpose in life and filled out a volunteer application prior to leaving the program to help in the therapeutic garden.

In conclusion, the horticultural therapist was successful in fulfilling a patient-centered and strength-based treatment plan. The interdisciplinary treatment team was hopeful that the patient had identified new coping skills to reduce self-harm after discharge. The patient wrote down gardening and cooking in his safety crisis plan as a leisure skill to embrace if feeling depressed.

Melissa Bierman, MS, HTR, MHT
Supervisor of Counseling and Therapy, Unity Center for Behavioral Health

Treatment

When considering the wide-ranging dynamics in approaching mental health from a horticultural therapy perspective, certain challenges and benefits are apparent. The horticultural therapist who leads the group must provide the structure, support, and supervision necessary to allow for exploration and change while simultaneously giving a sense of freedom. It is important to consider what items—plant or tool—could be triggering, used for harm, or lead to psychological degradation of participants. The therapeutic exercises will ideally be conducted to encourage but not force participation. Silence and a lack of sharing one's story only keep the patient locked up in his or her head, but therapists should understand that these steps may be intimidating for patients to process at first. Once these challenges are identified, an appropriate plan should be implemented by the treatment team. Working through challenges during a horticultural therapy session naturally leads to increased self-esteem, increased comfort and flexibility with materials (for example, if the client's challenge is being contaminated by "dirty" soil), increased problem-solving skills, and an even deeper understanding of how a mental health diagnosis affects life functioning. These outcomes may benefit all participants in the group. Understanding these challenges and benefits heightens the participant experience and informs the techniques and approach of the therapist who facilitates such groups (Figure 8.2).

The horticultural therapist is tasked with creating a proper therapeutic environment to ensure physical and emotional safety, while allowing room for emotional vulnerability and risk for clients in order to foment change. Purposeful metaphors may be drawn to help connect activities with the alleviation of psychological symptoms, while remaining sensitive to the experience of participants. Group activities can be adapted in order to focus on

Figure 8.2 The therapist and garden create a therapeutic environment in which to grow. (Courtesy of Pacific Quest.)

specific issues. Exhibit 8.4 discusses the ways that the horticultural therapy program at the Eating Disorders Center at Rogers Memorial Hospital in Oconomowoc, Wisconsin, provides sessions that are focused to meet the needs of clients with eating disorder diagnoses.

The participant in a horticultural therapy program who successfully meets these challenges will see great rewards. For both the client and therapist, it becomes enjoyable and meaningful to address issues as they arise using real-life horticultural concepts. Often, unexpected connections and personal growth come out of the more relaxed atmosphere and the purposeful activities integral to horticultural therapy groups. Clients may feel a

EXHIBIT 8.4

Program Example: Psychiatric, Rogers Memorial Hospital, Oconomowoc, Wisconsin

Rogers Memorial Hospital has a horticultural therapy program for patients with eating disorders in a residential setting. Patients live in the residential center for their entire stay, which is typically thirty to ninety days, spending some time off-site in groups, with day passes, and for outings. Residents are also encouraged to attend staffing meetings every other week. These meetings take a multidisciplinary approach and include the medical director, psychologist, nursing staff, clinical services manager, dietician, school staff for adolescents, primary therapist, behavioral specialist, and experiential therapists. Residents are encouraged to participate in all treatment-related activities and groups. They are not mandated to attend horticultural therapy sessions but may only advance to different system levels if they are actively engaged in all aspects of treatment offered. They are able to access gardens on site during appropriate groups, individual or group walks, or when addressing a particular concern in treatment. Supervision is required for use of garden tools, and tools are checked in and out as deemed appropriate by staff members. Treatment using garden spaces and horticulture elements is discussed and planned during team meetings, and staffing is customized to the specific needs of each resident. Goals are discussed and created during both group and individual meetings.

The hospital offers numerous programs with differing schedules and specialized staff. The following is an example taken from the eating disorder program at the residential level of care for sixteen adult patients and eight adolescent patients. The emphasis is on evidence-based programming utilizing cognitive behavioral therapy, dialectical behavioral therapy, and exposure-response prevention, among others. Residents work with a multidisciplinary team, which determines the most important areas of focus at the residential level of care. Rogers involves family whenever possible and seeks to stabilize residents and then refer them to the next level of care as soon as clinically appropriate.

Horticultural therapy is offered twice a week to both adults and adolescents. Sessions are designed to follow the seasons. An automated greenhouse for working with plants in the cold winter provides a warm, plant-rich setting for horticultural therapy groups and a year-round place for plant propagation. During the warmer

(Continued)

EXHIBIT 8.4 (Continued)

**Program Example: Psychiatric, Rogers Memorial
Hospital, Oconomowoc, Wisconsin**

months, therapeutic gardens and an edible landscape on campus are used for exploration, restoration, and a medium for horticultural therapy sessions.

Each session begins with a check-in that explores current patient-reported physical and emotional well-being. This is an open process and is typically tied to a question that allows for abstraction and gives an objective idea on which to place physical and mental states. For example, the therapist may give these directions: "Take five minutes to explore the greenhouse. Find one plant that represents how you feel physically and emotionally, and if possible, bring back for further discussion." The activity is then explained in a way that fits in with the overall purpose. For example, if the focus of the group is behavioral activation or something that is enjoyable to the patient, there may be very general guidelines given: "Today we are practicing self-care by eliminating weeds in the garden and will come back and talk at the end about what weeds in your life you want to eliminate." During the activity, the discussion and exploration of real-life parallels is encouraged, and appropriate support is given (finding tools, use of tools, weed identification, etc.). A group discussion is conducted at the end to summarize and process the challenges and successes of the group. A group geared specifically to explore what skills have been learned as the patient prepares for discharge would be centered around an individual task. A demonstration of how and why to transplant seedlings is given and the purpose of this activity is explained. Then the task is worked on with support and encouragement: "Today we are transplanting sage because it has grown all it can in this container without becoming root-bound or unhealthy. It needs room to grow. I will demonstrate how this is done and I want you all to work on this while considering what you have learned in treatment that you will use to successfully maintain the benefits gained while at Rogers. Experience the smells and textures of the sage. Which aspects of treatment have you appreciated, and which have been challenging?" This discussion concludes the session.

Jonathan Irish, HTR, MA, LPC

sense of well-being while being connected to nature, showing signs of body rhythm and thought regulation. A case example of this is portrayed in Exhibit 8.5.

Therapists must look for opportunities to point out cognitive distortions and readily use metaphors, analogies, or real-life examples. For example, a client may mention that depression feels like a dark time that will never end. The horticultural therapist may respond by introducing a discussion related to the ways plants weather and survive winter, emerging alive and thriving in the spring, in order to challenge the client's beliefs. Follow-up discussion will either confirm the benefit of this approach or suggest an alternative strategy to try in future sessions. These types of interactions are possible when the environment is physically conducive to connecting people and plants.

What then is physically necessary for a horticultural therapy program in a mental health setting?

EXHIBIT 8.5

Case Example: Veteran with PTSD

Tom, 44, homeless, and alcoholic, was referred to a veterans' service that ran a horticultural therapy program as a valuable adjunct to its three-phase clinical treatment for PTSD. Signing up with the army at the age of 17, Tom's basic training had reinforced his instinctive, automatic, "fight" response to threat, priming him for battle, but experiences of bullying in his unit meant that during his active tour of duty in Iraq and on his subsequent discharge from military service, he often felt alone and isolated. Postdischarge Tom began to drink heavily to numb his feelings of anger and despair, and to cope with flashbacks of burned bodies in Iraq, which haunted his dreams and prevented sleep. Fired from several civilian jobs as a security officer for erratic and risk-taking behaviors, he became long-term unemployed, moving from hostel to hostel owing to his explosive temper, and losing contact with his wife and two children.

When Tom arrived at the veterans' service, he described himself as "always on a short fuse," quick to irritation and anger; mistrustful; and highly critical of himself and others, especially civilians, and "the state of the world." Our first task involved engaging with Tom in order to help him understand these attitudes represented the trained "fight" state of his mind. He learned how his hypervigilant "threat brain" was perpetually on the alert for danger and that his historical way of managing this uncomfortable state of mind was by "flight" into addictions and risky behavior. Explaining and normalizing this vicious circle during phase one of trauma treatment helped Tom reduce his sense of shame and understand why breathing and grounding techniques were so effective in stabilizing his nervous system. Spending time in nature became a valuable support to this work, enhancing Tom's abilities to calm himself down, and providing a safe place to which he could retreat when feeling overwhelmed. This restorative aspect became particularly important when he moved to phase two of the clinical work, remembering and reworking his trauma memories. He would often come to the garden and sit on "Tom's bench," a secluded corner in the sun surrounded by butterfly- and bee-loving plants. As his interest and curiosity about the birds and plants he saw grew, Tom began to notice how this helped keep his "threat brain" from triggering his symptoms. Hours would pass as he became absorbed in the "flow" of potting up little seedlings, noticing too that caring for their needs enabled them to thrive. This gradually extended to Tom improving his own self-care and beginning to value his growing horticultural abilities and skills.

As Tom let go of his memories of the past, he began phase three, reintegration, with a renewed sense of hope and, to his surprise, a more compassionate understanding of his own and others' need for kindness and connection. He joined a local organic community growing scheme, producing fresh vegetables, which he took great pride in selling on the weekly farm stall, and putting on the table for his new partner and family.

Joanna Wise, DCounsPsych, MBPsS, HCPC, ACAT, EMDRIA, DipSTH

The treatment setting

Creating the proper environment for horticultural therapy in a psychological setting or any program where those issues are being addressed requires careful attention to small details. Horticultural therapy offers an experiential therapy working with plants in an environment with beneficial opportunities for connection with nature. A traditional psychotherapeutic approach involves a therapist and an individual, couple, or group of people sitting in a quiet room. This arrangement does not suit horticultural therapy programs, generally speaking, as they require physical engagement; as a result, horticultural therapists must create a unique environment for therapy. Within the proper environment, patients can practice balancing life and plant care at the same time, allowing growth to occur in multiple areas at once. This is a unique benefit of horticultural therapy, which will be fully realized only if the environment is conducive to both plant and people growth. Exhibit 8.6 gives a description of the whole-person therapeutic approach used in a carefully developed horticultural therapy program in place at Skyland Trail in the United States. The environment of a garden or the proper use of plants in an indoor space helps to create a judgment-free place to work on psychological concerns.

EXHIBIT 8.6

Program Example: Residential and Day-Treatment, Atlanta, Georgia

The Forward Spiraling Journey as a Model of Recovery in Mental Illness at Skyland Trail

When Skyland Trail first opened its doors in 1989, its founders understood the impact that horticultural therapy could have on a client's recovery. The horticultural therapy program was established the following year and was rapidly embraced as a transformative component of the whole-person therapeutic approach at Skyland Trail.

Located in Atlanta, Georgia, Skyland Trail is a nonprofit mental health organization offering residential and day treatment for adults age eighteen and older. Offering a continuum of care from residential care to job coaching and social opportunities for adults living in the community, Skyland Trail specializes in treating adults with bipolar illness, major depression, schizophrenia, schizoaffective disorder, and anxiety disorder, as well as patients who have a complex diagnosis involving substance misuse or personality disorders.

Horticultural therapy at Skyland Trail is under the collaborative umbrella of adjunctive therapies, including art, music, recreation, and woodworking. Often referred to as our 3D therapy, horticultural therapy complements the skills learned in verbal therapies, specifically cognitive behavioral therapy and dialectical behavioral therapy, and helps clients find new ways to express their thoughts and emotions, learn new skills, build confidence, and interact with peers and the community.

Integral to the horticultural therapy program at Skyland Trail is the mission of empowering the positive image of the recovery process in mental illness as a journey

(Continued)

EXHIBIT 8.6 (Continued)

Program Example: Residential and Day-Treatment, Atlanta, Georgia

that spirals forward, propelled by awe and the plant cycle. From the germination of vulnerability to the root growth of courage and the shoots of empowerment, to the full bloom of purposefulness and connection, the journey of recovery comes to fruition, parallel to the journey of a seed. Clients witness this parallel journey through weekly therapeutic groups called Inspired by Nature, Garden to Table, CAFÉ (Cognitive and First Episode) Gardens, Campus Caregivers, and Nature as Healer, and in the vocational groups that occur three mornings a week called the Green Team.

The threefold goals of both the therapeutic and vocational groups represent the backbone of programming, including:

- Instilling positive metaphors for recovery and relapse in mental illness by witnessing the plant cycle through hands-on, nature-related activities.
- Reaffirming self-esteem and self-worth through activities that engage exploration of nature, creative expression, and mindfulness.
- Fostering socialization, community reintegration, and nurturing through purposeful, community-focused group activities.

At Skyland Trail, our campuses are our canvases. A typical therapeutic group begins with harvesting and exploration of materials found in the gardens. It is part of the process. The main greenhouse functions as the Evergreen Group Room throughout the year where clients nurture seedlings and divisions that multiply and give forth opportunities to give back to the community, reinforcing a sense of purposefulness.

Over the decades, the campuses have benefitted from clients engaged in the design and implementation of garden rooms and features, which in turn serve as therapeutic spaces, tailored to their recovery. Examples include a Twelve-Step Deck, Butterfly Boardwalk, Metamorphosis Garden, Introspection Terrace, Golden Grotto, Labyrinth, Garden of Grace, and Interactive Mandala. These gardens embody a meaningfulness of recovery that is left as a legacy for others on their transformative, forward spiraling journey.

Libba Shortridge, HTR, MLA
Horticulture Department Coordinator, Skyland Trail

The horticultural therapist incorporates the garden and the people–plant response to influence outcomes. Horticultural therapy takes away transference and judgment by providing the neutral and objective media of plants with which to work. Horticultural therapy programs provide a nonjudgmental way of viewing psychological concerns as something that affects all humans and reduces the stigma associated with mental illness. Exhibit 8.6 describes how the gardens serve as creative treatment spaces that are tailored to aid in recovery in a whole-person, therapeutic approach to mental health care (Figure 8.3).

A carefully planned horticultural therapy environment includes safety precautions; program procedures and protocols, including practical processes such as how to

Figure 8.3 An ephemeral labyrinth is created at Skyland Trail in Atlanta, Georgia. (Courtesy of Libba Shortridge.)

get supplies; proper enclosures; and a variety of plants and activities. Safety precautions include but are not limited to monitoring where and when sharp tools are used, limiting group size, using nontoxic plants in order to reduce possible accidental poisoning or purposeful self-harm, awareness of possible triggering spaces within the garden (a dark, enclosed corner could trigger someone who has been traumatized), and clear communication with staff before and after each group.

Program protocols vary widely depending on the program and location of horticultural therapy sessions. In general, however, maintaining a clear and concise expectation of how to find, check out, maintain, use, and return horticultural tools is vital to safe programming. Knowing the uses of gardens and garden areas is also an important aspect to consider. Horticultural therapists have the responsibility of training other staff members and taking time during sessions to explain the use of tools and garden spaces within the program. If gardens and tools are available to clients outside horticultural therapy sessions, proper protocols and training procedures must be in place to maximize client safety as well garden care.

The enclosures and boundaries of a horticultural therapy garden space will also vary widely depending on the target population. Regardless, patient safety should be a top priority for all horticultural therapists. This can be achieved with a physical fence (decorative or functional, artistic or commercial), pathways, signs, or subtle biological markers (prickly shrubs, large trees, espaliered trees, etc.). Therapists may embrace the use of natural elements to metaphorically exemplify areas in which patients might need to use boundaries or increased awareness in their own mental health.

Physical, social, emotional, and spiritual safety is paramount in a psychological setting. Boundaries within a program help give voice and understanding to the concept of forming healthy boundaries in a person's life. A workable space for therapy groups can have fencing that physically illustrates diffuse, enmeshed, or rigid boundaries, which allows for purposeful or passive examples during sessions. Use stones, shrubs, fencing, or other natural borders to create a sense of privacy and security for clients. The physical environment must be cleared of hazards and potential dangers, such as icy walkways, loose steps and stones, buckling sidewalks, and low-hanging branches. Attention must be given to walkways and fences. Consider the entrances and exits of the garden space, and install signage and lighting to prevent slips, falls, and missteps. Tools necessary for the session should be available only as long as they are needed in order to reduce the chance of inappropriate use. When not in use, all tools and other necessary materials should be inventoried and stored in a locked structure. Plants used in the session and in the garden must be evaluated and labeled for toxicity in order to prevent accidental ingestion or exposure to toxins. Adhering to a defined schedule, set programming, and detailed goals adds to the feeling of security for clients because they know what to expect. For example, group sessions should begin and end on time. A group discussion needs to be carefully guided and remain focused on desired topics in order to build trust for a patient who has experienced trauma. Setting horticultural goals with clients gives them something to anticipate and can add to a client's overall commitment to the therapeutic process. These facets of a thriving horticultural therapy program require a dedication to adaptation, as described further in Chapter 11.

Program design

Horticultural therapists successfully implement horticultural therapy in a psychological setting by being intentional when designing programming around several factors unique to mental health. When working to combine traditional counseling approaches with horticultural therapy, one must have a clear picture of what is to be accomplished and how this is adapted for horticultural therapy. Horticultural therapists must be prepared to engage with a full treatment team made up of professionals from the mental health field, social workers, physicians, physical therapists, and others. Horticultural therapists must decide on group format. Decisions should be made on the following aspects of the group:

- Will the group be fluid, allowing for last-minute changes to the number of participants or the scope of the group, or will it be closed, retaining the same grouping of clients within a series of sessions for emotional safety and enhanced opportunities for continuity and group cohesion that may result?
- Will clients be permitted to leave or drop in at any time based on scheduling of other appointments?
- Should the group be open, allowing participation by anyone interested?
- Will the curriculum goals be ongoing?
- Will other staff members be needed for the group? Clients may benefit if additional staff are on hand to help with different needs of the group, such as behavioral support, assistance with tasks, and managing elements of client safety.

Therapeutic approaches

The horticultural therapist who practices within a mental health setting must have a good working knowledge of the following therapeutic approaches, which may be employed along with the people–plant connection made during horticultural therapy groups

(VandenBos 2007). If the horticultural therapist is not specifically trained in counseling or mental health treatment, it is advisable to co-treat or work alongside someone who does have the required knowledge and skills. Any of the following approaches are well-researched and effective in improving the mental health of people around the world:

- *Psychoanalytic psychotherapy*—interactions involving the one-on-one interaction between a therapist and a client that emphasizes the importance of unconscious motives and conflicts as determinants of human behavior.
- *Family systems theory*—model focusing on working with the entire family unit, with emphasis on the relationship between and among interacting individuals in the family.
- *Cognitive behavioral therapy*—a form of psychotherapy that integrates theories of cognition and learning with treatment techniques derived from cognitive therapy and behavior therapy. Treatment focuses on modifying the client's maladaptive thought processes and problematic behaviors.
- *Dialectical behavioral therapy*—a flexible, stage-based therapy combining principles of behavior therapy, cognitive therapy, and mindfulness, which promotes acceptance and change (Lineham 1980).
- *Eye movement desensitization and reprocessing*—a treatment methodology used to reduce the emotional impact of trauma-based symptomatology associated with anxiety, nightmares, flashbacks, or intrusive thought processes.
- *Existential*—treatment methodology used to reduce the emotional impact of trauma-based symptomatology associated with anxiety, nightmares, flashbacks, or intrusive thought processes.
- *Acceptance and commitment therapy*—a form of cognitive behavior therapy that helps clients to abandon ineffective control strategies, to accept difficult thoughts and feelings without taking them to be literally true, and to take actions in accordance with their own values and goals.
- *Exposure therapy*—a form of behavior therapy that is effective in treating anxiety disorders. It may encompass any of a number of behavioral interventions, including systematic desensitization, flooding, implosive therapy, and extinction-based techniques.
- *Neurosequential*—an approach that provides the clinician with a picture of the child's developmental trajectory to his or her present set of strengths or vulnerabilities (Mackinnon 2012).

With the proper training, horticultural therapists can incorporate any of the above modalities into their horticultural therapy sessions to address many of the common issues (identified by Weilert 2018) for which people seek mental health treatment, including: emotional, behavioral, relationship, spiritual, trauma, grief and loss, and life adjustment. Programs or sessions focused on the alleviation and treatment of psychological concerns are best achieved by creating a program that combines recognizable elements from gardens/plants and more traditional psychological treatment. Combining language from a cognitive behavioral therapy (CBT) or dialectical behavioral therapy (DBT) perspective, for example, with horticultural therapy elements can be highly effective.

The program examples in the next two exhibits illustrate ways in which horticultural therapy is combined with other mental health therapy techniques. Exhibit 8.7 describes therapists at Thrive in the United Kingdom using the various models alongside horticultural therapy with great success. Exhibit 8.8 shows how staff members at Mount Saint Vincent in Denver, Colorado, use horticultural therapy programming paired with a

EXHIBIT 8.7

Case Example: Personal Journey toward Mental Health, Thrive

At Thrive, we approach working with each individual on our core program in a person-centered way. This approach leaves us occasionally moving across the different model types and, even more frequently, combining different elements of the identified models in the way we enable personal journeys to take place. People coming into the program spend eight weeks with us, during which a two-way process of what we describe as being an assessment or an introduction takes place. They get to know us, as we get to know them. At the end of eight weeks, we then explore with the person how she or he thinks coming to our gardens can be valuable to her or him, and a goal is set in relation to this. One, two, or three goals are identified, and then our team of practitioners can be focused on managing their relationship with the person, facilitating a positive relationship between the person and the other members of the group, and enabling the positive relationship between the person and the natural environment—all shaped in relation to the personally identified goals. We then review the goals every three or six months, and then repeat this cycle for as long as it is objectively seen as valuable to the participant.

One of our most recognized success stories is John, who after a number of years accessing our gardens on the core program, moved on to volunteer and then be employed with Thrive. Our ongoing relationship enabled us to gain a deep understanding of his journey and our program through his own reflections. John made many different observations, some of them expected and others less perceived. Although his comments were only a small sample of the hundreds of people who access our programs across the United Kingdom, this reflection had great value. John observed, "This was the first time in twenty years that I got to experience ordinary human affection in a social group. The horticultural therapist talked to me on an equal level—a sense of common humanity." This statement demonstrates the importance of enabling as well as how important it is to encourage and facilitate positive conversation, not just between practitioner and client gardener but also between group members. Creating space for this to happen requires the therapeutic skills of empathy and unconditional positive regard, both identified as key elements within person-centered therapy approaches.

In addition, John identified his need for more responsibility and a chance to create something for himself. He stated, "In 2000, Thrive took over maintenance of the Battersea Park Herb Garden, and a sizeable bed became 'mine'—with responsibility for the planting. I decided on a medicinal theme and started to research medicinal herbs with benefits that could be interpreted as broadly relevant to mental health, such as helping with migraine or sleeplessness." Richard, who was his therapist at the time, said, "I feel his potential abilities were positively recognized, and he was trusted with additional responsibilities that I feel assisted with his positive development." This suggests that the program provides a space and opportunities for congruence, moving people from their current real self toward their ideal self as the core to the person-centered approach, along with the empathy and positive regard identified above.

(Continued)

EXHIBIT 8.7 (Continued)

Case Example: Personal Journey toward Mental Health, Thrive

This is a journey that John continues many years after leaving the program. "I believe that plant growth can be seen as a metaphor for the way a human personality can grow," he said. "The rhythm of physical work has a symbolic dimension that can benefit you even if you apprehend it only very dimly."

John continues to advocate for an improved perception of mental illness. His involvement in a network that focuses on that ensures that he continues to be focused on *direction*, not *destination*, which Carl Rogers would suggest keeps us happier and mentally healthier. His success demonstrates the capacity of the Thrive core program to enable people to benefit sustainably in their lives outside our garden and beyond their time with us.

Damien Newman

Training, Education and Consultancy Manager, Thrive, United Kingdom

EXHIBIT 8.8

Perspective: Childhood Trauma, Denver, Colorado

Mount Saint Vincent in Denver, Colorado, utilizes the neurosequential model of therapeutics to treat children who have experienced developmental trauma or neglect and applies this model to its horticultural therapy program. This trauma-informed care model developed by Bruce Perry (2009) focuses on brain development deficits due to trauma experienced at various developmental stages. The foundation of this model recognizes that prolonged stress and chronic adverse conditions, especially developmental trauma, cause major damage to children's neural system. Key areas of the brain are particularly implicated—the corpus callosum, linking the left and right hemispheres; the hippocampus, linking implicit memories to each other to form explicit memory; and the prefrontal cortex, linking the cortex, limbic area, and brainstem, which involve controlling the body, self-regulation, and relational capacities. Treatment is then focused on addressing these areas of the brain, from the bottom and working upward, targeting these areas with patterned, repetitive activation in the neural system. Therefore, treatment goals focus first on self-regulation and impulsivity before addressing emotional issues, such as anxiety, followed by cognitive challenges.

The horticultural therapy garden at Mount St. Vincent is used for both individual and group therapy sessions. These weekly sessions are led either by a clinician with

(Continued)

EXHIBIT 8.8 (Continued)

Perspective: Childhood Trauma, Denver, Colorado

a master's-level mental health degree and horticultural therapy training or co-led by a horticultural therapist and mental health therapist. The goals of this horticultural therapy program are to provide therapeutic interventions that offer physiological stress reduction through the self-regulatory benefits of nature as well as patterned, repetitive activities designed to help regulate the nervous system. For instance, specific interventions targeted at self-regulation through the use of repetition might look like slowly watering the plants at the beginning of each session in the same pattern utilizing a watering can, allowing extra time for shoveling regardless of its practicality, or turning the compost.

Philip Tedeschi, LCSW

Clinical Professor and Executive Director, Institute for Human-Animal Connection,

Graduate School of Social Work, University of Denver

Kristen Greenwald, MSW, LSW

Child and Family Therapist, Garden Resource Outreach

neurosequential therapeutic model to meet the physiological and psychological needs of children who have experienced developmental trauma or neglect.

Comprehensive programs in which horticultural therapy groups focus on alleviating psychological concerns involve communication with a multidisciplinary team, adherence to guidelines regarding patient confidentiality, a strong focus on safety and ethics, and adaptability to changing work environments and requirements. Information disclosed by the client must be treated with the utmost respect. On rare occasions, due to mandated reporting laws, disclosed information must be reported. If this is to occur, it must also be processed by therapist and patient. This process needs to be conducted within the policies and procedures of the program.

Considerations for those who have experienced trauma

Special attention should be paid to the work done with clients who have experienced trauma. Trauma-informed care dictates that therapists consider every aspect of a group or session from an understanding that one or more participants may have been touched by trauma. Balancing care and compassion with confrontation increases the overall feeling of safety and security within a group. Practitioners must realize that sometimes disruptive or distracting behavior can be rooted in trauma and is actually a reaction to trauma rather than defiance or purposeful choice. It is advisable to give participants the freedom to choose how invested or involved they want to be in a given session. This allows for a balance between challenge and autonomy and always gives people an out if needed. This helps to foster self-care. The skillful navigation of trauma influences on a session can add to group cohesion and safety. Collaboration among various team members is paramount to a living and dynamically diverse horticultural therapy program. Exhibit 8.9 outlines the horticultural therapy program at the Ann and Robert H. Lurie Children's Hospital of

EXHIBIT 8.9

Program Example: Goals and Methods at a Pediatric Hospital, Chicago

Nationally ranked in ten pediatric specialties, Ann and Robert H. Lurie Children's Hospital of Chicago is one of the nation's best children's hospitals. Since 1984, it has been home to a very special program called Garden Play, which is offered through Children's Services. Garden Play is a unique hybrid of the disciplines of horticultural therapy and child life. Child life is a profession that provides hospitalized children with age-appropriate preparation for medical procedures, pain management, and coping strategies. Garden Play helps pediatric patients and their families cope with the stress and trauma of hospitalization through the handling and nurturing of living plants, with an emphasis on play and self-expression.

Challenges faced by pediatric patients include separation anxiety, loss of control and autonomy, lengthy and/or multiple hospital stays, lack of sensory input, pain, fear, and fatigue. To counter these challenges, pediatric patients need to feel normalized by engaging in developmentally appropriate play and learning. They must be given effective coping strategies and diversions for dealing with pain and anxiety. They need a sensory-rich environment with choices to be made. Garden Play is an effective therapeutic modality that addresses these needs dramatically. Therapeutic treatment goals can be categorized as social, emotional, cognitive, and physical:

- Social
- Normalization
- Reversal of dependency roles
- Multigenerational social interaction
- Emotional
- Increased self-esteem through mastery
- Creative self-expression
- Coping with stress
- Cognitive
- Knowledge of the natural world
- Celebration of cultural diversity
- New hobby postdischarge
- Physical
- Mild physical activity
- Sensory stimulation
- Diversion from pain

Garden Play is offered by a registered horticultural therapist to inpatients and family members weekly in the Family Life Center of Lurie Children's for an hour and a half. It is also offered to outpatients and their families at special annual events. It is designed and adapted for children and teens, three years and older, who have approval from medical staff to come and participate. Certain aspects of Garden Play can be modified to include those younger than three.

(Continued)

EXHIBIT 8.9 (Continued)

Program Example: Goals and Methods at a Pediatric Hospital, Chicago

Precautions are taken to comply with the hospital's stringent infection control policies and procedures. Houseplants are hypoallergenic, nontoxic, disease- and pest-free, and in optimum health. A soil-free potting mix is moistened off-site the day of the program. Patients wear plastic smocks and any splints and bandages are covered in plastic to ensure they remain contaminant-free. Hands must be cleaned with sanitizing wipes or washed after participation, and all surfaces, containers, and utensils coming in contact with potting mix must be disinfected according to protocol.

Specially trained volunteers bring patients to the Family Life Center and support them in choosing from several activities, including dramatic play, medical play, art, games, and Garden Play. Patients regain a sense of control as they are able to choose an activity and participate at their own pace. Volunteers guide patients through the Garden Play activities, either individually or in small groups. Other staff members that may participate with patients include child life specialists, activity coordinators, art and music therapists, nurses, student nurses, occupational therapists and physical therapists.

Patients are allowed to handle living plants and learn about their care through play, art, and games. The process is as meaningful as the outcome as pots are decorated, fresh potting mix is handled, leaves and roots are examined, and plants are watered.

Reactions to Garden Play include squeals of joy from patients as water spills out the bottom of a pot into a saucer, sighs of relief from parents seeing their child distracted from pain, and laughter as patients and family members share a common experience. During the course of Garden Play, many metaphors for growth and healing emerge.

Roberta Yoss Hursthouse, BS, HTR
Garden Play Specialist, Ann and Robert H. Lurie Children's Hospital of Chicago

Chicago, where staff members have developed protocols and environmental safety procedures to enable children as young as three to participate in horticultural therapy. The horticultural therapy alleviates chronic stress and anxiety resulting from the trauma of childhood hospitalization.

Team treatment in psychologically focused programs

A patient is more likely to restore balance successfully if he or she is in a healthy and reciprocal relationship or has experienced reciprocal altruism. The balance between self and society often takes place and is in part orchestrated by appropriately trained professionals. Within the realm of mental health intervention, this is usually an individual or a team of social workers, therapists, caseworkers, psychologists, or psychiatrists in any combination. The addition of a horticultural therapist to the treatment team adds an element that may seem less intimidating or clinical to the client. This team member may have interdisciplinary training in mental health as well as horticultural therapy, or he or she may apply horticultural therapy treatment in cooperation with others in the mental health profession. In horticultural therapy, plants are an essential part of the treatment dynamics.

Multidisciplinary treatment teams

Factors considered by a multidisciplinary team when using horticultural therapy to treat psychological concerns include the level of care and the treatment location. In addition to the patient/client, treatment team members may include horticultural therapists, other licensed therapists, social workers, occupational therapists, physical therapists, doctors, psychiatrists, milieu staff members, outpatient providers, school personnel, and family members. The treatment team reviews the level of care on a continuum, from least restrictive to most restrictive, such as outpatient setting, intensive outpatient, day treatment, partial hospitalization, inpatient hospitalization, detoxification hospitalization, and incarceration. Treatment locations for horticultural therapy are assessed and considered along the continuum of care. They may include inpatient or outpatient clinics, partial hospitalization, locked or unlocked facilities, day treatment, live-in programs, long-term care, community settings, botanical gardens, or community gardens. The use of treatment teams varies depending on the type of program within the therapeutic model of horticultural therapy.

Treatment goals

Mental health treatment may offer many different outcomes. The end of treatment does not have to be the cessation of any psychological concerns. It can merely mean the successful resolution or management of a given situation. A person who seeks therapy for depression due to a recent death in the family on an outpatient basis may have a very different set of goals than a person who is depressed because his or her probation was revoked, and he or she can no longer expect the level of freedom previously thought possible. Both have depression as a main focus of treatment, but each has different levels of care and different desired outcomes. Regardless of the reason for treatment, goals must be made and can be realized. It is up to the horticultural therapist, treatment team, and individual to determine these goals together so that progress can be measured. Several examples of individualized goals that are addressed for the whole group pertaining to a psychologically focused horticultural therapy program are presented in the following lists.

Examples of goals for all group members (group goals):

- Follow the majority of instructions in order, to accomplish the task of the group.
- Demonstrate self-control and proper communication while working as a group.
- Understand and display healthy boundaries with other group members.
- Understand the relationship between psychological, physical, spiritual, and social health practice while in the group.
- Increase ability to work with and around others.
- Increase positive self-esteem.
- Develop and practice effective coping skills with other group members.

Examples of individualized goals:

- Increase ability to focus and follow a set of directions.
- Speak respectfully to others, even when upset.

- Attend horticultural therapy and participate actively.
- Speak assertively to get personal needs met.
- Examine the metaphors present while working with plants to better understand the reason for treatment.

Here are some examples of therapy objectives. Note that objectives may be written in the first person to encourage ownership, or in third person:

- Participant will report to the greenhouse each day at the agreed-upon time to water his or her assignment section, for three days per week for three consecutive weeks. (This objective addresses a goal to reduce depression through engagement in activity and task.)
- Participant will arrive at each horticultural therapy session appropriately groomed and in clean clothing for six consecutive sessions. (This objective addresses a goal to improve personal hygiene and self-care.)
- Between horticultural therapy sessions, I will write a journal entry daily about one positive thing I identified while working with plants, for five consecutive days. (This objective addresses a goal to improve positive thinking.)
- I will work in the outdoor flower garden for at least 30 minutes a day, two days a week, for four consecutive weeks. (This objective addresses a goal to increase physical activity in order to cope with stress.) (See Figure 8.4.)

Figure 8.4 A woman works on personal individualized goals while tending plants in a greenhouse. (Courtesy of Elizabeth Diehl.)

Plant use and connections

Healthy horticultural therapy programs do produce tangible evidence of the changes achieved by the patients involved in groups. The following questions then arise: "What to do with the plants and products made as a result of groups?" and "How can patients take with them some reminder of the knowledge gained during horticultural therapy sessions?" The process, not the product, is the focus of therapeutic horticultural therapy programs and is key to individual outcomes achieved through a program. Treating a psychological concern is neither quick nor easy. Working with plants must parallel the realities of the populations served in a mental health or psychologically focused setting. Are patients able to take plants with them when they leave? Do patients contribute to the ongoing programming by making plants or plant-based products? If plants are sold, who receives the proceeds? With creativity, a horticultural therapist can design transitional objects to be taken upon discharge, given to others, or left behind by the patients as a way to focus on the process of change that began during the course of treatment. This may depend on the level of care and can be worked into the program's foundation. For inpatient detox programs, patient stays are short, and plants would typically be tended by individuals for no more than five to seven days. In this situation, horticulture projects and tasks would need to be ongoing. Contrast strategies in these short-term patient stays with long-term residential programs where the patient is present long enough to participate in a full plant life cycle. Seeds can be started, tended, and harvested.

   In addition, the well-planned use of plants for different purposes—flowers that can be cut and used in horticultural therapy activities, for example—can add to the sustainability of the program and reduce materials costs from year to year. A program can work collaboratively with the environmental services department or garden staff members and volunteers to collect materials from the therapy garden, thereby building nature and human connections. Here are some examples of how to ensure that people–plant links extend beyond the duration of any group of sessions or series of horticultural therapy groups:

- Use metaphors to connect a traditional therapy skill to the natural world that involves the people–plant connection. Mindfulness, for example, can be taught and practiced outside in a garden where the wonderful health benefits can be explained. A patient's walk from the bus stop to a group treatment space can become an opportunity to practice mindfulness with the instruction to "notice all the animals that use the garden for their home or source of food, and practice using all five of your senses on the walk from the bus stop to group each day."
- Instead of drawing a traditional family genogram during a family therapy session, the horticultural therapist can take the family on a walk in the garden to work together to find elements in nature that represent different family members in some fashion—color, shape, emotional meaning, size, historical meaning, etc. These artifacts can then be placed on the genogram in lieu of a square or circle. This brings life, creativity, and reconnection to the process of understanding patterns within a family system.
- Materials from the garden may be collected during the fall, just before winter, and used to make a balanced mobile for which a stick is used to dangle different preserved dried plant elements. The discussion can then center on the need for balance and problem solving as patients transition to a lower level of care or back into the community to work and apply the skills used during therapy sessions.

- Use an acronym to teach skills that will be foundational to a horticultural therapy group throughout a series of sessions. For example, place the words PEOPLE PLANT inside a front cover or as section headers in a garden journal, with each letter corresponding to an intention or guide:

  - *Participate* appropriately more each session.
  - *Explore* the wonders of nature and the people–plant interaction.
  - *Observe* the patterns in nature and parallels to life while in the garden.
  - *Practice* skills or techniques presented during each session.
  - *Learn* about myself, about plants, and about life.
  - *Experience* a new way to be with acceptance, awe, and wonder for how things are, not the way I wish them to be.
  - *Plan* ahead for the next season purposefully.
  - *Listen* for the story underneath.
  - *Advocate* for myself respectfully.
  - *Narrate* my new chapter.
  - *Transition* effectively.

"… There are particular qualities of the plant-person relationship that promote people's interaction with their environment and hence their health, functional level, and subjective well-being" (Fieldhouse 2003).

Focus on physical outcome

Programs within the therapeutic model also focus on physical outcomes. This section highlights those programs and processes targeting physical outcomes that use a therapeutic model of treatment in a variety of treatment settings. Research dating back centuries has documented the benefits of gardening. Today, the practice of horticultural therapy is used for rehabilitation of lost skills, cognition and memory improvement, balance and muscle coordination, stress management, and providing a social connection. A therapeutic model within the context of a horticultural therapy program with a physical emphasis may be designed to address return of lost or impaired skills and abilities, maintain or improve strength or endurance, or help a participant adapt or modify physical activities to be successful (Gabaldo 2003). These goals are measurable and clearly defined. A vocational model teaches specific job skills as well as the social and physical skills that may lead to employment outcomes. A social or wellness model helps to engage participants' aims to improve overall quality of life and may offer a recreational outlet that connects them to the community in which they live (Haller 2008). Although these three models are clearly defined, horticultural therapy programs often combine models, and horticultural therapy practitioners treat within a variety of settings. Programs that emphasize vocational or wellness outcomes may also target physical goals for individual participants.

Horticultural therapy in physical rehabilitation

Participants of a therapeutic model in a physical rehabilitation setting are treated for a wide range of conditions, including brain injury, spinal cord injury, concussions, sports injuries, stroke, and other neurological diagnoses. The treatment team for horticultural therapy programs in this setting often includes a physician, case manager, physical therapist, occupational therapist, speech therapist, recreational therapist, psychologist, and horticultural therapist. Assessments completed by the team, as well as the needs and desires of the patient, are the basis for establishing the treatment goals. Goals define the outcomes to be achieved and a plan of care is developed to achieve them. Team meetings or case conferences are held to review goals, assessing and reassessing to determine action plans. Accommodating patient preferences is consistent with current definitions of evidence-based practice. The link between therapy activities and functional outcome may not be obvious to the patient. It is often up to the horticultural therapist to describe the purpose of engaging in horticulture activities to the patient in order to help her or him understand the therapeutic aspect of what may seem to be a diversional pursuit as well as to motivate the patient to be fully involved in sessions. When goals are practical in their application and are relevant and challenging, clients are more likely to work toward achieving them (Scobbie 2011; Sivaraman 2003).

This is validated in Exhibit 8.10, which shares the treatment of an inpatient recovering from a brain injury. The co-treatment sessions between speech therapy and horticultural therapy proved to be especially meaningful, encouraged carryover to other sessions, and contributed to the patient's recovery. The exhibit also highlights the importance of

EXHIBIT 8.10

Case Example: The Importance Communicating with Families

The following is a note from a patient's speech therapist to his horticultural therapist in a rehabilitation hospital:

> I spoke with Joseph's mom today and she mentioned that Joseph had told her that his horticulture project and the related co-treats were the most meaningful things he did at Craig. I think this is a big compliment to you for coming up with the idea. The effort you put into it made it a success! I also think it was a great idea to wrap it up today with actual fertilization! Thank you again for your hard work and dedication to Joseph!

Early in his inpatient stay, Joseph's speech therapist contacted the horticultural therapist to inquire about setting up a co-treatment session for the patient with his speech therapist and horticultural therapist. She stated that the patient's brain injury resulted in his indecisiveness, lack of motivation, poor initiation, and apathy. She wanted to get him engaged in an activity that he would enjoy while still working on cognitive and language skills.

Communication Goals

Joseph's goals included conciseness in verbal communication, not interrupting other people, and sticking to only the important information. The speech therapist described that he was doing fairly well cognitively, except for his initiation and time management deficits. Also, it would be great for him to work on flexibility of thinking, planning, problem solving, organizing, and presenting information

(Continued)

EXHIBIT 8.10 (Continued)

Case Example: The Importance Communicating with Families

clearly and concisely. The activity used to work on these skills would need to be a complex task of some sort.

Joseph majored in environmental sciences and enjoys anything outdoors. His parents shared that he likes growing his own food and planted a vegetable garden each year. Sustainability is also important to him.

Horticultural Therapy Intervention

During Joseph's first co-treatment session, the horticultural therapist talked with Joseph about his background and gardening interests. Then the horticultural therapist explained a current need of the horticultural therapy program and asked if Joseph would be willing to help. This need involved creating a fertilization handout chart that would be used by horticultural therapy program volunteers and patients who assist in fertilizing indoor and outdoor plants in the program. Joseph willingly accepted the request to help.

Additional meetings with Joseph and his speech therapist were spent creating a handout and chart as requested. According to his speech therapist, he showed up on time to sessions, worked the entire hour, and looked forward to continuing work on the project. Joseph emailed a copy of his work to the horticultural therapist and requested a meeting to review his work in person. His rough draft was also shared with horticultural therapy program volunteers, who gave input about layout and detail.

During the second co-treatment session, they met to review the rough draft and discussed suggested edits. Joseph took notes and willingly made changes. He was pleased and excited that his work would be useful and would help clarify questions others may have when fertilizing.

At the final co-treatment session, Joseph led others and assisted with fertilizing flowers and vegetables in the outdoor gardens.

Joseph's speech therapist noted that initially Joseph was unmotivated, did not initiate, and was apathetic with most therapy treatments. Once he was participating in something that he was actually passionate about, his entire affect changed, and his positive attitude carried over into other treatment sessions.

Susie Hall, HTR, CTRS
Horticultural Therapy Program Coordinator, Craig Hospital

effective communication with family members who are often available to provide insight into interests and hobbies, which can be motivational during the therapy process.

The garden and horticultural therapy setting provide this practical application to achieve goals in the therapeutic model. An example is a garden-to-table experience offered to an individual recovering in a physical rehabilitation setting. The garden-to-table experience starts when the participant is an inpatient and, depending on the length of stay, can expand to the continuum of care as an outpatient. From seed to harvest, the analogy of the cycle of life in the garden often replicates the road to recovery for individuals. The first step (which can be literal, as in ambulation for an individual) is replicated in planting a seed. Hand function and fine and gross motor skills are used through planting seeds; eye-hand coordination is addressed through repotting cuttings or harvesting produce from the garden. As the new seedlings emerge or the cutting flourishes, responsibility and accountability are fostered, increasing self-esteem as the client becomes the caregiver of the plant.

Opportunities for improving the client's self-worth are provided by giving clients a sense of purpose, by demonstrating that the clients are needed as they assist with plant cultivation that will benefit future participants in horticultural therapy. Dynamic standing balance and proper body mechanics are improved through navigation of uneven terrain and through movement such as bending, lifting, and twisting. This prepares individuals for the challenges they may face in the community by utilizing the varying surfaces and design features of the garden. Practical use of adaptive tools for operation both in the garden and in the home are taught, as are safety behaviors. Cognition, memory, and understanding of abstract concepts such as time, change, and growth are witnessed in the life cycle of plants in the garden. The final step is the harvest, and the cycle is complete where activities of daily life are achieved by preparing a meal, reaping the benefits of a season of growth in the garden. Cognition is addressed through culinary activities utilizing the harvest, which might include following a recipe, noting necessary supplies, and measuring required ingredients. Word retrieval, speech, compensating for field cuts, and visual tracking can also be incorporated into the meal preparation activity. Nutrition and cooking with fresh ingredients are encouraged, and gardening as a healthy leisure activity is introduced. Note that throughout the treatment process, the client works on practical, meaningful activities of daily living. The therapy considers the whole person and addresses social-emotional needs while focusing on physical outcomes (Figure 8.5).

Figure 8.5 Harvesting produce as well as enhanced physical, social, and emotional health. (Courtesy of Kirk Hines.)

Measuring physical health benefits

Research validates the efficacy of horticultural activity interventions, and it is important to recognize the therapeutic mechanisms that help measure physical health benefits. Park et al. (2016) reviewed more than 500 papers to determine the physical health benefits of horticultural activity interventions. The articles reported the following physical health benefits: muscle strength, flexibility, range of motion, eye-hand coordination, hand function ability, agility, endurance, movement ability, physical functional ability, weight control, immunity, autonomic nerves, cortisol levels, blood pressure, improved quality of sleep, less pain, and increases in activities of daily living. As noted above, the garden-to-table experience incorporated these goals in a fluid and familiar way, as described in terms correlated to nature and its cycle.

Researcher and horticultural therapist Sin-ae Park describes the specific muscular and other physical benefits of gardening activities in Exhibit 8.11. This type of research may inform practice by helping horticultural therapists to understand the physical mechanics of horticulture tasks These interventions engage the participant by activating a wide range of muscles, which lead to positive physical health outcomes in a meaningful and practical way.

Further verification and analysis are documented by A. Y. Lee (2017), noting the rehabilitative effectiveness of horticulture activity. In addition, a study using gardening as an

EXHIBIT 8.11

Research Example: Muscle Activation by Gardening Activities

In order to further understand the potential physical benefits of various gardening activities, it is helpful to look at specific muscles that are affected and the range of physical health benefits that stem from gardening activities. The external stress produced by weight-bearing exercises stimulates new bone formation (Turner and Robling 2003) and improves muscle strength and physical functional ability (Olivetti et al. 2007). Weight-bearing gardening activities such as digging or weeding can improve muscle strength and bone mineral density (Restuccio 1992). Gardening activities include weight-bearing body movements and working against gravity by using the upper and lower limb muscles (Park et al. 2013, 2014).

Another mechanism to measure physical outcomes of therapy is electromyography, which measures muscle activation while a subject performs an activity (Bolgla and Uhl 2007). A study was conducted to measure muscle activation associated with common indoor horticultural activities by electromyography (Park et al. 2013). Subjects used eight muscles on upper limbs and hands. The major muscle used in the indoor horticultural activities was the upper trapezius because most indoor horticultural activities use the shoulder and arms. Specifically identified hand muscles were also actively used for the horticultural activities.

Another study measured the muscle activation of upper and lower limb muscles for five common gardening activities such as digging, raking, hoeing, weeding, and using a trowel. The subjects used the upper limb muscles, such as the bilateral anterior

(Continued)

EXHIBIT 8.11 (Continued)

Research Example: Muscle Activation by Gardening Activities

deltoid, biceps brachialis, brachioradialis, and flexor carpi ulnaris, as well as lower limb muscles, such as the bilateral vastus lateralis, vastus medialis, biceps femoris, and gastrocnemius (Park et al. 2014). Specific upper limb muscles were mostly used during the gardening activities, and the lower limb muscles played a role in supporting the body.

Sin-ae Park, PhD

Chair of International People Plant Council, Professor, Department of Environmental Health Science, Konkuk University, South Korea

intervention to increase physical activity in women over the age of 70 improved blood lipid profiles, blood pressure, and inflammation and stress levels (Park et al. 2017).

Exhibit 8.12 further documents this research in practice as it applies to the use of therapeutic mechanisms and to a task-oriented approach to measure muscle function with patients recovering from stroke at a rehabilitation facility.

EXHIBIT 8.12

Program Example: A Horticultural Therapy Program for Stroke Rehabilitation

To apply and test research in practice, a horticultural therapy program was designed to include horticultural activities, including motions such as reaching, grasping, squatting, stepping, and stooping (Lee 2017). Fourteen patients who were in rehabilitation with stroke diagnosis participated in horticultural therapy sessions for sixty minutes, three times per week, for six weeks. Seventeen patients with the same diagnosis were in a control group that did not participate in horticultural therapy. The patients in the horticultural therapy group showed significant improvements in upper limb function, hand force, balance, fall efficacy, and activities of daily living as well as decreased symptoms of depression. In contrast, there was no significant change in the control group. The horticultural therapy sessions, based on task-oriented training, improved the physical and psychological function during rehabilitation following a stroke.

A horticultural therapy program as a task-oriented training provides repetitive motions in horticultural activities. The horticultural therapist can correct the posture one-on-one based on kinematic and kinetic characteristics of horticultural activities. Moreover, the horticultural therapist can do progressive training with horticulture activities according to improvements in a client.

Sin-ae Park, PhD

Chair of International People Plant Council, Professor, Department of Environmental Health Science, Konkuk University, South Korea

In addition to physical rehabilitation, horticultural therapy may seek physical improvement outcomes in programs and settings that have a different or broader focus. While a therapeutic model may be used, physical issues may also be addressed in vocational or wellness models and are common in a wide range of programs. The following describe a few of these applications with examples of programs addressing increased activity for people with developmental disabilities, elders aging in place, and children coping with obesity.

School, residential, or day program setting

For individuals with developmental disabilities, increasing physical activity is often a big challenge. Horticultural therapy can be the catalyst for treating such individuals within a therapeutic model, whether it occurs in a school setting, group or residential home, or day program. In schools, horticultural therapy often focuses on the needs of students with learning difficulties and challenging behaviors that are secondary to a broad range of intellectual disabilities, including autism, brain injuries, and other genetic and neurological disorders. In the United States, an Individualized Education Plan (IEP) or *individualized support plan* (ISP) is integrated into the classroom treatment plan and is assessed on an individual basis. The treatment team varies depending on a student's needs and might consist of a board-certified behavior analyst; clinical social worker; speech, physical and occupational therapists; certified special education teacher; vocational coordinator; registered nurse; and other complementary therapies such as a horticultural therapy, music therapy, or art therapy. Family members are also an integral part of the treatment team. The program may be based on a therapeutic model and can emphasize physical movement, motor skills, or endurance. Sometimes a transitional vocational program is available and may be funded through a school district. Goals of the program vary and may include promotion of independence, fostering an increase in self-awareness and confidence, offering participants opportunities to make friends and develop personal interests, or providing vocational training opportunities. Horticultural therapy can be an integral part of the treatment plan to develop these interests and also to provide vocational training in the field of floriculture or horticulture. Often programs include opportunities for individuals to create horticulture-related products and share in the overall profits from the sale of their products. This approach combines the therapeutic, vocational, and social models and incorporates physical goals, vocational activities that provide meaningful work for individuals, and opportunities for social interaction through the use of creative arts.

Aging in place

Aging in place is a term used to describe the ability to remain in one's own residence as an individual ages for as long as the individual is able. This includes the ability to have any services or other support an individual might need over time as his or her needs change. Aging brings changes such as reduced vision; decreased muscle strength and endurance, which can lead to decreased mobility; reduced mental processing capabilities; an increase in the risk of falling; and reduced hearing. The impact of these changes can often be seen in the daily life of an elder. The goal is to maintain one's independence, quality of life, and dignity. Gardening is a meaningful opportunity for exercise. Planting, pruning, watering and taking a garden walk can help an individual stay physically fit and promote muscle movement, coordination, and flexibility. Overall, the aim is to maintain a range of physical abilities and slow the progression of age-related decline.

In the United States, continuing care retirement communities (CCRCs) provide individuals with the opportunity to age in place by offering independent living and a transition of care, with accommodations made for assisted living and nursing home care when necessary. Horticultural therapy programs exist in CCRCs, and the garden can provide programming for all levels of residents and address their needs and goals. As noted in other examples, in this setting, horticultural therapists take a holistic approach by attending to psychosocial as well as physical concerns. Gardens often have a deep-seated familiarity to elders, which is calming to an individual with dementia who feels as though nothing can be relied on to stay the same. Often the rote memory of gardening comes to life through a horticultural therapy session. Memory is connected to the sense of smell; using fragrant plants or aromatherapy provides sensory stimulation for those confronted by short- and/or long-term memory loss. Elders may retreat into themselves and feel hopeless or cut off from their surroundings. Caring for plants encourages bonding and socialization, stimulates communication, and may evoke memories of previous gardening experiences. At times those living in CCRCs are unable to make decisions about their own care; this loss of independence can lead to feelings of helplessness and powerlessness. Evidence of the positive change that a horticultural therapy session can make through the use of a mobile horticultural therapy program is highlighted in Exhibit 8.13. Other applications for using a mobile horticultural treatment strategy include situations where patients have extremely limited mobility and/or medical fragility, or are in secure (locked) environments without access to a greenhouse or garden space. These circumstances might be found in a hospital or other institutional setting (Figure 8.6).

EXHIBIT 8.13

Case Example: Horticultural Therapy Plant Cart

Working on a residential assisted-living campus for people with Alzheimer's disease and dementia sometimes presents gardening challenges. Indoor horticulture is preferred during any time of warm or cold weather, not necessarily only in extreme temperatures. Participants with limited mobility or understanding are sometimes best served when the activity can be brought to them indoors. This program has the ability to improve attention, reduce stress, and increase socialization.

Use of a rolling two-tier cart has allowed for horticultural therapy to go to the resident. The therapist transports the tools and materials to activity tables, small or private meeting rooms, or even into an individual residential room or suite. The cart provides an activity surface with a raised edge to contain spills and plenty of storage space below.

The therapist uses life histories provided by family members to gather materials needed for individualized experiences. A basic setup includes the following list of supplies: a two-tier horticultural cart, plant materials, small hand tools, small trash bags, spray bottles or watering can, plastic pots, writing utensils, labels, and hand wipes or a wet towel. Horticultural materials such as soil, sand, pebbles, etc., are stored in shoebox-sized plastic boxes with lids so they may be stacked to save space. Any sharp tools are also kept in a box to be out of view of the residents for their safety.

A successful example of this use is with a resident of an assisted-living community who is living with Alzheimer's disease. Lydia spends a lot of time pacing

(Continued)

EXHIBIT 8.13 (Continued)

Case Example: Horticultural Therapy Plant Cart

from area to area, shows symptoms of being anxious, and stays to herself most of the time, sometimes only leaving her room for meals. To help address Lydia's unmet needs, she is engaged with the horticultural cart. To empower Lydia, she is given a choice of seeds to plant and how she will plant them. She is encouraged to either reach her bare hands into the container of soil or use tools and gloves. Together with the horticultural therapist, she fills small containers with soil and plants the seeds. For 10–15 minutes, Lydia and the therapist stay actively engaged with smiles on their faces. Lydia ends most interactions with a proud smile and a celebratory high-five. This simple interaction has Lydia happily engaging, socializing, and feeling productive and accomplished!

By using the horticultural cart, the horticultural therapist is able to provide purposeful activities to the residents. They express feelings of accomplishment after planting seeds or taking cuttings for propagation. Participants with more advanced dementia often appreciate a sensory experience, such as holding a flower, feeling pleasant textures, or smelling the fragrance of herbs. The use of this engagement continues to show benefits to other residents like Lydia to capture resident attention, increase happiness, and provide an opportunity to engage with other people.

Debra Edwards, HTR, BLA

Life Engagement Coordinator and Horticultural Therapist

Figure 8.6 Mobile horticultural therapy plant cart. (Courtesy of Debra Edwards.)

Childhood obesity

Another opportunity for horticultural therapy within the context of a therapeutic model addresses childhood obesity. While physical in nature, obesity in often intertwined with many other challenges such as self-confidence, impulse control, motivation, and depression. This example again demonstrates that horticultural therapy often affects participants on many dimensions, even when the focus may be primarily physical in nature. Engaging in gardening activities for children offers a hands-on approach to learning, which can include designing, planting, and maintaining gardens; harvesting, preparing, and sharing food; collaborating together in small groups; learning about science and nutrition; and creating art and stories, which gardens inspire. Facts published by the American Heart Association (2017) note that one in three American children are overweight or obese and fewer than one in ten high school students receive the recommended amounts of fruits and vegetables daily. The horticultural therapist, facilitating garden-based, nutrition-intervention programs, can use gardening to improve physical health by promoting increased fruit and vegetable intake among children and at the same time improve physical skills such as strength and flexibility (Blair 2009). Thorp and Townsend (2001) note, "Gardening changes the status of food for all involved. When one gardens, food can no longer be viewed as a mere commodity for consumption; we are brought into the ritual of communal goodness that is found at the intersection of people and plants. Food that we grow with our own hands becomes a portal for personal transformation."

Research shows that children who grow vegetables may be more likely to eat them (McAleese and Rankin 2007) and people eat more fruits and vegetables if they participate in community gardens that also "grow" community (Glover 2004). Studies also indicate that those living in close proximity to a garden in childhood were more likely to show an interest in gardening later in life, which has lasting benefits for better health (Lohr and Pearson-Mims 2005). For urban dwellers, an introduction to the garden and how things grow allows children to become explorers in a world to which they are often unaccustomed. Multicultural school gardens provide an additional avenue for cultural diversity, providing a space where children can share cultural heritages and feel a sense of belonging. Since food disparities are often prevalent in many communities, associations with food banks are a natural connection for horticultural therapy programs. This collaboration not only provides meaningful programming for clients, who may be experiencing their own difficult time of recovery, but it also helps them feel they are contributing to something larger than themselves. This connection also serves the community and those that are often underserved in neighborhoods in which horticultural therapy programs exist or should exist.

Continuing care in a community-based setting

Horticultural therapy programs also provide a wide range of benefits for those who are recovering from illness or injuries that affect brain function and other ongoing physical issues after leaving intensive hospital-based rehabilitation programs. Exhibit 8.14 highlights the progression of adults recovering from stroke and traumatic brain injury (TBI) in a facility near Paris, France. It describes the impact that horticultural therapy has made on those individuals participating in the program and the difference the program is making to the surrounding community (Figure 8.7).

EXHIBIT 8.14

**Program Example: Patients Recovering from Stroke and
Brain Injury—A Garden by Request, Paris, France**

In 2011, a patient at the Maison des Aulnes, a then-brand-new, long-term home near
Paris housing forty-five people recovering from strokes and brain injuries, started
clandestinely planting flowers and herbs around the grounds. As a recreational
therapist, Stéphane Lanel recognized the therapeutic value of this initiative. Because
horticultural therapy is not a well-known practice in France, Stéphane had to be creative for the *Jardin d'Epi Cure* to come to life—the name is a play on word referring
to Epicurus, injections and cure. Lanel sought the help of Anne Ribes, a pioneer in
horticultural therapy, who lives nearby. He also educated the management and the
staff about this uncommon activity and approached donors while co-creating a more
than two-thousand-square-foot garden inside a larger park with a core group of residents. Because no training in horticultural therapy was offered in France, Lanel, in
his spare time, received several horticultural degrees to complement his initial training in helping people with disabilities.

In this program based on therapeutic and wellness models, residents were
instrumental in designing the garden; building willow planter boxes; preparing the soil; and planting fruit trees, vegetables, and flowers. Beehives were also
added, and three other residents now take care of them. Stéphane leads individual
weekly sessions with eight volunteer participants. Every Monday, Anne Ribes
comes for a visit, honoring expected rituals such as herbal tea and homemade jam
tasting under the pergola. In the past six years, participants have taken ownership
of the garden. They will often drop by to weed, water, or just sit under the pergola.
Residents who don't wish to participate in the activity also visit the garden as a
restorative space and as a place to entertain visitors. Occupational and psychomo-
tor therapists routinely use the garden for their sessions, and a Qi Gong class is
taught there.

The benefits for participating patients are both cognitive and physical. More
recently, the program, which attracts many visitors and is a great source of pride
for the participants, also developed a strong psychosocial component. Two residents
tend a special plot where they grow vegetables for a local food bank, Les Restos du
Cœur, while two others visit a nearby preschool for regular gardening sessions with
the children. These ties to the local community continue to deepen with an incredible edible project of planter boxes that will eventually lead from the residence to a
school in an underprivileged neighborhood. In addition, the crops are used for a
therapeutic cooking activity, which will be extended as part of a strong demand from
residents for healthier, more nutritious meals.

In a recent pilot study conducted at the residence, participants in the horticultural therapy activity had higher self-compassion and higher optimism than non-
gardeners. On the edge of the garden runs a path along a river, a favorite walking

(Continued)

EXHIBIT 8.14 (Continued)

**Program Example: Patients Recovering from Stroke and
Brain Injury—A Garden by Request, Paris, France**

spot for the locals. "We used to look at people walking there with envy," said the
resident who sparked the horticultural therapy program. "Now they look at us
with envy."

Isabelle Boucq

Clinical Psychologist, President, Fédération Française Jardins, Nature et Santé, France

Stephane Lanel

*Certified Medical/Psychological Assistant, Certification in Organic
Farming, Recreational Therapist, La Maison des Aulnes, France*

Figure 8.7 Anne Ribes interacts with a resident in the Jardin d'Epi Cure. (Courtesy of Stephane
Lanel.)

Use of space within the therapeutic model

In the United States, guidelines for characteristics of therapeutic garden spaces were initially developed by the AHTA (Kavanaugh 1995), and more recently by the American Society
of Landscape Architects, based on best practices and evidence-based design principles. As
defined by the AHTA (2013), a therapeutic garden is "a plant-dominated environment purposefully designed to facilitate interaction with the healing elements of nature." There is a
high degree of correlation between the intended programming of a garden and the physical design, which contributes to improving the care for patients, residents, and clients.

Therapeutic benefits from garden environments provide relief from physical symptoms, an opportunity to reduce stress, and an overall sense of well-being (Barnes and Cooper Marcus 1999). Understanding the characteristics of therapeutic gardens is an essential skill set for the practicing horticultural therapist, and all models—therapeutic, vocational, and social—use a therapeutic garden as a tool to engage a participant in horticultural activities. Horticultural therapy sessions often start at the garden gate, and there are distinct users of indoor and outdoor horticultural therapy gardens and spaces. They are patients, clients or participants, employees, and visitors (Neducin et al. 2010). Disability does not define someone. Rather their age, physical skills, and psychological needs determines how they might use an indoor or outdoor space. Are they children, elders, people with injuries, those who are recovering from surgery, patients in physical rehabilitation, individuals recovering from a concussion, those with psychiatric issues, or people who are undergoing chemotherapy with suppressed immune systems? As previously mentioned in this Chapter under *The treatment setting*, providing a supportive horticultural therapy environment is paramount. This includes focusing on the safety of participants, providing accessible areas of quiet, establishing well-defined perimeters, using plants safe for people–plant interactions, and creating areas that are unified and easy to comprehend or navigate as participants will use the spaces in unique and various ways, depending on their circumstances, capabilities, and desires.

One of the goals of an effective horticultural therapy session is to enhance the experience for everyone by making accommodations for a variety of uses and activities. In the outdoor space, according to a therapeutic model, features include easily accessible entrances and walking paths that are clearly defined, and raised beds with established, sensory-oriented plants in a variety of colors, textures, and fragrances. (Additional discussion about accommodations and design of the horticultural therapy garden space is found in Chapter 11.)

It's imperative to note the often-undocumented benefits of access to outdoor spaces. For those going through a difficult time of recovery, the garden provides an area for respite and a welcome change of pace for family members, caregivers, visitors, and staff members. Outdoor spaces provide an interactive area for individuals to address therapy goals in a unique, clinical setting that does not feel clinical to the participant, which lends itself to restoration and provides an opportunity to be in the moment. Gardens in horticultural therapy programs encourage active participation and are rewarding for several reasons: Humans are dependent on plants for existence; gardens and the observation of them can distract from pain or problems; through cultivation of plants within the garden setting, a person may develop attachment; and horticultural activities facilitate integration into society through a practical application or learned skill set (Relf 1999; Stigsdotter et al. 2011).

Guidelines for outdoor spaces

Further recommendations for specific guidelines for outdoor spaces are found in Roger Ulrich's (1999) theory of supportive garden design, which suggests the need to consider the following as key components for a successful space for therapeutic benefits or gardens used for horticultural therapy practice: visibility, accessibility, familiarity, quiet and comfort, and unambiguously positive art. When considering the first component, *visibility*, the garden space should be placed in a prominent area, noticed by clients, visitors, caregivers, and staff members. Signage should be provided to direct clients for horticultural therapy programming.

Accessibility is also a key component and begins with the ability to get to the garden area. Once there, navigation within the space should promote ease of use for all abilities within the garden setting and also provide clients the opportunity to practice real-life skills and activities of daily life. Planned design elements might also include suitable or appropriate challenges for ambulation or wheelchair mobility because either one would be a goal in a rehabilitation setting and encountered in the community postdischarge. Provide a variety of spaces to accommodate multiple activities that might take place simultaneously. Raised beds within the garden setting should vary in height because one size (or height) does not fit all.

Familiarity should be taken into consideration for supportive garden design and may be applied through the use of familiar plants, such as those historically found in home gardens, in gardens in the country, or in locations where patients previously lived. The introduction of gardening as a healthy leisure activity might be a new skill or hobby for participants. Since horticultural therapy programs often use the garden space for a defined functional purpose or an intentional activity, familiarity of space can increase active participation, engagement, and transference of learned skills. An opportunity to provide connections for those from diverse cultures through horticultural therapy programming can be achieved by plant selection or specific design features. Also, garden features that tell a story are interesting and engaging, so the therapy garden might be enhanced by displaying particular themes such as edible flowers, butterfly food sources, rainbow colors, plants in literature, or other topics.

Supportive garden design also provides *quiet and comfort*. Gardening is universally a mindful, meaningful, and quiet activity. Attending to the garden is a powerful elixir for the stressors of daily life. Creating a quiet space allows participants to focus on new perspectives and provides an opportunity to encourage healing in a nonthreatening environment.

The final element of supportive garden design, as noted by Roger Ulrich, is the use of *unambiguously positive art*, which may be soothing or uplifting and easily interpreted. Art can provide a long axis view or sight line to pull clients or visitors through the garden.

The design should include a variety of forms, textures, and colors to provide sensory stimulation. Indoor spaces are equally important, particularly when weather and temperatures prohibit use of an outdoor space. Designs should provide transition areas between outdoor and indoor spaces to allow participants to adjust to changes of light, guard against the elements, and welcome participants to the outdoor horticultural therapy environment.

Collaboration with community and allied health professionals

In horticultural therapy programs that focus on a physical outcome, a common setting is a hospital or other rehabilitation setting. Like spokes on the wheel, opportunities for community collaboration exist within the framework of many horticultural therapy programs. Viable horticultural therapy programs remain sustainable because they operate on more than one dimension. In a landscape or industry that continues to change and evolve, particularly in the field of medicine and health care insurance reimbursement, community connections help to promote the rationale behind funding for horticultural therapy programs. (See Chapter 13 for more on community connections and external communications in managing horticultural therapy programs.) A variety of factors affect the feasibility and sustainability of community connections. Close proximity to an existing horticultural therapy program is required; adequate staffing to manage and sustain these networks is imperative; and a strong commitment from administrators, funding sources, and facilitators of horticultural therapy programs to pledge and support their community is essential. An example is a horticultural therapy program aligning with a local food bank. Factors that make the relationship viable are close proximity for delivery of produce; assurance of adequate staffing to maintain, harvest, and deliver goods in a timely fashion; and a commitment to work together to maintain and sustain the relationship, which might include a signed agreement or contract.

Allied health professionals provide an extensive network and a platform to promote horticultural therapy as a profession. It is possible to share a common vision and collaborate with music, art, mental health, and recreational therapists, and the challenges often faced by allied health professionals are analogous to horticultural therapy. An example of a horticultural therapy program that collaborates with allied health professionals working with children who are hospitalized was highlighted earlier in the chapter in Exhibit 8.9.

Other opportunities for collaboration might include organizations to which former clients have transitioned postdischarge from a medical facility. This includes clients who have progressed through intensive inpatient and outpatient programs. When therapy goals are no longer being met in a timely manner or progress and gains have plateaued or slowed, funding sources for therapy typically end, and the patient is discharged. Individuals in this stage of recovery, who are unable to go home and live independently, have developed additional insight into their own abilities yet need to understand that progressing through the phases of recovery can be a lengthy process. This poses a challenge for them because the desire to return home is often paramount, yet that return to independence may be preceded by treatment in day programs, group homes, or transitional living arrangements. Opportunities exist for horticultural therapy programming to be provided in any of these settings on a continuum of care for community integration. It may fill a niche for the encouragement of gardening as a healthy leisure activity and continue to address therapy goals. This is important to this population, where services have usually been discontinued or diminished, yet progress and opportunities for further recovery still exist. Such collaboration is shown in Exhibit 8.15, which describes the journey of an individual recovering from TBI and the transition from inpatient rehabilitation to community integration through collaboration with a public garden.

Other strategies that horticultural therapists may use to support this continuum of rehabilitation after hospital discharge include forming partnerships with local parks and recreation departments, wildlife refuges, or state and federal parks that are creating

EXHIBIT 8.15

Case Example: Traumatic Brain Injury, Physical Rehabilitation

The following describes a prevocational experience for a woman with a brain injury, and experience that was arranged through a partnership between horticultural therapists at a public garden and a rehabilitation hospital.

Assessment

Demographics: The patient is an 18-year-old female from a small rural town in a western state of the Rocky Mountains. She is a recent high school graduate and was active in several extracurricular activities at school and in her community. She excelled in school and was valedictorian of her class. She is an only child with loving, supportive parents.

Case History: The client was in a motor vehicle accident that resulted in a traumatic brain injury. Computerized tomography (CT) scans showed swelling, multiple intracranial hemorrhages, and bruising of the frontal lobes. Once medically stable, she was transported to Craig Hospital for intensive neurological rehabilitation.

(Continued)

EXHIBIT 8.15 (Continued)

Case Example: Traumatic Brain Injury, Physical Rehabilitation

Existing Behavior, Symptoms, Diagnosis: Poor memory, decreased attention, and a lack of social awareness and insight into her cognitive deficits.

Planning

Long-Term Goal: The client will implement learned cognitive and social strategies to assist with her independence and return to an active community lifestyle.

More specific goals addressed in horticultural therapy include:

- Increase memory with organizational tools.
- Stay organized and on task.
- Complete tasks during allotted time.
- Participate in appropriate social conversation and behavior.

Methods (Horticultural Therapy Techniques, Tasks, Activities, etc., Including Any Co-treatment)

As a patient in the hospital, the client was involved in the horticultural therapy program. Because of an internship collaboration between Craig Hospital and Denver Botanic Gardens (DBG), the opportunity to volunteer at DBG was presented to the client. She was excited about the idea of giving back to the community. She was also comforted by the fact that the horticultural therapy intern would be working alongside her in a supportive role in both settings.

Horticultural therapy treatment sessions occurred twice a week for ten weeks. On Mondays, a co-treatment therapy session between horticultural therapy and various therapists (physical, occupational, and speech) occurred in the hospital and addressed the client's cognitive and social goals. On Wednesdays, strategies that were introduced during these co-treatment sessions were then implemented in a real-life setting when she volunteered at the DBG Sensory Garden.

Client Goals

- Practicing functional cognition (memory, attention, time management).
- Maintaining appropriate social interactions.
- Promoting independent work habits.
- Tracking/managing fatigue.

Results and Conclusion

Documentation of Progress on Objectives: horticultural therapy documented activities, progress, and challenges after each treatment. Attendance in weekly patient rounds meetings provided opportunities for the horticultural therapist to exchange information with the treatment team. Additional emails and discussions occurred,

(Continued)

EXHIBIT 8.15 (Continued)
Case Example: Traumatic Brain Injury, Physical Rehabilitation

when necessary, to relay information. Subsequent sessions were modified based on noted progress and challenges.

Outcomes and/or Long-Range Expectations:

- Impaired memory interfered with daily activities; the client benefited from cueing to take notes and make written lists.
- Difficulty with multi-tasking, needed to listen until complete directions were given.
- Lack of insight prevented independence.
- Extra time was needed to complete tasks.

In conclusion, the client will continue to benefit from support and structure to address these areas of challenge. Discussions occurred with client and parents to seek similar opportunities in home community.

Susie Hall, HTR, CTRS
Horticultural Therapy Program Coordinator, Craig Hospital

Kristina Gehrer
Horticultural Therapy Intern, Denver Botanic Gardens and Craig Hospital

Angie Andrade, HTR
Senior Horticulturist, Denver Botanic Gardens

outdoor spaces for therapeutic purposes. The expertise of a horticultural therapist as a consultant in the design process may lead to future horticultural therapy programming. Other possible avenues for therapeutic horticultural therapy in rehabilitation or other programs seeking physical outcomes might include summer camps, after-school programs, and collaboration for hands-on science programs.

Summary

This chapter explored the therapeutic model of horticultural therapy practice, with particular focus on those programs that seek psychological or physical outcomes for participants. As noted, a wide range of programs exists within this model that focus on an extensive array of life skills and address capacities that go beyond a singular goal of addressing physical and psychological improvement. The aim of treatment in horticultural therapy, as in other therapies, is to help each participant advance functioning and improve quality of life (Haller 2017). To be an effective horticultural therapy practitioner in the therapeutic model, one must understand the unique characteristics and circumstances of the individual receiving services, establish objectives and goals as they relate to the participant, and introduce challenging therapeutic interventions using evidence-based practice.

Key terms

Ambulation The ability to walk from place to place, usually noted by distance during therapy sessions in documentation.

Cognitive behavioral therapy (CBT) A type of psychotherapy based on the model of understanding cognition, that is, a model that considers and analyzes the ways individuals perceive a situation and react based on thoughts rather than the situation itself. It focuses on helping clients change their unhelpful patterns of thinking and behavior that lead to lasting improvement in their mood and functioning (Beck Institute 2018).

Continuing care retirement community (CCRC) A retirement community where individuals can live independently, with assistance, and with nursing home care. This type of living offers residents the ability to move between levels of care as health care needs change.

Continuum of care A broad range system that guides and directs clients through the health care spectrum and includes all levels of services.

Electromyography Technique used to measure muscle activation. This chapter refers to electromyography as it relates to an individual performing a horticulture activity.

Enmeshed boundaries In family therapy, the concept whereby the established boundaries between individual members are blurred and overly flexible, and it is unclear who is responsible for roles within the family.

Individual family service plan (IFSP) A plan for special services for young children with developmental delays. An IFSP applies to children only from birth to three years of age. Once a child turns three years old, an individualized education program (IEP) is put into place.

Individualized education program (IEP) A document that outlines goals identified during an evaluation process that meet the needs of individual students. The IEP defines the individualized objectives of a child who has been determined to have a disability or requires specialized accommodation, as defined by federal regulations.

Kinematics The study of the motion of objects without reference to the masses or forces involved in it. This chapter refers to bodily movement as it relates to horticulture activities.

Kinetic energy The energy a body possesses by being in motion. This chapter refers to energy stemming from movement as it relates to horticulture activities.

Kokedama Literally translates from *koke*, meaning "moss," and *dama*, meaning "ball." It is considered a modern art form that provides a unique presentation of a plant wrapped in a moss ball.

Mandala An abstract design that is usually circular in form. It is a Sanskrit word that means "circle." Mandalas generally have one, identifiable center point from which emanates an array of symbols, shapes, and forms. Mandalas can contain both geometric and organic forms.

Rote learning A memorization technique based on repetition. The idea is that one can quickly recall the meaning of the material the more one repeats it. In the case of a horticulture activity, repetition of an activity is a strategy used to help with memory.

Transference The projection of one's own thoughts or feelings onto another person.

Treatment team Team consisting of those who provide care as it relates to treatment of an individual. The treatment team collaborates to accomplish shared goals within and across settings. The treatment team varies according to the facility, school, or program where services are provided. This chapter refers to the horticulture therapist as part of the treatment team.

References

American Horticultural Therapy Association (AHTA). 2013. Definitions and positions. Accessed October 10, 2017. www/ahta.org.

Barnes, M., and C. C. Marcus. 1999. *Healing Gardens: Therapeutic Benefits and Design Recommendations.* New York: Wiley.

Beck Institute for Cognitive Behavior Therapy. "What Is Cognitive Behavior Therapy (CBT)?" Accessed February 28, 2018. https://beckinstitute.org/get-informed/what-is-cognitive-therapy/.

Blair, D. 2009. "The Child in the Garden: An Evaluative Review of the Benefits of School Gardening." *The Journal of Environmental Education* 40(2): 18.

Bolgla, L. A., and T. L. Uhl. 2007. "Reliability of Electromyographic Normalization Methods for Evaluating the Hip Musculature." *Journal of Electromyography and Kinesiology* 17: 102–11.

Detweiler, M. B. et al. 2012. "What Is the Evidence to Support the Use of Therapeutic Gardens for the Elderly?" *Psychiatry Investigation* 9(2): 100–110. *PMC.* October 7, 2017.

Fieldhouse, J. 2003. "The Impact of an Allotment Group on Mental Health Clients Health, Well Being and Social Networking." *British Journal of Occupational Therapy* 66(7): 286–296.

Gabaldo, M. M., M. D. King, and E. A. Rothert. 2003. *Health through Horticulture: A Guide for Using the Outdoor Garden for Therapeutic Outcomes.* Glencoe, IL: Chicago Botanic Garden.

Glover, T. D. 2004. "Social Capital in the Lived Experiences of Community Gardeners." *Journal of Leisure Sciences* 26(2): 143–162. Accessed September 4, 2017.

Haller, R. L. 2008. "Vocational, Social, and Therapeutic Programs." In *Horticulture as Therapy: Principles and Practice,* ed. S. Simson, and M. Straus. 43–44. Binghamton, NY: Haworth Press.

Haller, R. L. 2017. "Goals and Treatment Planning." In *Horticultural Therapy Methods: Connecting People and Plants in Health Care, Human Services, and Therapeutic Programs,* 2nd ed., ed. R. L. Haller, and C. L. Capra. Boca Raton, FL: CRC Press, pp. 2–27.

Kavanaugh, J. S. 1995. "Therapeutic Landscapes: Gardens for Horticultural Therapy Coming of Age." *HortTechnology* 5(2): 104–107.

Lee, A. Y. 2017. "Analysis of Kinematic and Kinetic Characteristics of Horticultural Therapy Activity and Verification of Its Rehabilitative Effectiveness." PhD thesis, Konkuk University, Seoul, Korea.

Linehan, M. 1980. "What Is Dialectical Behavior Therapy (DBT)?" Behavioral Technologies. Accessed February 28, 2018. https://behavioraltech.org/resources/faqs/dialectical-behavior-therapy-dbt/.

Lohr, V., and C. H. Pearson-Mims. 2005. "Children's Active and Passive Interactions with Plants Influence their Attitudes and Actions as Adults." *HortTechnology* 15: 472–476.

Mackinnon, L. 2012. "The Neurosequential Model of Therapeutics: An Interview with Bruce Perry." *Australian and New Zealand Journal of Family Therapy* 33(03): 210–218. doi:10.1017/aft.2012.26.

McAleese, J. D., and L. L. Rankin. 2007. "Garden-Based Nutrition Education Affects Fruit and Vegetable Consumption in Sixth Grade Adolescence." *American Dietetic Association* 107: 662–665.

Nedučin, D., M. Krklješ, and N. Kurtović-Folić. 2010 "Hospital Outdoor Spaces: Therapeutic Benefits and Design Considerations." *Architecture and Civil Engineering* 8(3): 293–305.

Olivetti, L., et al. 2007. "A Novel Weight-Bearing Strengthening Program during Rehabilitation of Older People Is Feasible and Improves Standing up More Than a Non-Weight-Bearing Strengthening Program: A Randomized Trial." *Australian Journal of Physiotherapy* 53: 147–153.

Park, S.-A., A.-Y. Lee, J.-J. Kim, K.-S. Lee, J.-M. So, and K.-C. Son. 2014. "Electromyographic Analysis of Upper and Lower Limbs Muscles during Gardening Tasks." *Korean Journal of Horticultural Science and Technology* 32: 710–720.

Park, S.-A., S.-R. Oh, K.-S. Lee, and K.-C. Son. 2013. "Electromyographic Analysis of Upper Limb and Hand Muscles during Horticultural Activity Motions." *HortTechnology* 23: 51–56.

Park, S.-A., C. Song, J.-Y. Choi, K.-C. Son, and Y. Miyazaki. 2016. "Foliage Plants Cause Physiological and Psychological Relaxation as Evidenced by Measurements of Prefrontal Cortex Activity and Profile of Mood States." *HortScience* 51: 1308–1312.

Park, S.-A., A.-Y. Lee, H.-G. Park, K.-C. Son, D.-S. Kim, and W.-L. Lee. 2017. "Gardening Intervention as a Low- to Moderate-Intensity Physical Activity for Improving Blood Lipid Profiles, Blood Pressure, Inflammation, and Oxidative Stress in Women Over the Age of 70: A Pilot Study." *HortScience* 52(1): 200–205.

Perry, B. D. 2009. "Examining Child Maltreatment through a Neurodevelopmental Lens: Clinical Application of the Neurosequential Model of Therapeutics." *Journal of Loss Trauma* 14: 240–255.

Relf, P. D. 1999. "The Role of Horticulture in Human Well-Being and Quality of Life." *Journal of Therapeutic Horticulture* 10: 10–14.

Restuccio, J. P. 1992. *Fitness the Dynamic Gardening Way: A Health and Wellness Lifestyle.* Cordova, TN: Balance of Nature Publications.

Scobbie, L., D. Dixon, and S. Wyke. 2011. "Goals Setting and Action Planning in the Rehabilitation Setting: Development of a Theoretically Informed Practice Framework." *Clinical Rehabilitation* 25: 468–482.

Sivaraman Nair, K. P. 2003. "Life Goals: The Concept and Its Relevance to Rehabilitation." *Clinical Rehabilitation* 17: 192–202.

Stigsdotter, U., A. M. Palsdottir, A. Burls, A. Chermez, F. Ferrini, and P. Grahn. 2011. "Nature-Based Therapeutic Interventions." In *Forests, Tress and Human Health*, ed. K. Nilsson and M. Sangster. New York: Springer, pp. 309–342.

Thorp, L., and C. Townsend. 2001. Agriculture education in an elementary school: An ethnographic study of a school garden. *Paper presented at the 28th Annual National Agricultural Education Research Conference*, New Orleans, LA. Accessed September 9, 2017. http://www.aaaeonline.org/Resources/Documents/National/ResearchProceedings, National2001.pdf.

Turner, C. H., and A. G. Robling. 2003. "Designing Exercise Regimens to Increase Bone Strength." *Exercise and Sport Sciences Reviews* 31: 45–50.

Ulrich, R. 1999. "Effects of Gardens in Health Outcomes: Theory and Research." In *Healing Gardens: Therapeutic Benefits and Design Recommendations*, ed. C. C. Marcus and M. Barnes. New York: John Wiley & Sons, pp. 27–86.

VandenBos, G. R. 2007. *APA Dictionary of Psychology.* Washington, DC: American Psychological Association.

Weilert, L. "Ten Most Common Reasons for Seeking Help from a Therapist or Counselor." Accessed February 28, 2018. http://kchealthandwellness.com.

chapter nine

Vocational model

Gwenn Fried and Rebecca L. Haller

Contents

This chapter examines the use of a vocational model in horticultural therapy programming. Characteristics of the model are described, and background information on career development, particularly for people with disabilities, is reviewed. Varied types of vocational programs and placements are presented, along with considerations for space needs, program development, professional collaboration, and treatment processes. Examples include how to set up and manage the effectiveness of a vocational rehabilitation approach as well as descriptions of wide-ranging programs, individualized goals, and outcomes.

Characteristics of a vocational model in horticultural therapy

Vocational horticultural therapy uses a program model that emphasizes work-related skills and outcomes. In these programs, horticultural therapists apply principles of vocational training, vocational horticulture, and horticultural therapy to practice.

Vocational training refers to training in the skills and knowledge to work in a particular trade. This type of training is typically undertaken during or after high school and is focused on gaining specific skills in order to prepare for a desired job (The Law Dictionary 2018).

Vocational horticulture was defined by the American Horticultural Therapy Association (AHTA) in 2007 as follows:

> A vocational horticulture program, which is often a major component of a horticultural therapy program, focuses on providing training that enables individuals to work in the horticulture industry professionally, either independently or semi-independently. These individuals may or may not have some type of disability. Vocational horticulture programs may be found in schools, residential facilities, or rehabilitation facilities, among others. (AHTA 2007)

A program in *vocational horticultural therapy*, on the other hand, incorporates the horticultural therapy elements described in Chapter 1, including a trained professional, a client who is in some form of treatment with defined goals, and the cultivation of plants. Note that the term *treatment* includes procedures beyond those found in a medical model, and thus may emphasize a full spectrum of care, including skills and knowledge related to employment. In vocational horticultural therapy, the therapist documents assessments, goals and objectives, and progress notes. (See Chapter 12 for more information on documentation.)

In contrast, *vocational horticulture* programs (without the term *therapy*) usually teach horticulture skills and may or may not apply horticultural therapy processes; it is a broader category of training. This chapter predominately illustrates programs and processes specifically applicable to a vocational model of horticultural therapy services, in which the outcomes sought are related to employment. "The desired outcome for individuals in the program is placement in the least restrictive environment in which the individual functions successfully" (Haller 1998). This may be competitive independent job attainment, one of the many types of supported employment, or the acquisition of skills that are related to work and independent functioning. The program may offer paid employment and/or training in the skills to obtain employment in the community. In many vocational horticultural therapy programs, the emphasis is on teaching work readiness and job-related skills, although programs that train workers for employment in horticulture are also offered. Work-related objectives and benefits of horticultural therapy have been recognized since the early days of the profession. In 1973, "vocational and prevocational training" was listed as one of the intellectual benefits that may be stressed (Hefley 1973). Throughout the development of the profession, vocationally oriented programs have been prevalent and have evolved along with changes to the foci of vocational services for people with disabilities.

Efforts to promote employment of people with disabilities in community-based jobs are common worldwide. The United Nations 2030 Agenda for Sustainable Development

Figure 9.1 Preparing tomatoes for sale in a vocational program for addiction recovery. (Courtesy of Eugene Jones.)

Goal 8 is to "promote sustained, inclusive, and sustainable economic growth; full and productive employment; and decent work for all." One of the targets is expressed as "By 2030, achieve full and productive employment and decent work for all women and men, including for young people and persons with disabilities, and equal pay for work of equal value" (United Nations 2018). Vocational services that apply horticultural therapy principles and processes seek to offer an effective strategy for meeting this target (Figure 9.1).

Perspective on horticulture-based vocational training

Employment in horticulture is a viable option for many people who face economic or social exclusion. Horticulture also offers a particularly effective modality for teaching work-related skills and behaviors. Vocational programming that uses horticulture has an advantage due to the motivating characteristic of working with plants; clients are motivated to overcome difficulties to succeed. Some of the components of using horticulture for training also illustrate advantages that this modality offers. The first component is practicing select actions that lead the trainee to gain the expertise necessary to complete the task successfully. Horticultural therapists are uniquely qualified to provide accommodations, training techniques, and supports to foster achievement (see more about these topics in Chapter 11). Allowing the time for acquiring individual skills and combining them to become competent in a specific job has a direct effect on the success of the individual. Another component of the vocational model of horticultural therapy is task identity. In horticulture, a worker can create a whole, identifiable piece of work, such as trays of seedlings or potted planters, in many ways. The products of the work are tangible and valued. This is unlike putting a computer chip onto a motherboard time and time again, never seeing a recognizable end product. See Exhibit 9.1 for an example of a man who benefitted vocationally and behaviorally from the use of a well-structured horticulture task.

Task significance is an asset of using the vocational model of horticultural therapy. Witnessing the impact that the job a client has completed has on the work or life of others

EXHIBIT 9.1

Autism Spectrum Disorder

Horticultural Therapy Facilitates Sensory Integration for Those on the Autism Spectrum: An Individual Case: Parker Picks Produce

Assessment: Parker is a 27-year-old male diagnosed with autism spectrum disorder. He lives in a group home with three other individuals. Parker has significant antisocial behaviors, self-injurious tendencies, and an intense drive to manipulate items repetitively with his hands. Parker maintains intense focus when prompted to complete sorting or packaging type tasks, and his observed inappropriate/ aggressive behaviors during these skill sessions are noted to be markedly decreased. Parker would like to get a job so that he can work to earn money to purchase meals out with his peers as well as movies to add to his collection.

Planning: Parker's clinical support team met to discuss options for his vocational training. Based on his predisposition for repetitive activities and tactile sensory stimulation preferences, recommendations were made for Parker, with assistance from his one-to-one support staff, and the oversight of a trained horticultural therapy professional, to sample job duties of picking blueberries at a local produce farm.

The long-term goal for Parker was to develop skills that would enhance his ability to become gainfully employed. Treatment also included assistance with meaningful sensory stimulation activities to aid in decreasing maladaptive self-injurious or aggressive behaviors.

Objectives utilized to assist Parker with learning a job skill included maintaining positive interactions with the support staff trainer, following a visually structured schedule of horticultural tasks, utilizing assisted decision-making processes while completing horticultural tasks, completing multistep task assignments while engaging in horticultural tasks, and actively participating in gross motor activities while completing horticultural tasks.

Methods: Parker was given a visual, five-step schedule with pictures and words, set up on a letter-sized laminated piece of cardstock. The schedule's visual instructions outlined the process for picking a pint of blueberries. Parker's one-to-one support staff person visually and verbally reviewed the steps in the blueberry picking process with him, before his arrival at the farm. Necessary materials were set out prior to Parker's arrival, and identically matched those he had been shown on his visual schedule. Parker was directed to follow his schedule using verbal prompting, job skills modeling by his support staff, and hand-over-hand assistance. He was able to successfully pick a half pint of blueberries during his initial visit to the farm, without any negative behavioral outbursts. Parker enjoyed this experience, and a set of specific objectives were developed for his continued vocational training.

Results and Conclusions: Over the course of the next six weeks, Parker was prompted to gradually increase his harvest production. His objectives were increased in frequency, volume of production, and time worked, and the prompting/support required for him to complete tasks as well as inappropriate social aggression was shown to decrease.

(Continued)

EXHIBIT 9.1 (Continued)
Autism Spectrum Disorder

Parker was given a job opportunity to continue his work at the produce farm pick-ing other summer vegetables. Through the implementation of this horticultural therapy program, Parker has become employed, his self-esteem has improved, and his obsessive-compulsive tactile manipulations and maladaptive behavior rates have decreased.

John Fields
Certificate in Horticultural Therapy, Director of Operations, GHA Autism Supports

offers a comparable level of satisfaction. Seeing a flat of seedlings growing or being used in the garden is rewarding and motivational for trainees. These tangible rewards, along with feedback given by the vocational trainer or horticultural therapist, increases performance and satisfaction. Engagement in horticulture tasks often provides a focus that effectively replaces disruptive behaviors with more positive ones, as we saw in Exhibit 9.1. Exhibit 9.2 describes a similar result in a practice in Taiwan. In addition, the significance of the task itself helps to motivate trainees and employees to come to work each day, reducing absen-teeism and employee turnover.

EXHIBIT 9.2
Case Example: Developmental Disability, Taiwan

This exhibit describes a client in a horticultural therapy program that uses a voca-tional model to serve people with: Down syndrome, intellectual disability, cerebral palsy, autism spectrum disorder, and Prader–Willi syndrome. Lee is a 22-year-old man who is diagnosed with Prader–Willi syndrome. This syndrome is genetically based and causes him to constantly look for and eat food without self-control. Teachers usually need to keep a close watch on him so that he won't steal food.

Lee has been involved in a prevocational horticultural therapy program for at least two years. He started knowing nothing about plants. During the horticultural therapy program, he is in charge of watering vegetables in a greenhouse. This task has shown to be highly motivating to him. He is so interested in watching for the time that the vegetables can be harvested and when they may be eaten, of course. Therefore, Lee works very hard and focuses on his job. This engagement and focus can efficiently decrease the chance of going out to steal food.

By participation in the horticultural therapy program, Lee has learned a lot of knowledge of plants to increase his cognitive skills and has engaged in physical activity by working in the greenhouse. When he sells vegetables to clients, he learns how to communicate with people, increasing his social skills and confidence. Horticultural therapy turns his disability into motivation to know, learn, plant, harvest, and share with others in a positive way.

Chin-Yung Wung, HTR
Horticultural Therapist, Daniel A. Poling Memorial Babies' Home, Taiwan

Options and opportunities for horticultural therapists

Vocational horticultural therapy programs are designed and implemented in a myriad of places and systems, and they serve clientele of diverse ages, abilities, and social circumstances. Even children may be served, as schools adopt work-based models of teaching students with disabilities. One such school-based program begins training during elementary school as part of the curriculum, transitions students into an in-house cottage industry, and eventually moves the student into a community setting and paid employment. (See Exhibit 9.3.) Programs that incorporate vocational training into rehabilitation in facilities for individuals with mental health, substance abuse, or adverse legal issues also have been established in many regions. Programs for individuals with developmental disabilities, acquired cognitive deficits, and social or physical limitations can be found as freestanding programs or programs that use community resources to place individuals in employment. Programs designed specifically to train youth at risk (see Exhibit 9.4) provide underserved urban students training in an afterschool setting. Some programs gainfully employ people at various levels of support, either permanently or with the goal of transitioning into a mainstream horticulture career. In addition to providing training and job placement, many vocational horticultural therapy programs provide ongoing employment for people

---

EXHIBIT 9.3

Program Example: Work-Based Learning and Vocational Horticulture for Students with Developmental Disabilities

Mason smiles as he enters the greenhouse, his favorite spot in school. He has come to do his school job, being a garden helper to the horticultural therapist. Mason is a seven-year-old, nonverbal student with autism at the Monarch School of New England (MSNE) in Rochester, New Hampshire. Every morning when Mason sets up his daily schedule, he places the garden icon on his schedule book; it is his way of communicating that he wants to work in the greenhouse. His teacher reports that horticultural therapy is the only activity in which Mason will focus for fifteen minutes without attempting to leave. In the greenhouse, he works on his individual education program (IEP) goals, which will someday help him succeed in a community job, with skills such as following directions, sequencing tasks, manipulating tools, and communicating with others.

Recognized for excellence in academic, vocational, and therapeutic programs, MSNE is a private, nonprofit day school for students, ages five to twenty-one, with significant developmental disabilities. MSNE offers an array of prevocational/vocational training opportunities in different career sectors. Students of all ages participate in what is now called work-based learning, progressing from jobs in school to jobs in the community.

Horticultural therapy at MSNE integrates individual therapy and work-based learning goals within an educational theme. Horticultural therapy groups focus on reducing stress while teaching specific work skills in a meaningful context, skills such as propagating plants, sorting and packaging seeds, and cooking with garden produce. Students sell their garden products at the school store and farmstand, practicing customer service and money skills.

(Continued)

EXHIBIT 9.3 (Continued)

Program Example: Work-Based Learning and Vocational Horticulture for Students with Developmental Disabilities

All students participate in horticultural therapy groups, led by the horticultural therapist. Teachers, educational technicians, nurses, and other therapists work with individual students during horticultural therapy, which enhances the holistic approach. The students who show a special interest in gardening are given additional school jobs as garden helpers, helping the horticultural therapist with plant care in the greenhouse and therapeutic garden.

By the age of fourteen, a student's team, comprised of the student, the family, and the educational staff member, start planning for life after high school. A transition plan is written into the student's IEP, with student-centered goals related to job training, employment, activities of daily living, and community access. Through formal and informal vocational assessment, a student identifies his or her interests and skills, which helps in career planning. Situational assessments, such as observing the student in different work settings, helps to determine which types of jobs are the best fit.

After the age of sixteen, students are matched to volunteer jobs in the community, consistent with their interests and skills. For example, students who have a special interest in horticulture may volunteer at a garden center. Before landing the job, students prepare a resume, interview with the on-site supervisor, and train for the job. If successful, the job becomes a regular part of the student's vocational curriculum. The student's goal is to work with increasing independence and learn skills that will support meaningful employment after graduation.

Kathryn B. Perry, MA, OTR/L, HTR
Horticultural Therapist, Monarch School of New England

EXHIBIT 9.4

Program Example: At-Risk Youth

Urban Roots is a vocational horticultural therapy program that provides agricultural training to at-risk youth in Greensboro, North Carolina. The program is a collaboration between local and national nonprofit organizations and was funded through a $15,000 grant from the Campaign for Black Male Achievement. Through this program, ten young men in Greensboro gained employment, developed vocational skills, and built a new school garden for a local elementary school. Urban Roots seeks to elevate existing black male achievement work in Greensboro and open new doors for the youth who participate in the program.

In the program's initial year, Urban Roots consisted of eighteen, two-hour after-school sessions in January through May. Participants were required to apply, attend

(Continued)

EXHIBIT 9.4 (Continued)

Program Example: At-Risk Youth

interviews, and obtain work permits as part of their vocational training. Following program completion, each participant received a $400 stipend. The program participants were ten black and Latino high school youth who were part of a local program called the African American Male Initiative. Urban Roots was designed and led by horticultural therapist Hailey Moses, garden educator at the Greensboro Children's Museum, with support from Rashard Jones, director of the African American Male Initiative. Urban Roots participants had three primary group goals:

- To gain vocational skills that can translate to future employment.
- To gain knowledge and skills in small-scale agriculture.
- To develop increased self-esteem and self-efficacy.

During the sessions, participants built eight raised garden beds at their neighborhood elementary school, planted and learned about seasonal crops, and provided weekly maintenance for the garden. The sessions also included youth community organizing training from a local nonprofit organization, as well as lessons by guest educators related to vocational skills. The program culminated in a youth-led farmers market, in which Urban Roots participants sold the crops they produced to fellow residents of their community. Throughout the program's initial year, steps were also taken to support future sustainability. Social media promotion occurred throughout the year, potential funding partners for subsequent years were invited to the farmer's market, and records were kept that detailed what happened during each session.

Vocational horticultural therapy is an important model for many populations but is a particularly powerful tool for work with at-risk youth. Paying Urban Roots participants communicated to the young men that their time was valuable, allowed youth to participate for whom employment was a financial necessity, and provided a source of motivation for hard work and dedication to the project. In addition, working within the framework of an organization that already serves at-risk youth and has an existing relationship with the participants helped to ensure program success. Urban Roots will seek additional grant funding for following years and hopes to establish a sustained vocational horticultural therapy program for young men for years to come.

Haley Moses
Certificate in Horticultural Therapy, Garden Educator, The Edible Schoolyard

who are disabled or disenfranchised. These include programs that employ people with disabilities, or those in recovery or reintegration programs, in horticulture businesses. For example, a community-supported agriculture business, a production greenhouse, or a plant nursery may exist solely to train and employ people with disabilities or those who have been displaced due to conflicts or natural disasters. Similarly, greenhouse production may offer vocationally based services that also serve as a key component of recovery from addictions (as seen later in this chapter in Exhibit 9.8) (Figures 9.2 and 9.3).

Horticultural therapists may be employed in vocational training programs as therapists, teachers, trainers, job coaches, corrections counselors, crew leaders, or rehabilitation

Figure 9.2 Student at Monarch School of New England. (Courtesy of Cynthia Tokos.)

Figure 9.3 Urban Roots participants building a raised bed. (Courtesy of Hailey Moses.)

specialists. Many times, the job title of the horticultural therapist is not horticultural therapist; instead, it fits the system in which the program exists. Professionals may apply a vocational approach in many types of different organizations. The interdisciplinary skills of horticultural therapists uniquely qualify them to offer excellent vocational training. It is important that the therapist is well versed in gardening, landscaping, and greenhouse or nursery production in order to be effective in using a vocational model of practice. It is also incumbent upon therapists to inform potential employers that they have the added skills as human service professionals to successfully accommodate and address the individual needs of trainees.

Vocational horticultural therapy programs offer a distinct budgetary advantage to organizations that apply it and offer financial sustainability along with program endurance. A natural product of work training is horticulture production and services. These may take the form of greenhouse growing, food or ornamental crops production, or indoor and outdoor plant maintenance. In most programs, significant income is generated from the sales of plants or food, or by charging customers for services provided. Sales are often included as a key part of tasks available to participants and provide opportunities for social contacts and community integration. Many vocational programs use a business model to provide the best training or supported employment possible by blending therapy and training with a connection to real work conditions. The Melwood Horticultural Training Center (1980) promoted the idea of using a horticulture business model for serving people with developmental disabilities. The model described in 1980 has since evolved to focus on outcomes that emphasize individual-supported employment in regular workplaces in the community, but it still utilizes horticulture as a key to success.

Horticulture career opportunities for the people served

A worldwide trend in employment for people with disabilities is to provide supports that enable individual placement in jobs or self-owned businesses. Note that the term *people with disabilities* includes those who have intellectual, physical, psychiatric, and sensory disabilities. The broad range of options in horticulture facilitates job acquisition and retainment. Horticulture provides a comprehensive group of challenging and rewarding careers in everything from production to research. There are opportunities in interior urban landscaping companies; design, build, and maintenance firms; nurseries; and greenhouses.

Although many employers hire unskilled laborers to fill positions, vocational training gives individuals with disabilities an advantage in the competitive marketplace. A well-trained, career-minded employee is desirable to business owners because such an employee cuts training costs and reduces turnover.

Horticulture jobs are uniquely suited to a wide range of differently abled individuals. Jobs require a specific set of skills and abilities and can often be adapted to accommodate exceptional needs without difficulty, for example:

- Production jobs offer tasks that can be repeated again and again, which may be a good job match for someone who prefers very structured and familiar situations.
- Public gardens often have both public and behind-the-scenes opportunities to meet individual preferences for levels of interaction.
- Research can provide field and desk work, which may be suitable for those with physical limitations.
- Marketing runs the gamut from sales of seeds and plugs to growers through end-user retail sales.

Unlike some other industries, there is a need for full-time, part-time, and seasonal employees in the horticulture industry due to the ebb and flow of nature. This creates opportunities for employees who need to gradually build endurance and skills over a longer period of time or may need a planned period of time off to take care of medical issues.

A large number of individuals who become disabled during their productive years due to accidents, military service, or disease often need to revise or rethink their original career aspirations. As they prepare to reenter the workforce, horticultural job training programs can provide prevocational as well as vocational training. Issues of job tolerance, interpersonal appropriateness, multi-tasking, and other areas can be assessed and improved. Decisions can be made regarding whether an individual can work independently or may need coaching or a more sheltered opportunity.

See Exhibit 8.15 in Chapter 8 for an example of a vocational emphasis in a horticultural therapy program that primarily uses a therapeutic model. The vocational approach was an appropriate client-centered response to the needs and desires of a patient who had experienced a brain injury. It provided an opportunity to observe the patient outside the hospital; allowed for observation of the individual's functioning in a simulated work setting; and identified a possible postdischarge outcome, which may include work as a volunteer.

Benefits of using horticulture as a training modality

Horticulture offers unique advantages as a medium for vocational and prevocational training. Regardless of the intended job placement, the environment and tasks themselves offer authentic feedback and consequences to trainees. When proper care is given to plants, they flourish. If tasks are not completed with accuracy and attention, they may wither or die. For example, watering must be sufficient, seedlings must be handled gently, and lawn mowing must be methodical so that all sections of grass are cut. The results of tasks that are well done are tangible and visible to the participant as well as others who encounter the program. This authenticity offers tremendous advantage for training.

In addition, participants in vocational horticultural therapy programs have the opportunity to produce or care for plants that are valued by others. In fact, in many such programs, plants or maintenance services are sold to outside buyers. When the result of one's work is valued, this helps to build self-esteem and a sense of efficacy. In order for this personal worth to be maximized, quality standards in horticultural therapy programs with a vocational emphasis must be in alignment with those found in the horticulture industry (Figure 9.4).

Figure 9.4 The sale of valuable microgreens provides personal worth to a program participant. (Courtesy of Eugene Jones.)

The therapeutic benefits of working with plants and nature are not lost on those choosing horticulture as a vocation. Studies that find improved concentration, performance, and productivity in work areas that contain plants or a view of nature might indicate that the benefit would exist or even be stronger for those working directly with plants and nature. More quantitative research is needed to further understand or substantiate this idea. Other benefits of plants in the workplace include lowered perceived stress and reduced use of sick leave (Bringslimark 2007). Exhibit 9.5 describes an individual who thrived and managed difficult behaviors successfully when working in a garden environment.

EXHIBIT 9.5

Case Example: High School Student with Multiple Disabilities

Alexei is a teenager in an exceptional children's class in high school. Adopted at an early age from a Russian orphanage by a remarkable family, he faces several challenges in life. Alexei is deaf and uses American Sign Language (ASL) and a combination of gestures and voicing to communicate. During school hours, he has a dedicated sign language interpreter. He has also been diagnosed with autism and fetal alcohol syndrome.

Alexei is placed in a life skills class, a class in which he struggles to understand his role as a student and in which he can often get frustrated. He frequently tries to assume the role of teacher, being bossy and telling students what to do. When he is asked to tend to himself, he can get quite angry. This struggle with his proper role has devolved into violent meltdowns in the past. Alexei takes medication to keep his emotions under control, and it does keep him calmer. Without this medication, his mother says he would not be able to go to school or even live at home.

For the most part though, Alexei is a happy seventeen-year-old who is extremely helpful. He seems to know when something needs to be done before it's even mentioned. He assists with pushing the wheelchairs of other students, jumps up to get a trash can or a broom, and volunteers for any chore. He enjoys being helpful. Other than his hearing impairment, Alexei has no physical limitations and is a strong, athletic young man who enjoys being active and is capable of any physical activity.

With his class, Alexei has been part of the horticultural therapy program at Bullington Gardens for two years. Due to his high physical capabilities he is being transitioned to a program that is more work-based at the gardens. He thrives on the physical labor and is quite comfortable in the role of worker. He easily understands what needs to be done and wholeheartedly contributes to the task.

Alexei absolutely loves coming to Bullington. He enjoys the small setting, the appropriateness of horticultural therapy activities, and especially being able to make something that he can give as a gift. Giving is very important to him and it brings him pride. Bullington has a very calming influence on Alexei. If he is in the throes of an emotionally rough day, he snaps out of it once he is at the gardens.

(Continued)

EXHIBIT 9.5 (Continued)

Case Example: High School Student with Multiple Disabilities

A primary goal for Alexei during his time at Bullington is to become more independent and reduce his reliance on his interpreter. When he comes, he will be given a task and shown what tools he needs. His interpreter will then retreat to a designated place where he could find her if needed. Given his intuition of knowing what to do and his love for working outdoors, it is expected he will be quite successful working at Bullington. This will prepare him for the time when he won't have the support he now has and can function more independently.

John H. Murphy, MS, HTR
Director, Bullington Gardens

Concepts and theories

Basic constructs of work and disability

Vocational training first appeared in the United States in the early 1900s, in the work of Frank Parsons. Parsons proposed a three-step process of choosing a vocation, consisting of (1) gaining information about the person, (2) gaining information about the world of work, and (3) matching these two to arrive at an appropriate occupation for that person. Over the years, theories changed and developed to make career development more of a lifelong process than a one-time event. Note that the process did not consider any adjustments for working with people with disabilities.

Work may ease, cause, or exacerbate psychological or physical problems (Neff 1985; Quick et al. 1992). The organization of many businesses leads to added and unnecessary stress. Specially designed horticulture settings utilize natural tendencies to be restorative and calming. When looking to place individuals with disabilities in ideal employment situations, mitigating negative or stress-reducing factors and matching the job to personality goes far in job retention. When considering what motivates an individual with disabilities to want to work, job satisfaction and self-image are important factors because government aid often provides a similar income without any of the stress and effort involved in employment.

Disability has many different faces, and those with differing abilities cannot be grouped together due to the diverse nature of the individuals. Individuals with disabilities could prove to be the single largest minority in many countries; for example, in the United States 19% of Americans with disabilities were identified in the 2010 census (Brault 2012). In tough economic times, when jobs are scarce, it is even more difficult for those with disabilities to secure employment. In the United States, concerns about losing benefits needed for medical treatment and potential changes in coverage for preexisting conditions make many individuals fearful of making the leap to paid employment. Even with all the negatives and unknowns, the majority of people with disabilities report that they would like to work. Vocational rehabilitation counselors can help trainees navigate the intricate system to transition to an employment situation that takes all factors into account. Horticultural therapists who take a strengths-based approach can offer the training and support to help individuals to be vocationally successful.

Career development theories

"Person-environment interaction approaches have been the foundation of much of vocational rehabilitation" (Szymanski et al. 1996). The theory that individuals seek environments that fit their personality and that the success of the fit directly affects the person and her or his success is especially poignant for the individual with disabilities. People with disabilities as a whole have far less choice in the companies and types of jobs that they can attain. It is the responsibility of the therapist or trainer to understand the importance of the fit of the job in client satisfaction and to take this into consideration.

John Holland spent forty-two years devoted to career counseling. His theory develops person-environment theory, adding developmental aspects. Holland (1997) says, "Both people and environments can be categorized into six types: realistic, investigative, artistic, social, enterprising, and conventional," and "[p]airing of personality types with certain environmental models will lead to predictable outcomes including vocational choice, stability and achievement." Holland's work reinforces the importance of including individuals in career planning and encouraging them not to limit their choice based on perception of barriers created by disability. These barriers may be resolved with accommodation. In fact, in the United States, barrier removal is required under the Americans with Disabilities Act of 1990, and similar laws in the European Union and other countries.

Social Cognitive Career Theory focuses on self-efficacy, outcome expectation, and personal goals as the key variables in career success (Lent et al. 2000) "In this theory the contextual importance of culture is vital. It not only influences how career development and work are perceived but also affects how disability is viewed" (Trevino and Szymanski 1996). Today with the diversity in cultures, in many countries around the world, the view of disability is varied. Often the vocational counselor must consider long-held beliefs of family and caregivers about the ability and need for the individual to work. Understanding these cultural differences aids in identifying reasons for and overcoming compliance issues, such as a belief that the person with a disability is incapable of work, in vocational training settings.

Types and characteristics of programs with vocational emphasis

Prevocational training

The most basic type of training is prevocational training. Prevocational training is not necessarily focused on building skills in specific horticultural tasks, which is the cornerstone of vocational training. Prevocational skills are work-readiness skills and may address a wide range of behaviors, attitudes, and capacities. For people with disabilities, building endurance is often one of the first areas addressed. Individuals need to be able to physically and cognitively tolerate the duration of a work training experience—perhaps starting with one to two hours. Being able to accept and understand direction, restrain emotions, and interact with superiors and coworkers are also areas of focus. Self-checking before moving on to a new task or location is also an important skill. Navigating to and from the job site, being on time, and dressing appropriately for work and weather are addressed. Horticultural therapy prevocational programs are appropriate for trainees who are interested in horticulture careers as well as those who may ultimately seek other types of work. In the community, prevocational training programs are found in programs for people with developmental disabilities, day treatment sites for individuals with brain injuries, veteran's hospitals, rehabilitation ambulatory care centers, and schools.

EXHIBIT 9.6

Program Example: Public Garden and School Partnership

Bullington Gardens is a nonprofit public garden in Hendersonville, North Carolina. It functions primarily as a center where a variety of horticultural educational opportunities are offered to the community. Horticultural therapy is a central part of its mission.

The horticultural therapy programs at Bullington are conducted chiefly through its close partnership with the Exceptional Children's department of the Henderson County school system. One program, Bullington Onsite Occupational Student Training (BOOST), dovetails with the Occupational Course of Study (OCS). OCS is a statewide high school course of study for students with special needs. It balances a standard curriculum with job training and competitive work hours. The goal is for students to learn the basic skills necessary to obtain and retain employment and ultimately live independently.

Since 2003, Bullington Gardens has worked with sophomores in OCS helping them to acquire some of their required work hours. Students from the county high schools work each week at Bullington doing a variety of jobs, including growing plants in the greenhouse for future sales and maintaining the gardens, grounds, and nature trails.

The highlight of the BOOST program is a garden competition among the schools where students design a garden, grow all the plants from seed, and install it, all within a budget. The garden competition is a project that challenges the students in many ways. Few of the students have ever previously grown plants or planned for such a long-term project. The garden themes chosen by the students often reflect the interests and issues facing them. Themes in past years have included antibullying, acceptance of Mexican culture, saving bees, Appalachia, peace and harmony, student memorials, and school pride. As the process develops over a semester, ideas and leaders emerge, and the garden gradually takes shape. The result is something of which the students are usually very proud.

During their year at Bullington, students certainly gain horticultural knowledge; however, the focus of BOOST is to impart the basic prevocational skills that are valuable in any job setting, such as staying on task, being productive, working as a team, working safely, and taking care of tools and equipment. Students have the opportunity to learn these values at Bullington before they go on to do internships in their junior year and go on to employment as seniors. At Bullington, they have a chance to get away from the traditional school environment and work with new people, although teachers are almost always present. This unique hybrid program combines school with an outdoor workplace and provides students with an opportunity to experience success that will hopefully set them on the right path for the rest of their lives.

John Murphy, MS, HTR
Director, Bullington Gardens

Gardening and horticulture can provide a supportive, restorative environment for individuals to be assessed for and to transition into a more formal vocational training environment, as described in Exhibit 9.6. Note that each of these prevocational skills is also addressed in vocational training but typically requires less focus due to functioning levels (Figures 9.5 and 9.6).

Figure 9.5 Students in the BOOST program at Bullington Gardens learn work skills by designing and creating gardens. (Courtesy of John Murphy.)

Figure 9.6 Skill building in the Adaptive Horticulture Program at Solano Community College. (Courtesy of Rebecca Haller.)

Sheltered training and employment

Sheltered training and employment are used for individuals who prefer this environment to a competitive workplace and for those with disabilities severe enough that they are unable to succeed in the competitive job marketplace. The sheltered workshop employs a group of individuals who complete tasks within their cognitive and physical ability range. The support level remains consistently high, with the size of the group determined by the requirements of the job and the ability of the participants. People can transition out of this model into a program with a less intensive level of support. Sheltered employment has been replaced in many regions by supported employment due to a preference for individual job placements, which occur in the community. The drawback of this model is considered to be a lack of integration with the typical (nondisabled) population. Some sheltered or work activity settings reduce this limitation by actively accomplishing integration within a sheltered environment. See Exhibit 9.7 for an example of a long-running vocational horticultural therapy program that has taken advantage of community integration opportunities and successfully navigated changing needs and program emphases, with a constant focus on the provision of client-centered services.

EXHIBIT 9.7

Program Example: Greenhouse for People with Developmental Disabilities

At Mountain Valley Developmental Services in Glenwood Springs, Colorado, the vocational services for adults with developmental disabilities have included horticulture since 1981. Developed by a horticultural therapist, the program began as a viable component of employment and work skills training. From its roots in a small attached greenroom with limited growing and work space, the facility expanded to a 9,000 square-foot greenhouse complex in 1983. Designed to provide accessible and climate-controlled work space, storage, and two separate growing areas, the greenhouse has been used to produce a variety of crops, including bedding plants, native perennials, culinary herbs, sunflower sprouts, ornamental potted plants, and salad greens. It has served clients in an evolving array of programs and emphases, such as supported employment, job skills training, and community participation.

Operating as a commercial greenhouse, the center sells directly to consumers as well as to businesses who resell the products or use them in landscaping projects. Participants who work in the greenhouse also receive individualized supports, which are designed to maximize their growth and development, vocationally as well as personally. Clients have community contact through people who come to the center and by going into the community for a variety of jobs. Interactions with the community occur through plant sales, contracts with downtown businesses to plant and maintain annual flower containers, and partnerships with a local school and its students to grow salad greens as a component of greenhouse production.

Throughout its history, the program has maintained accredited programs that focus on individual needs and desires. It offers exemplary person-centered services that include treatment teams and individualized goals and objectives. Through vocational horticultural therapy, participants may work on improving skills such as communication, anger management, fine motor, following directions, or staying on task (Mountain Valley Developmental Services 2018). They also earn a paycheck for their work while building self-esteem, self-efficacy, and a sense of purpose in life, along with learning that they contribute to their local community.

Rebecca L. Haller, HTM

Employment of groups

Other options for ongoing employment have involved the employment of groups of people in the community. These programs provide a lower level of supervision than the sheltered model, yet they do have a supervisor present at the job site. One of these, an enclave model of support, employs a group of people in a community-based workplace to perform specific job responsibilities with the support of a trained supervisor. The ratio can vary but ideally would not exceed eight workers to each supervisor. The individuals can be all in one area or dispersed throughout the business, but they are still supervised. The mobile crew model of supported employment is similar to the enclave model, but the supported group moves from site to site completing a specified contracted job at each location with support and supervision. Jobs in uncomplicated garden or lawn maintenance are typical in the mobile horticulture crew. Both enclave and mobile crew workers are usually paid a stipend based on the number of employee slots they fill. For example, two supported employees may be taking the place of one competitive employee. They can be paid by the employer or the agency based on contract nuances. Like sheltered employment, these group employment schemes are less prevalent today because individual employment is considered a more integrated, and therefore a more desirable approach.

Supported employment

Supported employment involves individual coaching and may be referred to as supported competitive employment. Supported employment means "competitive employment in an integrated setting with ongoing support services" (Wehman 2012). The World Association for Supported Employment describes the attributes as follows:

> Supported employment has been shown to be an effective way for persons with disabilities to get and keep a job in the open labor market. Supported employment does this by its focus on ability and not disability, by its provision of individualized support to the person and advice to employers. The principles of supported employment can be applied in all parts of the world, provided that they are adapted to the cultural context and labor market trends of any given region (Kamp 2018).

Typically, an individual has a job coach through the interview, orientation, and job training process. As the individual becomes more comfortable and competent with the job, the coach is present less and less often. The goal may be for the coach not to support the employee eventually unless the employee has a severe disability and requires intensive and ongoing support. In this model, the employee is paid the current competitive wage for all hours worked.

Competitive employment training

Competitive employment training programs are designed to give trainees an advantage when they are applying for entry-level jobs. Often individuals in vocational rehabilitation programs have resumes that have gaps or are nonexistent. Having documented training and professional references often compensates for the spotty resume or needed accommodations. Vocational rehabilitation programs typically work with clients who seek competitive employment. See Table 9.1 for a summary of these types of vocational services.

Table 9.1 Vocational services

Service type	Focus	Support or Staffing
Prevocational training	Acquisition of work readiness skills	Support and teaching
Sheltered training and employment	Training and supervised employment that takes place in a private setting (outside industry)	Continuous supervision and support of staff
Group employment (enclaves or work crews)	Paid supported employment in an integrated setting with a group that functions to complete particular job tasks	Continuous work supervisor on site
Supported employment	Individual employment in an integrated setting in the community with support	Job coaching or other support prior to and during employment
Competitive employment training	Acquisition of specific task skills that are needed for independent employment	Training for specific tasks

Source: Rebecca L. Haller, HTM.

The process of programming for vocational rehabilitation

This section describes how a horticultural therapist might work with vocational rehabilitation professionals and organizations to develop training programs for individuals who desire employment in horticulture. A training program could be added to an existing horticultural therapy program or be autonomous. Included in the following subsections are examples of the systems and processes found in the United States.

Identifying potential trainees

Locating individuals to participate in a horticultural therapy-based vocational or pre-vocational training program can be challenging when beginning a new program. In the United States, each state and most big cities have a governmental agency that provides vocational rehabilitation services. In other countries, systems and organizations vary, but frequently there is some governmental agency that addresses employment training. In addition to career training and counseling for adults, vocational rehabilitation services often provide literacy education, transition, and youth services to help young people with disabilities move from a school to work setting; career training; and high school equivalency diploma programs. These local or regional programs are good places to start. They are connected to a wide range of agencies that provide services and can refer individuals to horticultural therapy programs. Ambulatory rehabilitation facilities, which are often affiliated with hospitals or health care networks and private treatment programs, are also able to refer clients to a horticultural training program. They may not require the level of certification that a state program requires and could provide clients while a new program awaits certification from the state agency. Another avenue for locating potential participants is in correctional programs because vocational rehabilitation is often provided for incarcerated individuals and those on parole.

Horticultural therapists can seek employment in any of these agencies. Social workers and rehabilitation counselors understand the rehabilitation process but lack the horticultural knowledge necessary to train competitive job seekers effectively. Greenhouse staff members understand horticulture but lack the knowledge of the special needs that must be addressed when working with these populations. The horticultural therapist is the expert in bridging the gap and forming a successful training program.

Working with vocational rehabilitation counselors

Each of these agencies has vocational rehabilitation counselors with whom to work closely when selecting individuals for the program. This relationship continues as the horticultural therapist updates the counselor on the successes and difficulties that each client is having as she or he navigates the training provided.

Prior to a client being introduced or suggested for the horticulture training program, the vocational rehabilitation counselor will have completed a series of assessments. She or he will have information on physical, cognitive, and psychosocial performance of the individual. The counselor will be able to provide information regarding strengths and weaknesses and offer strategies to help compensate for the deficits. This groundwork with the vocational counselor guides the horticultural therapist in setting up the necessary supports to enable the client to have a positive initial experience and address any safety concerns.

Vocational counselors sometimes accompany a trainee through an interview and "hiring" process, if it is deemed appropriate. Higher-functioning individuals may navigate this procedure with less support. Once the client is accepted into the vocational horticultural therapy program and a start date is agreed upon, communication with the vocational counselor is necessary to ensure success. For the first two or three days, a daily report, formal or informal, should be generated. After the initial encounter, the therapist/trainer and the counselor may decide on a schedule of weekly or biweekly reports and periodic site visits, with time for meeting with all parties.

If at any time the client is not functioning successfully in the program, this should be communicated to the counselor immediately, with a meeting scheduled to add supports, make adjustments, or terminate the relationship if there is no other recourse.

Most programs have a specific number of days to completion. Milestones and markers should be agreed upon with the counselor so that, as the trainee progresses, all parties will know if the trainee is making adequate progress. For example, in a forty-day program, practical performance evaluations are given at ten, twenty, and thirty days. If the trainee is making the markers, the last ten days of the program should see the client performing at the level she or he will need for employment with increasingly less supervision.

At this time, the vocational counselor begins working with the trainee on a resume, interview, and job search skills. Because horticulture is a unique industry, it is often helpful for the horticultural therapist to look over the resume to be sure it highlights the skills that the trainee has acquired. This is also a good time to assist trainees and/or counselors with connections to any horticultural professional with whom the program is connected.

As a final step, the horticultural therapist develops a program to follow up on recent graduates of the training program. The vocational counselor also often follows them for a prescribed period of time. Input on the performance of the client in the real world offers tools to improve on the training program and its success rate.

In a facility with a horticultural therapy program that accepts vocational trainees from a governmental or nonprofit program, the horticultural therapy program often receives funding for services rendered to the individual. There is usually a set rate per day for the specific type of training program.

Program development: Competitive employment in horticulture

When developing a horticulture-based vocational training program, the first thing to look at is the necessity for services. Are there horticulture positions available in the area? Do potential trainees exist? Are the needs unmet? If the area has jobs to fill, what are the specific requirements for the jobs? Relationships with employers who are looking for employees must be fostered because they will inform many decisions regarding programming and goals and be an invaluable resource when looking to place trainees in jobs.

Locations for potential job placement

One type of employment in horticulture is in retail establishments. Some businesses have programs to welcome adults with disabilities as employees. For example, the US home improvement store, Home Depot, has a program that facilitates accommodation and training needed for individuals to be successful. Family-owned and -run retail nurseries especially have difficulty with the high employee turnover in the horticulture industry and welcome the potential for a long-term employee who is part of the community. People with disabilities often stay with the same employer for multiple years, offering a stable workforce to these businesses. Retail horticulture businesses often provide year-round employment opportunities as well as part-time seasonal jobs, meeting the varied needs of trainees.

In cities or urban areas, there is a market for interior landscaping. Large office buildings often have green space mandated by local ordinances or sustainable development standards such as Leadership in Energy and Environmental Design (LEED) or other certification standards. The development of biophilic design theory also entices corporations to incorporate plants into outdoor planters, lobbies, waiting rooms, boardrooms, and offices. Companies or building managers then hire companies to design, install, and maintain these plants. The companies in turn hire technicians to water and clean the plants. Employees can work independently or as part of a team. There are also desk jobs for individuals who have an understanding of plants and are able to do desktop design and ordering.

Public parks and public gardens hire both seasonal and regular employees. Entry-level positions are usually part of a team and can be simple raking, sweeping, shoveling, and mulching jobs. Trainees who benefit from a supervised, team environment will thrive here. Instructing trainees to self-check after completing an assignment is key for employment in this setting. Seasonal employees are brought in each spring to bolster the staff during peak gardening months. These part- or full-time positions are usually assigned tasks such as clean-up, planting, watering, weeding, and fall plant removal. Although all-weather work is required, seasonal employees do not have to cope with the winter cold in temperate climates. Another benefit is that these organizations often select their year-round full-time employees from seasonal staff members who have proven themselves worthy and have shown interest in a long-term position.

Production businesses, such as greenhouses and nurseries, offer yet another viable type of work in the horticulture industry. While public gardens often require a considerable amount of interactions with visitors, in production work there is much less interaction with the general public, and jobs can often be repetitive and relatively consistent in their demands. For the right trainee who thrives on routine and structure, this could be the perfect match. In some areas there are horticulture production businesses that are staffed primarily by individuals with disabilities.

The training site

Once the employment needs and opportunities have been determined, the next step is to select a location for training. It may be possible to add this type of program to an existing horticultural therapy site. If a horticultural space already exists, it will need to be outfitted to maximize the training opportunity. Retail training could be satisfied by a cottage market that sells to staff members or visitors within the facility. Alternatively, a role-playing set up to orient and assess trainees provides an initiation into the retail world and could be followed by a coached experience.

Interior landscaping training can be facilitated by beginning interior plant training focusing on identifying and understanding the cultural requirements of commonly used interior plants and a hands-on immersion in the industry. Trainees can assume maintenance of plants in the training facility, or the program may contract to take care of plants in a nearby business. Any income generated can be returned to the program and eventually could pay for a staff position to expand the training program. Once trainees are oriented and ready for additional responsibility, a partnership with a local interior landscaping company can provide an internship site for the training program and a steady supply of employees for the local company.

Training for jobs in parks or botanical gardens requires outdoor space. If the training facility has outdoor space, the program may be able to assume responsibility for a specific area that is not maintained by others. Another option for securing outdoor space includes partnering with a local park to adopt a space in the park and maintain it during the year.

Designing a training program for jobs in greenhouse or nursery production requires access to and knowledge of each step in the production process. Greenhouses offer a unique set of benefits and challenges. A greenhouse with space for seeding and repotting through wholesale or retail sale offers an opportunity for income generation and multimodal training. Greenhouses require a high level of competency in pest management as well as climate control. A horticultural therapist who specializes in greenhouse production or a combination of a therapist with a production specialist can spell success. Another way to look at this opportunity is to foster collaboration with a production greenhouse. The horticultural therapist could do preliminary training and then coach the trainee as she or he transitions to the work environment. Exhibit 9.8 describes a program that utilizes commercial greenhouse production as training and therapy for people recovering from addictions.

Curriculum

A curriculum for the training program needs to be very basic. Many of the vocational training discussed is designed to give an individual an advantage in a competitive interview for an entry-level position. Very often the business owner who has the potential to hire the trainee offers expertise on skills and other characteristics that would be helpful for someone to secure a position. Often included in the curriculum is a plant identification component, which contains a comprehensive list of the most used plants, cultural requirements, and pictures. It is helpful to have the trainee fill out a workbook. The trainee enters each plant as she or he encounters it and, by the end of the training period, she or he has a book to refer to when on the job. In addition, hands-on checklists of the trainee's daily duties can gradually grow to encompass the entire job and thus can promote autonomy. Training also includes appropriate behaviors and basic job skills, such as punctuality. Another effective approach to curriculum is to train the person at an actual job site, integrating the work and skills unique to a particular setting.

EXHIBIT 9.8

Program Example: Vocational Training in a Therapeutic Residential Program for Adults with Addictions

PROGRAM OVERVIEW

Demographics and admission

Recovery Ventures Corporation is a two-year residential therapeutic community that combines job training with counseling to serve approximately one hundred adults who are addicted to drugs or alcohol. The program model provides peer-based programming, and comprehensive services to promote sobriety, personal accountability, and productive work skills for self-sufficiency. All clients (referred to as "associates") accepted into this program have an Axis I, II, or III diagnosis, which includes attention deficit hyperactivity disorder, oppositional defiant disorder, anxiety disorders, depressive disorders, bipolar disorders, cognitive deficits, substance use disorders, and post-traumatic stress disorder. They also may have a criminal history and antisocial behaviors. Approximately 40% of the associates are on probation and are court-ordered to seek residential substance abuse treatment. Prior to admittance, potential associates are assessed by the clinical director for cognitive, mental, physical, educational, and criminal history status. Minimum standards and criteria regarding self-care, mobility, and behavioral skills are set for those accepted into the program.

Associates typically enter the program with a variety of antisocial behaviors such as poor life choices, anger issues, and lack of social skills. In addition, they are often homeless; lack job skills or training; have achieved only a low level of education; and exhibit dysfunctional behaviors, low self-esteem, and low self-efficacy.

Individual vocational assessment

The initial assessments also identify the individual's vocational skill set. Work experience and skills are varied, with some having very limited work experience or skills and others who are trained professionals. The program makes an effort to place the associates in jobs similar to their skill sets. The program has construction, landscaping, and maintenance crews, as well as work placement in hospitality jobs at various hotels and resorts.

Upon admittance, an on-the-job evaluation is conducted while the associate spends several days or weeks participating in job training in the community settings. This period provides staff members with observation of the applicant's abilities while the associate gains knowledge of the training program and its opportunities.

Horticultural therapy program

Theoretical framework

The approach is to use horticultural therapy as a modality in addition to other therapeutic components used in the program, such as dialectical behavior therapy, group therapy, and individual therapy. Horticulture adds a unique therapeutic

(Continued)

EXHIBIT 9.8 (Continued)

**Program Example: Vocational Training in a Therapeutic
Residential Program for Adults with Addictions**

element that can enhance associates' growth and at the same time provide vocational skills, which can increase their employment opportunities. Thus, horticultural therapy bridges all aspects of the agency's mission, using both therapeutic and vocational models, and introduces associates to the field of horticulture.

Program goals

Program goals include the promotion of individual well-being, as well as production goals that offer real-life work experiences and training to promote self-sufficiency along with improvements in mental health. Production is inextricably tied to overall treatment; thus, program goals include:

- Generating a volume of production of hydroponic lettuce, microgreens, tomatoes, kale, and collard greens to support the operation and to provide horticultural therapy and provide vocational training to associates.
- Using horticulture/hydroponics growing methods suitable for health care, human service, and educational settings.
- Selecting appropriate plants for successful growing and to support program goals.
- Planning indoor and outdoor growing spaces to fit the physical site, population served, and program goals.
- Preparing plants for retail sales, managing greenhouse/hydroponics operations, and successfully producing plants using hydroponics growing methods.

Individual treatment plans

Each associate has a treatment plan that is evaluated each week by the treatment teams. Each associate treatment plan is reviewed to determine adaptability to the entire treatment process and to make changes as necessary. This process is conducted by the clinical director.

Methods

Throughout the treatment process, horticultural therapy techniques, tasks, and activities are employed to provide the associates an opportunity to observe greenhouse (hot houses) and hydroponics practices first-hand. Each associate is involved with hands-on activities each day, such as conducting maintenance in the greenhouses, harvesting, preparing harvest for the consumers, planting seeds, and transplanting various plants. Associates are rated each day on their understanding and knowledge of horticulture tasks completed. In addition, each day there is a teaching moment and a walk-and-talk session to afford each associate an opportunity to discuss any concerns or problems she or he might have, to express her or his interest, and express what she or he has learned.

Documentation

In order to record individual progress and outcomes, written documentation occurs throughout the program. Verbal one-on-one reviews between staff members and associates happen each day to provide discussion and feedback. Long-term outcomes

(Continued)

EXHIBIT 9.8 (Continued)

**Program Example: Vocational Training in a Therapeutic
Residential Program for Adults with Addictions**

are determined by observations and follow-up with graduates. Many of the graduates are employed in various aspects of horticulture in the private sector, such as garden centers, greenhouse operations, and landscaping companies.

Facilities used

The horticultural therapy program is conducted on a fifteen-acre farm that has two hot houses for hydroponic crops, such as tomatoes, edible flowers, collard greens, and kale to be sold to consumers. A state-of-the-art hydroponics greenhouse is used to grow lettuce and microgreens. Fully engaged in all aspects of the operation, the associates constructed all the greenhouses at the farm and conduct maintenance on the facilities, as well as acquire skills in the proper use of farm and garden tools, equipment, and machinery. Staff members use various hands-on, one-to-one techniques throughout the horticultural therapy process, including demonstrations with various tasks, to include pruning, transplanting, cleaning equipment, providing nutrients to plants, harvesting, and preparing product for delivery to consumers.

Program success

In operation since 2011, experience with horticultural therapy in this setting has been very successful with the community and the associates as a sustainable and innovative approach to addiction recovery. Following is an example of an individual associate in the program.

Stan, a fifty-six-year-old man who was homeless and addicted to alcohol and cocaine, was a skilled factory worker, has a minimum high school education, and is a divorced father of two adult sons. He has been in the program for seventeen months. He was homeless for several years after he lost his job due to his addiction. Upon introduction to the horticultural therapy program, he demonstrated an interest in working with plants and the various concepts involving horticulture. After successfully completing the first four phases of treatment, he is working with the horticultural therapy program as an intern. Internship is the fifth phase of treatment, whereby associates begin to transition back into society. Upon graduation, it is anticipated that he will be employed as a full-time assistant to the horticulture staff.

In addition to successful vocational and recovery outcomes, it is important to note that the organization is self-sustaining, without the receipt of any government funding. The horticultural therapy program operates as a business called Farm Fresh Ventures to provide real-life vocational training as well as therapy. The monies earned from contracts and sales from the farm operation provide sustainability and operational income to support the full range of services.

Eugene Jones, MS Ed, HTR
Horticultural Therapist, Farm Fresh Ventures

Program implementation

Once an individual is referred, a formal interview is the next step. Although the potential trainee has been assessed, an interview provides an opportunity to meet and observe the trainee in person. In a competitive employment training program, it is appropriate to require the trainee to call and set up an appointment and to arrive to the interview on time and dressed appropriately. The vocational counselors involved should understand the training program requirements. For coached or sheltered employment training, a less strenuous set of rules should be communicated to the referring agency.

Once a trainee has been approved for the program, the next step is to set up a concrete training schedule, including days and hours of work, responsibilities, chain of command, sick-leave or days off procedures, dress code, and what to do if the trainee has a problem.

Time should be scheduled for a review of the trainee after the first or second day, with a report back to the referring agency. After the initial review, there should be a regular schedule of documented assessments with the referring agency at predetermined intervals. If at any time the individual experiences difficulty, it is necessary to accommodate or adjust the training style or pace to individualize the training. Communication with the referring agency is key in properly serving the trainee and improving the training program.

Evaluation of the trainee at predetermined times ensures timely completion or extension of the training program. Evaluation should include assessment of knowledge as well as job performance. Accommodation of testing method should be based on the learning style of the trainee. Verbal, written, pictorial, or hands-on testing should be used to get an accurate assessment of the trainee.

Logistics for vocational horticultural therapy

Every state in the United States has different rules concerning approved vocational training programs. And systems and policies vary widely internationally. Before beginning the process, it is necessary to obtain information about it for approval through the appropriate governmental agency, develop the curriculum and parameters for the training program, obtain letters of approval from potential employers, and submit them for approval. Approved programs in many areas receive government-funded payment for training days.

In evaluating a vocational training program, the goals of the program must be considered. If the goal is for participants to obtain competitive employment, success may be gauged by placement data. If the population in training is not seeking competitive employment or is a mixed group of trainees, evaluation of completion of curriculum, changes in an individual's self-concept or skill development, or overall census can be indications of the level of achievement of programs goals.

Collaboration is key in the success of any vocational training program. Stakeholders in the program are valuable resources. The agency that offers the program can provide support in various ways, such as space and functional training opportunities as well as the work of allied professionals. For example, speech therapists can help with cognitive and communication issues. This could make a positive difference for individuals who need to use augmentative adaptive communication devices to talk to employers or coworkers. Occupational therapists can help with adapting tools or equipment. Mental health counselors and others may also provide support.

For horticultural professionals, it is important to assist the trainees to identify jobs that are the best fit for each. For employees, that will require job coaching or other supports, a call to the employer prior to the candidate's application to explain the situation, and the benefits to the employer. In addition, this communication helps introduce the program concept and can even add a new collaborating employer.

As previously described, training programs are also often embedded in businesses or agencies like prisons, day treatment facilities, production greenhouses, and community-supported agriculture farms. Although the processes need to be adjusted to fit the individual goals and objectives of the program, using the vocational rehabilitation model as a loose guideline provides a foundation for successful outcomes.

The horticultural therapy site: Components for vocational programs

Based on the individuals served, the elements necessary in a program site will vary, but a discussion of some of the horticulture-specific considerations will benefit program developers and clients alike. In contrast to other models of horticultural therapy, one of the essential components of a site used for vocational programming is that there is enough space to sustain a significant amount of production or other related horticultural tasks. It may also be important that the treatment space and methods used take a commercial approach, simulating systems and methods used in the horticulture industry. In order to reach a wide variety of users, this approach is balanced with design and techniques that accommodate and enable independent access and task performance. To practice work skills, the trainees benefit from opportunities to engage in the practice of horticulture, so sites must be designed to facilitate full engagement and work.

When developing a site for a horticultural therapy vocational training program, accessibility must be addressed as a primary issue. If the clients are to arrive by vehicular transport, there should be a safe loading and unloading area. This area can also serve for deliveries or another area that can accommodate soil and other messy items that must be included. The next issue is the surface in the production space. An approved surface must be able to absorb or wick away water, avoid glare, and be easy to clean. Something like a porous concrete dyed to a medium dark color is ideal.

A defined entry area, where individuals can observe the space before entering, allows for a feeling of security. Seating provided in this area should serve not only those needing a resting area but also as a place to observe what is going on before entering the area. Once the individuals enter the training site, a climate controlled, shaded work area would be ideal. Shade can be provided by a structure or by plant material organized to define space, reduce glare, and shelter workers from intense sun and bright light. Plant material in the work area not only provides a holistic work environment but also adds opportunities for plants to be pruned, groomed, and maintained as part of the learning experience. The space should be organized to resemble the space that trainees would work in after completing the program. The more familiar the work environment, the greater the chance the trainee will make a successful transition.

Aspects of a training site that uses horticulture that should be considered include noise, temperature, and sun. Because greenhouses are inherently noisy, it is best to find fans that minimize noise especially when turning them on and off. Temperature is another issue specific to horticulture. The program should provide climate-controlled areas for respite from high temperatures. Sunlight is not tolerated by some individuals on psychotropic or other medications. Easy access to shaded areas for work and breaks is necessary.

Programming

Types of treatment goals, objectives, and desired outcomes

Because work is a social interaction, it is important for those designing a vocational training program for individuals with a disability to include organizational socialization into the training. Organizational socialization is "the process through which individual are transformed from outsiders to participating, effective members of organizations" (Baron and Greenberg 1990). It is especially important for the individual with a disability to feel confident that they are a contributing member of the work team in all types of employment.

In developing a program for trainees who seek competitive employment in the community, the horticultural therapy practitioner not only provides training in task skills but also must also address prevocational issues such as endurance, interpersonal relationships, communication skills, and general job performance. If the goal is supported employment with a job coach, this serves as a jumping-off point that can also train the vocational coach on the nuances of horticulture work. In some programs, the desired outcome or goal is not necessarily immediate job placement but rather combines a focus on personal growth with career education. One such program at a correctional facility is described in Exhibit 9.9 (Figure 9.7).

EXHIBIT 9.9

Program Example: Vocational Horticultural Therapy Correctional Program for Adolescent Girls

The Growing Opportunities Gardening Program is a vocational horticultural therapy program at Girls Rehabilitation Facility (GRF), a minimum-security detention center in San Diego, California, for girls thirteen to seventeen years old. GRF is a highly structured environment and serves up to fifty girls placed there by the court system. The girls attend an on-site, state-certified high school and participate in a wide array of personal-growth programs.

The gardening program was initiated in 2013 as a collaborative undertaking of master gardeners and the county probation and public health departments to help address the complex developmental and mental health needs of the girls incarcerated at GRF. Studies show that girls in the juvenile justice system are more likely than their peers to have experienced violence and abuse in their lives, and most have more than one psychiatric disorder, such as substance use disorder and post-traumatic stress

(Continued)

EXHIBIT 9.9 (Continued)

Program Example: Vocational Horticultural Therapy Correctional Program for Adolescent Girls

disorder (US Department of Justice 2018). The facility has a spacious, onsite garden that is dense with flowers, native plants, and seasonal fruits and vegetables, which was designed and built by girls under the guidance of master gardener volunteers. Its diversity of plant and wildlife, shaded benches and workspaces, and flexibility of use combine to create an inviting, highly functional educational environment that also serves as a place for solace and inspiration for staff members, visitors, and girls.

Girls assigned to the Growing Opportunities Gardening Program are enrolled in an introductory Career Technical Education (CTE) horticulture course taught by a credentialed CTE instructor, with assistance from master gardener volunteers under the guidance of a person trained in horticultural therapy. The curriculum meets California content standards for high school students. Instructional methods, which are delivered in the classroom and the garden, include lectures, readings and responses, group discussions, visual demonstrations, presentations by industry professionals, field trips, and hands-on activities. In the garden, the girls are partnered one-on-one with master gardeners, and together they engage in the seasonal ebb and flow of maintaining a garden: prepping garden beds, thinning seedlings, building trellises, managing pests, and harvesting. San Diego has a year-round growing climate so there is always something to do.

The low-ratio, intergenerational partnering of girls with master gardeners offers many benefits, both practical and therapeutic. It enhances safety by minimizing distractions and easing the task of monitoring girls' use of tools, provides opportunities for modeling positive behaviors, and builds trust between teacher and student. It also ensures a less stressful, more satisfying experience for volunteers, and this in turn keeps them coming back. Retaining trained, seasoned volunteers is crucial to the program's sustainability.

Once the girls have worked in the garden for a week or two, they become less anxious and more curious, their self-confidence grows, and they begin developing skills in collaborating and problem solving. "I learned how to cope with people I didn't get along with," said one fifteen-year-old girl. Another girl observed, "Being in the garden has taught me many new things: how to work with others, how to make healthier choices, how to be responsible."

Girls' progress in the course is assessed through hands-on practice, activities in the garden, and group discussions. They receive a written summative assessment at the end of each unit and a formal grade at the end of the course. They earn high school credit for completing the course, and they earn community service hours and a certificate for the time they spend maintaining the garden.

Joni Gabriel, BA
Certificate in Horticultural Therapy

Figure 9.7 Collaborating on projects in a vocational horticultural therapy program strengthens participants' communication and problem-solving skills. (Courtesy of Joni Gabriel.)

The goals of participants should be clear and agreed upon by all members of the treatment team. For each goal, objectives are written to describe measurable steps to achieve it. These should be clearly written, specific, and achievable, and should span all spheres of the human experience. Physical objectives that challenge the trainee yet fall within the scope of his or her abilities are commonly addressed in vocational horticultural therapy programs. They may also provide information about what practical accommodations a competitive employer might need to make in defining potential jobs for the individual. Endurance can be addressed and enhanced with an understanding for the amount and duration of physical activity that will ultimately be tolerated by the trainee. Cognitive goals and objectives address new learning, habituation, as well as short- and long-term memory. Addressing these issues in training enhances the skills of the client and also informs vocational coaches of the amount and type of support the trainee may require in a new environment. Behavioral goals and objectives for some clients include attention to task, interpersonal interactions with coworkers and supervisors, acceptance of constructive criticism or correction, and anger management. Psychosocial goals and objectives overlap the other areas of focus and address the day-to-day interactions and productivity of the trainee.

By assessing these areas of functioning and by using the horticultural therapy vocational training program to address areas that need improvement, trainees will ultimately be able to utilize their abilities to be productive, contributing members of the workforce and the community.

Summary

The use of the vocational model of horticultural therapy programming provides invaluable benefits to those served, addressing a key component of overall wellness and also social inclusion. Well-designed vocational training equips the employee with tools that enhance abilities and minimize disabilities. Trainees who find employment situations that are well matched to their skills and personal traits continue to develop and grow. They feel empowered by the prestige that having appropriate work brings. As contributing members of society, they add to the economy, garner accolades from family and friends, and have the satisfaction and identity of an individual holding a productive place in society.

References

American Horticultural Therapy Association. 2007. *AHTA Definitions and Positions: Vocational Horticulture.* http://www.ahta.org/documents/Final_HT_Position_Paper_updated_updated.pdf.

Baron, R. A., and J. Greenberg. 1990. *Behavior in Organizations:Understanding and Managing the Human Side of Work* (3rd ed.). Boston, MA: Allyn & Bacon.

Brault, M. W. 2012. "Americans with Disabilities: 2010." In United States Census Report Number P70-131.

Bringslimark, T., T. Hartig, and G. G. Patil. 2007. "Psychological Benefits of Indoor Plants in Workplaces: Putting Experimental Results into Context." *HortScience* 42(3): 581–587.

Haller, R. 1998. "Vocational, Social, and Therapeutic Programs." In *Horticulture as Therapy: Principles and Practice,* ed. S. P. Simson and M. C. Straus. Binghamton, NY: Haworth Press, pp. 43–68.

Hefley, P. D. 1973. "Horticulture: A Therapeutic Tool." *Journal of Rehabilitation* 39(1): 27–29.

Holland, J. L. 1997. *Making Vocational Choices: A Theory of Careers.* Englewood Cliffs, NJ: Prentice Hall.

"Issues." United Nations Division for Social Policy and Development: Disability. Accessed April 2, 2018. https://www.un.org/development/desa/disabilities/issues.html.

Kamp, M., C. Lynch, and R. Haccou. *Handbook Supported Employment*, World Association for Supported Employment and International Labor Organization. Accessed April 2, 2018. http://www.wase.net/handbookSE.pdf.

Lent, R. W., S. D. Brown, and G. Hackett. 2000. "Contextual Supports and Barriers to Career Choice: A Social Cognitive Analysis." *Journal of Counseling Psychology* 47: 36–49.

Melwood Horticultural Training Center. 1980. *The Melwood Manual: A Planning and Operations Manual for Horticultural Training and Work Co-op Programs.* Accessed May 10. 2018. https://eric.ed.gov/?q=Melwood+Manual&id=ED200793.

Neff, W. S. 1985. *Work and Human Behavior.* New York: Aldine.

Office of Juvenile Justice and Delinquency Prevention. "Girls and the Juvenile Justice System." US Department of Justice. Accessed February 8, 2018. https://rights4girls.org/wp-content/uploads/r4g/2016/08/OJJDP-Policy-Guidance-on-Girls.pdf.

"Our Greenhouse." Mountain Valley Developmental Services. Accessed March 30, 2018. https://www.mtnvalley.org/mountain-valley-greenhouse.

Parsons, F. 1909. *Choosing a Vocation.* Boston, MA: Houghton Mifflin.

Quick, J. C., L. R. Murphy, and J. J. Hurrell, Jr., Eds. 1992. *Stress and Well-Being at Work: Assessments and Interventions for Occupational Mental Health*. Washington, DC: American Psychological Association.

Szymanski, E. M., D. B. Hershenson, J. Ettinger, and M. S. Enright. 1996. "Career Development Interventions for People with Disabilities." In *Work and Disability: Issues and Strategies in Career Development and Job Placement*, Eds. E. M. Szymanski and R. M. Parker. Austin, TX: PRO-ED, pp. 255–276.

Trevino, B., and E. M. Szymanski. 1996. "A Qualitative Study of the Career Development of Hispanics with Disabilities." *Journal of Rehabilitation* 62(3): 5–9.

United Nations. 2018. "Sustainable Development Goal 8." https://sustainabledevelopment.un.org/sdg8.

"Vocational Training?" The Law Dictionary. Accessed March 30, 2018. https://thelawdictionary.org/vocational-training.

Wehman, P. 2012. "Supported Employment: What Is It?" *Journal of Vocational Rehabilitation* 37: 139–142.

chapter ten

Horticultural therapy grounded in wellness models
Theory and practice

Jane Saiers

Contents

This chapter discusses horticultural therapy grounded in wellness models of care. In the sections that follow, the definition of wellness as related to horticultural therapy is considered, and wellness-model-based horticultural therapy is compared with approaches based on other models. The theoretical underpinnings and the research-based evidence for the usefulness of wellness-model-based horticultural therapy are discussed. Characteristics of wellness-model-based horticultural therapy are outlined, and example programs are described. Finally, challenges and opportunities for wellness-model-based horticultural therapy are considered.

Wellness defined

The World Health Organization (WHO) defines health (wellness) as a state of complete physical, emotional, mental, and social well-being and not merely the absence of disease or infirmity (WHO 2005). This definition, established in 1948, was proposed in part to counteract the disease-centric conception of health—still pervasive in Western medicine today—as simply the absence of disease or illness. In the holistic conception put forth by WHO, health subsumes not only amelioration or absence of disease symptoms but also the presence of positive emotions, general well-being and quality of life, and the ability to perform normal daily activities. Health also encompasses the development of coping skills and perspectives useful in living life to its fullest in the presence of acute and/or chronic diseases or challenges.

Wellness models of care

The concept of health as encompassing physical, emotional, mental, and social aspects of being and functioning underpins wellness models of care. Approaches to care guided by wellness models help participants to achieve their potential, address the whole person (body, mind, spirit), and capitalize on participants' strengths. Based on the recognition that well-being and quality of life critically depend on interpersonal and community relationships and on the quality of the physical environment, wellness models also emphasize the importance of relationships with others and with the ecosystem.

Several specific wellness models have been developed (Hettler 1976, RHIhub 2017). Their relative merits have not been studied. The models share the conceptions of wellness as (1) an intentional, continuous process of striving to realize full potential, and (2) a multifaceted, holistic concept that includes physical, mental, social, and ecological aspects. All arguably offer useful frameworks for care. One popular model—the Six Dimensions of Wellness—defines wellness based on occupational, physical, emotional, social, spiritual, and intellectual dimensions (Hettler 1976). In Hettler's words, the occupational dimension "recognizes personal satisfaction and enrichment in one's life through work;" the physical dimension "recognizes the need for regular physical activity;" the social dimension "encourages contributing to one's environment and community;" the intellectual dimension "recognizes one's creative, stimulating mental activities;" the spiritual dimension "recognizes our search for meaning and purpose in human existence;" and the emotional dimension "recognizes awareness and acceptance of one's feelings" (Hettler 1976). Hettler conceptualizes the dimensions as interconnected determinants of healthy living.

Care based on wellness models differ from that associated with the traditional medical model in several respects (Table 10.1). Wellness models focus on development and well-being of the whole person rather than management or treatment of a specific disease or injury. Wellness models employ multimodal approaches—physical exercise, meditation, and healthy eating, for example—directed at body, mind, and spirit, whereas the traditional medical model relies heavily on pharmaceuticals and surgery directed at disease or injury. In wellness models, the participant is encouraged to assume responsibility for his or her health and actively partners with the health care provider in self-improvement and self-healing. In the traditional medical model, the patient's primary role is a recipient of care. Social interaction, community engagement, and environmental stewardship are encouraged in wellness models. In the traditional medical model, community, society, and the environment are seldom considered.

Table 10.1 Wellness model vis-à-vis conventional medical model

Aspect	Wellness model	Conventional medical model
Conception of health	State of physical, mental, emotional, and social well-being	Absence of disease
Framework	Focus on well-being of whole person	Focus on managing or treating the disease
Approaches	Multimodal, affecting body, mind, spirit	Pharmaceuticals, surgery, other interventions directed at disease or injury
Role of the participant	Active partner in self-improvement and self-healing	Recipient of care
Community/ society	Social interaction and development of community encouraged	Not generally considered
Environment	Focus is a harmonious relationship	Not generally considered

Horticultural therapy based in wellness models of care

Horticultural therapy is uniquely suited to support wellness-based programming. Working with plants is a holistic endeavor that involves multiple domains of function, including physical activity, cognitive/intellectual engagement, emotional investment, and social interaction. It engages body, mind, and spirit. By definition, horticultural therapy is an active process in which the participant, aided by the therapist, engages with the plant world to improve physical, cognitive, emotional, and/or social aspects of well-being. The participant assumes a degree of responsibility in cultivating and maintaining plants. Horticultural therapy also supports social interaction and community engagement. Many wellness model horticultural therapy programs, for example, are conducted in groups and/or in community settings, such as community gardens. Horticultural therapy also fosters environmental stewardship. Working with plants helps participants experience interconnectedness with other life forms and can awaken and sustain concern for and care of the ecosystem. Charles A. Lewis in *Green Nature/Human Nature* describes further:

> Many other craft activities can keep patients busy and give them a sense of achievement. Why grow plants rather than knit an afghan or build a ship in a bottle? Plants and people share the rhythm of life. They both evolve and change, respond to nurture and climate, and live and die. This biological link allows a patient to make an emotional investment in a plant; however, it is a safe, non-threatening investment.... Should the patient choose to withdraw, there will be no recriminations. In severely damaged patients, such a relationship can signify the first willingness to reach out to another living being. (Lewis 1996)

In his 2008 *New York Times Magazine* article "Why Bother?," journalist Michael Pollan touches on many of the multifaceted rewards that are the basis of horticultural therapy programs grounded in wellness models (although his focus is growing a vegetable garden rather than horticultural therapy per se) (Pollan 2008). Grow some of your own food, advocates Pollan:

- You will set an example for other people: viral social change.
- You can grow the proverbial free lunch: CO_2-free and dollar-free.
- You will get a good workout in your garden, burning calories without having to drive to the gym.

- You engage both body and mind and subtract time and energy from electronic forms of entertainment.
- You develop new habits of mind.
- You learn that you need not depend on specialists to provide for you.
- You will re-engage with your neighbors to give away produce and to borrow their tools.
- You will have begun to heal the split between what you think and what you do, to commingle your identities as consumer and producer and citizen.

> "The single greatest lesson that the garden teaches," writes Pollan, "is that our relationship to the planet need not be zero-sum, and that as long as the sun still shines and people still can plan and plant, think and do, we can, if we bother to try, find ways to provide for ourselves without diminishing the world."

Horticultural therapy practiced in the context of wellness models differs in goals and focus from horticultural therapy based on therapeutic models (discussed in Chapter 8) and vocational models (discussed in Chapter 9). In wellness-model-based horticultural therapy, improvement in general well-being and quality of life as opposed to specific vocational skills and capacities (vocational models) or clinical improvements (therapeutic models) is sought. Further, wellness-model-based horticultural therapy focuses on the growth of the whole person rather than the development of specific skills (vocational models) or remediation or cure of disease or illness (therapeutic models).

Although therapeutic goals and foci vary by model, many horticultural therapy programs incorporate aspects of multiple models. For example, while improvement in physical and mental health and general well-being is a primary goal of wellness-model-based programs, they might also incorporate elements of therapeutic and/or vocational programs (Figure 10.1).

Figure 10.1 Interesting and engaging activities improve mood and increase a sense of well-being in someone living with dementia. (Courtesy of Beth Bruno.)

Theoretical framework for horticultural therapy grounded in wellness models

The *biophilia hypothesis* and tenets of *ecopsychology*, described in detail in Chapter 5, form a theoretical framework for horticultural therapy grounded in wellness models. Ecopsychology is the study of the relationship between humans and nature. Ecopsychology aims to understand and develop the emotional connection between people and the natural world. The biophilia hypothesis, introduced by sociobiologist Edward O Wilson (1984) in his book *Biophilia*, is a central tenet of ecopsychology. The term *biophilia* literally means "love of life" and has been described as the urge to affiliate with other life forms. According to the biophilia hypothesis, humans have an innate, biologically based affinity for the natural world. They instinctively seek connection with nature because the human body and mind evolved in adaptive response to the natural environment.

In the ecopsychological framework, connection with nature is essential for physical and mental health. The beneficial mental and physical effects of a walk in the woods or gardening arise partly from the fact that humans evolved to engage in activities such as these. Conversely, disconnection from nature contributes to stress and illness. (See Chapter 5 for more discussion on these concepts.)

Many modern human societies are disconnected from nature. Stephen R. Kellert, during his keynote address to the American Horticultural Therapy Association's (AHTA) annual conference in 2014, said, "The experience of Nature is not a luxury or an amenity but an anvil on which human health is forged." For the first time in evolutionary history, modern humans in developed countries spend more than 95% of their lives indoors and in settings increasingly dominated by electronic technology and synthetic materials. In the context of the centrality in ecopsychology of humans' experience of nature to mental and physical health, modern humans' disconnection from nature and growing things is generally viewed as pathological. In his book *Last Child in the Woods: Saving Our Children from Nature-Deficit Disorder*, Richard Louv (2005) argues that society teaches youth to avoid direct experience with nature. The resulting *nature-deficit disorder*, in Louv's view, is associated with a host of behavioral, cognitive, and emotional problems best remedied by direct contact with aspects of the natural world:

> In the most nature-deprived corners of our world we can see the rise of … cultural autism. The symptoms? Tunneled senses, feelings of isolation and containment. Experience, including physical risk, is narrowing to about the size of a cathode tube, or flat panel if you prefer. Atrophy of the senses was occurring long before we came to be bombarded with the latest generation of computers, high-definition TV, and wireless phones … But the new technology accelerates the phenomenon (Louv 2005).

In advocating contact with nature, Louv advances *ecotherapy*, which is the participatory outgrowth of ecopsychology. Ecotherapy seeks to reestablish connections between people and nature. Also known as green therapy, ecotherapy encompasses approaches including wilderness excursions and some types of animal-assisted therapy (Chalquist 2009).

Evidence-based benefits of horticultural therapy grounded in wellness models

Numerous studies support the benefits of ecotherapeutic approaches to health and wellness (Chalquist 2009). Improvements in mood, coping ability, self-efficacy, self-esteem, and feeling of belonging and connection, and reduction in stress have been documented in association with ecotherapy. To date, the ecotherapeutic research most relevant to horticultural therapy has involved gardening interventions. Gardening, including work in school gardens and community gardens, is associated with an array of positive outcomes, including increases in fruit and vegetable consumption and physical activity, and improvement in mental health and well-being (Ohly 2016; Shiue 2016; van Lier et al. 2016; Savoie-Roskos et al. 2017).

The majority of publications specifically documenting the impact of horticultural therapy describe studies conducted in the context of therapeutic models (see Chapter 8). The body of research on the impact of horticultural therapy grounded in wellness models on health outcomes is smaller but growing. Much of the latter research was included in a meta-analysis reported by Soga and colleagues (2017). These authors analyzed published research that investigated the effects of gardening, including horticultural therapy, on various health outcomes measures. To be included in the analysis, studies needed to report data on health outcomes in the context of gardening, be published in peer-reviewed scientific journals after 2001, and be written in English. The authors identified twenty-two studies for inclusion in the analysis. Twelve of the twenty-two studies involved horticultural therapy conducted in the context of wellness models or therapeutic models. Of the studies that did not involve horticultural therapy, the majority involved a daily gardening intervention. The meta-analysis revealed a strong positive association between gardening, including horticultural therapy, and diverse health outcomes, including alleviation of symptoms of depression, symptoms of anxiety, stress, and disturbed mood; reduction of body mass index; enhancement of quality of life, sense of community, and cognitive function; and increase in physical activity.

In subgroup analyses, both those participating in gardening through horticultural therapy and those participating in gardening outside the context of horticultural therapy experienced beneficial health outcomes. However, the positive association between gardening and health outcomes was significantly more robust among participants exposed to horticultural therapy. This result arguably points to the importance of the therapeutic program and the role of the horticultural therapist in optimizing outcomes (Figure 10.2).

The studies in the meta-analysis involved diverse frequencies and durations of gardening and horticultural therapy. Examination of outcomes as a function of frequency and duration of gardening intervention revealed that wellness benefits were both immediate and persistent. The three studies that assessed outcomes both before and after a few hours of gardening that did not involve horticultural therapy found positive associations with gardening and outcomes such as reductions in depression and anxiety. Whether these effects persisted in these studies was not assessed. In twelve studies involving horticultural therapy that assessed health outcomes over weeks to months, horticultural therapy–associated positive health outcomes were found to persist for the duration of the follow-up periods. Soga and colleagues concluded:

> Our meta-analysis has provided robust evidence for the positive effects of gardening on health. With an increasing demand for reduction of health care costs worldwide, our findings have important policy implications. The results presented here suggest that gardening can improve physical, psychological, and social health, which can, from a long-term perspective, alleviate and prevent various health issues facing today's society. We therefore suggest that government and health organizations should consider gardening as a beneficial health intervention and encourage people to participate in regular exercise in gardens. (Soga et al. 2017)

Figure 10.2 Active participation in garden tasks provides therapeutic outcomes. (Courtesy of Rebecca Haller.)

Horticultural therapy programs based in wellness models: Characteristics and examples

Settings and participants

Horticultural therapy programs based in wellness models are conducted in varied settings, including botanical gardens, farms, community gardens, assisted-living centers, senior centers, retirement homes, residential group homes, prisons, churches, schools, and colleges. The client populations are as varied as the settings and include people with diagnoses of mental or physical illness, those who could benefit from programming directed at the whole person (as opposed to specific symptoms or challenges), and those who seek help in the ongoing process of realizing their potential. Participants in wellness-model-based horticultural therapy include but are not limited to older adults, people with developmental disabilities, youth at risk of delinquency, the homeless, and people who are institutionalized. For older adults, horticultural therapy can provide a sense of purpose, help revive pleasurable memories, and reduce feelings of isolation (Detweiler et al. 2012). For people with developmental disabilities, horticultural therapy can provide the opportunities for exercise, social interaction and integration, and stress reduction. For at-risk youth, horticultural therapy can enhance self-esteem, encourage self-reflection, and build life skills. For institutionalized people or those who are homeless, horticultural therapy can help overcome the losses of sense of identity and control (Niklasson 2007). Nelson Mandela, the

South African anti-apartheid revolutionary, politician, and philanthropist who served as president of South Africa from 1994 to 1999, spent twenty-seven years in prison. He later spoke of the importance gardening had become to him during his incarceration: "A garden was one of the few things in prison that one could control. To plant a seed, watch it grow, to tend it, and then harvest it offered a simple but enduring satisfaction. The sense of being custodian of this small patch of earth offered a small taste of freedom" (Mandela 1994).

Program examples

The following three examples of horticultural therapy programs in the United States—the GreenHouse program at Rikers Island, and the Herbal Essence program and the Transplanting Traditions Community Farm program in North Carolina—illustrate the diverse settings, populations, and foci possible in wellness-model-based horticultural therapy.

GreenHouse program

GreenHouse, based at New York City's Rikers Island jail complex, is a horticultural therapy program that helps to prepare incarcerated men and women for release and reentry into society. GreenHouse is a program of the Horticultural Society of New York, which is a nonprofit, nongovernmental organization (NGO) that relies on grants and contributions for funding. The program offers the opportunity for socialization, education, physical activity, and vocational training in horticulture. Activities include garden design, installation, and maintenance as well as design and construction of garden-related infrastructure such as benches and trellises. A greenhouse, a classroom, and nearly three acres of outdoor gardens that are designed and maintained by inmates provide the setting for this program. After participating in the program, inmates have positive feelings about the experience, as described by GreenHouse program participants. "You get a lot of frustration out when you're digging a hole—it's like 'this person made me mad, let me dig this hole' and as time goes by, you forget what you were frustrated and mad about." Another said, "It's about learning to care for things … for living things, not just plants…. If you can care for a plant, you can care for a person."

GreenHouse incorporates bridged programming that provides participants with the option of progressing from the horticultural therapy program while incarcerated to a postrelease vocational internship program. The horticultural therapy program, in connection with the bridged programming, is associated with good outcomes, including low rates of recidivism, postrelease employment opportunities, and enhanced personal well-being. More information about the program is provided in Exhibit 10.1.

EXHIBIT 10.1

Program Example: Rikers Island Jail, New York

Wellness and corrections seem like a contradiction. The primary focus of jail or prison is not the wellness of people in its custody. However, 95% of people who are incarcerated in the United States return to the community. How individuals return is highly relevant for successful reentry—defeated, angry, without changed perspective,

(Continued)

EXHIBIT 10.1 (Continued)

Program Example: Rikers Island Jail, New York

practiced in negative behaviors, with low self-esteem, and stressed? Or physically recovered, with positive self-image, coping skills, tools to appropriately address challenging situations, and plans for a meaningful life outside correctional institutions?

Jail creates high levels of stress for inmates and staff members. This program example focuses on inmates, though we recently also started exploring staff wellness. Jail is an environment deprived of cognitive stimulation and positive sensory stimulation. Inmates have no control over their personal space. They often lack access to healthy food and cannot get undisturbed sleep. They need to keep their guard up constantly. Skills that help inmates survive often are negative, such as fighting to settle conflicts or becoming a bully. The ability to actively participate in the lives of loved ones outside is diminished or gone. Participants mention loss of control, loneliness, and feelings of abandonment, which are often preexisting. Their lives lack purpose and sense of accomplishment.

Horticultural therapy, in corrections, plays a significant role in improving participants' wellness in all aspects of their being. Gardens provide an environment different from the setting inside. Our gardens focus on sensory stimulation: fragrant herbs and flowers, plants interesting to touch, cut flowers, water features and ornamental grasses for sound, and a large variety of edibles chosen by participants. There are sheltered areas, providing personal space for reflection and to take a deep breath. Sharing unrushed meals and developing access to healthy food outside play a crucial role. We eat what we grow, welcome family recipes from our students' varied backgrounds, and cultivate joyful table traditions. Many students struggle with substance abuse. Cultivating plants and patiently waiting for crops to ripen gently teaches about instant and late gratification. Participants are intentionally and gradually given the role of caretakers, asked for input on plant care, learn to make choices regarding landscape design, and help plan daily tasks or long-term projects. A sense of control, accomplishment, and purpose is fostered. Students share crops or flower arrangements with staff members inside the facility, reversing the usual roles and allowing for moments of simple positive human interaction. All staff members, civilian and uniform, model respectful work relationships. While caring for plants together, healthy and safe relationships are built. For the students, being kind and supportive of each other doesn't backfire. Students mention that in the garden, they "feel like a person, not an inmate," empowered to share who they are and reminded of their value. Many students report improved sleep, reduced anxiety, and lowered blood pressure and sugar levels. We can't aim to solve the structural and personal issues leading to our students' incarceration. But through horticultural therapy, we can help students to find moments of reprieve; experience that the beauty of life is accessible for them, too; and foster the sense that their person and actions matter.

Hilda Mechthild Krus, MSW, HTR

Director GreenHouse and Horticultural Therapy Programs,

The Horticultural Society of New York

Herbal Essence program

Herbal Essence, based at the University of North Carolina's North Carolina Botanical Garden in Chapel Hill, is a horticultural therapy program that offers community-dwelling people with mental illnesses such as depression, schizophrenia, and anxiety the opportunity for socialization, physical activity, and education as they work side-by-side with other garden volunteers from the community. Participants come to the botanical garden once weekly to tend program-dedicated vegetable and flower gardens and to work on garden-related crafts. Participants with mental illness are not differentiated from other garden volunteers during the work sessions so that participants do not know who among them has a diagnosis of mental illness. Herbal Essence is run by a registered horticultural therapist with help from horticultural therapy interns and volunteers from the community. The program is supported by donations to the North Carolina Botanical Garden, which is a nonprofit NGO.

Transplanting Traditions Community Farm

Transplanting Traditions Community Farm, based in Pittsboro, North Carolina, is a horticultural therapy program that provides Burmese refugees with access to land for farming and gardening, vocational training and education, and a place for preserving their cultural traditions and strengthening community while transitioning to living in the United States. Open to both individuals and families, it is a program of the Orange County Partnership for Young Children, a nonprofit dedicated to enhancing children's health and school readiness. The farm includes several acres of vegetable gardens, and a vegetable-packing facility where participants cultivate and process foods from their native country as well as crops commonly grown in North Carolina. Participants have the opportunity to sell the vegetables they grow at the farm through a community-supported agriculture (CSA) program or at a local farmer's market. Exhibit 10.2 further describes this program (Figure 10.3).

EXHIBIT 10.2

**Program Example: Transplanting Traditions
Community Farm, North Carolina**

Program Type and Focus

This is a wellness program with vocational aspects.

Population served

Refugees

The Mission of Transplanting Traditions Community Farm (TTCF) is to provide refugee adults and youth access to land, healthy food, and agricultural and entrepreneurial opportunities. The farm provides a cultural community space for families to come together, build healthy communities, and continue agricultural traditions in the Piedmont of North Carolina.

TTCF is an eight-acre working farm in a rural buffer zone just outside the city. Currently thirty-five refugee families, approximately 163 refugee individuals from Myanmar (Burma), farm on the site. Each family involved has its own set of goals

(Continued)

EXHIBIT 10.2 (Continued)

**Program Example: Transplanting Traditions
Community Farm, North Carolina**

for involvement. Some families simply grow for themselves, while other families are involved in various levels of the Marketing Program, which is a vocational agriculture program that helps participants transfer their already advanced agricultural skills to a new environment and business climate. Programming focuses on vocational skills; social integration; community support; and a space that provides mental, physical, and emotional support.

The farm is a community gathering space for the local refugee community from Burma and is well connected with local refugee support services, providing referrals for participants on a regular basis. This cross-agency support system is an innovative model in providing participant services, creating a network of support. The project is funded through local, state, and federal grants; donations; and project income.

Community

One of the main reasons that participants farm at TTCF is to maintain a sense of community in a space that is familiar and self-generated. The farm also provides a gathering space to maintain multigenerational cultural knowledge. Passing on traditions to future generations is an important part of the survival of refugee culture in the United States. Many older generations worry these things will be lost amid the demands of adapting to American life and culture. Program participants expressed the importance of maintaining culture, community, and sharing.

Participant A: "This is part of Karen culture, families coming together, working together and talking together and getting to know one another. It reminds me of my motherland. We the Karen people daily depended on growing crops and raising animals to live."

Participant B: "When I work in my garden, it gives me a good exercise and I breathe fresh air too. I am so happy when I see my vegetables are germinating and growing. Here I can get many friends talking to each other, sharing with each other and also listening to a sweet singing of birds."

Human agency and mental health

Participants come to the United States with generations of knowledge as well as a deep love of farming. The farm provides a space to use existing skills in a culturally meaningful way and opportunity to build skills through educational workshops and interparticipant skill share. Participants have created a space where they are the experts among otherwise discouraging work opportunities and challenging life situations. Further comments reflect the personal programmatic impact.

Participant C: "I think when I come to the farm and I look around, it looks perfect, I think, 'Oh my family will survive, and I don't need to buy things, and my kitchen will be filled with vegetables,' and this makes me feel less stressed. Also, when I grow vegetables and have enough to share this makes me feel good."

(Continued)

EXHIBIT 10.2 (Continued)

Program Example: Transplanting Traditions Community Farm, North Carolina

Participant D: "I am less stressed since I have started coming to the farm. When you see your friends and plants growing and then have fresh vegetables to put on the table, you are less stressed. In our house, we feel that having fresh vegetables available is more important and better than buying vegetables from the store."

Nicole Accordino, HTR

Program Coordinator, Transplanting Traditions Community Farm

Figure 10.3 Growing culturally relevant plants, such as taro (pictured here), contributes to the overall well-being of individuals who have been displaced, helping maintain a connection to traditions and culture. (Courtesy of Natalie Ross.)

Personnel

Horticultural therapy programs based in wellness models are often initiated and sustained by horticultural therapists, who sometimes work in conjunction with community organizers, master gardeners, activity directors, and therapeutic recreation specialists. These individuals often work with collaborators in the health care setting, academia, and the community at large. Members of the treatment team in horticultural therapy programs based in wellness models may include the horticultural therapist, health care providers, allied health care providers, and people in the community, depending on the nature of the program. In wellness-model-based horticultural therapy programs, as in therapeutic and vocational programs, volunteers and interns play crucial roles with functions such as handling administration, publicity and communications, logistical support and garden maintenance tasks as well as providing assistance to participants during sessions.

Group size and session schedules

Horticultural therapy programs based in wellness models are conducted both with individuals and groups. However, because socialization, community building, and/or reduction in social isolation are common goals of these programs, they are more often conducted with groups than with individuals. Group size varies depending on program goals and characteristics of the participants. To provide participants with the optimum attention and guidance, a therapist-to-participant or therapist assistant-to-participant ratio of one to ten or fewer seems to be ideal.

Horticultural therapy programs based in wellness models vary in frequency and duration. A common schedule entails once- or twice-weekly sessions of a few hours each. Duration of programs can be defined or open-ended. Generally speaking, programs of longer duration (months to years) are best suited for fully realizing the lifestyle changes often sought in wellness-model-based horticultural therapy.

Horticultural therapy programming in the wellness model: Developing goals, objectives, and activities and documenting outcomes

Goals and objectives

Defining goals and objectives is as important in wellness-model-based horticultural therapy programs as it is in programs with therapeutic or vocational emphases. Goals and objectives of wellness-model-based horticultural therapy programs pertain to specific aspects of general health and well-being that the participant, working with the therapist, seeks to enhance, and activities are chosen to support participants' goals and objectives. Often, wellness-model-based horticultural therapy programs incorporate multidimensional goals involving the interconnected physical, emotional, social, spiritual, and intellectual determinants of healthy living. Horticultural therapy programs based in wellness models may also incorporate a prominent educational component, including topics such as nutrition, food preparation, or plant biology (Figure 10.4).

Figure 10.4 Pea shoots help a participant reach a personal goal to adapt to home in a new country at the Transplanting Traditions Community Farm. (Courtesy of Natalie Ross.)

Horticultural therapy programs based in wellness models also may incorporate therapeutic and/or vocational elements depending on the client population and goals. The GreenHouse program at Rikers Island in New York and the Herbal Essence and Transplanting Traditions Community Farm programs in North Carolina described above illustrate the overlap among therapeutic, vocational, and wellness-model-based programs. Each of these programs is designed to support and develop multiple dimensions of wellness—including occupational, physical, emotional, social, spiritual, and intellectual—although their emphases differ. For example, the horticultural therapy program at Transplanting Traditions Community Farm emphasizes development of the social, occupational, and spiritual dimensions of wellness and illustrates the importance of incorporating social and cultural aspects into wellness-based programming. It also includes educational and occupational emphases characteristic of vocational horticultural therapy programs. The GreenHouse program at Rikers Island focuses on developing the occupational dimension of wellness as it helps participants transition from prison to postrelease employment.

Goals and objectives and their relation to activities

Goals, objectives, and activities supporting the intended outcomes in wellness-model-based programming are illustrated in the following example of a potential horticultural therapy program involving children transitioning from foster care. The program would be conducted at a farm, where participants engage in at least two, two-hour group horticultural therapy sessions per week for a minimum total weekly time commitment of four hours.

The transition to independent living can be fraught with challenges for youth in foster care, many of whom lack family and social support networks, financial resources, education, and vocational skills. Compared with their peers who are not in foster care, youth leaving foster care are at greater risk of homelessness, reliance on social assistance, substance abuse, single parenthood, criminal activity, and mental illness (Tweddle 2007). The need to enhance services directed at improving outcomes for youth transitioning from foster care is recognized at the federal, state, and local levels. Target outcomes for youth transitioning from foster care include development of a sense of well-being with a positive sense of personal identity; establishment of connections to and emotional support from a variety of adults outside the public child welfare system; avoidance of high-risk behaviors; and training in keeping with the youth's goals, interests, and abilities. Horticultural therapy can be useful in helping to achieve several of the desired outcomes for youth transitioning from foster care. Horticultural therapy can provide opportunities for physical, social, emotional, and cognitive development; vocational training; education; being mentored; developing relationships with caring adults; and community involvement.

A wellness-model-based horticultural therapy program conducted in the setting of a farm could help youth transitioning from foster care to (1) improve specific aspects of health and well-being—physical, cognitive, emotional, social—through gardening and other plant-related activities; (2) develop self-esteem and self-reliance by confronting challenges and rewards inherent in working on the farm; (3) gain practical, transferable experience in growing, preparing, and storing food for self and others and in marketing it locally; and (4) develop a sense of connection with other people and with the natural world.

The program focuses on the following three group goals, which support the target outcomes identified by federal and state governments for youth transitioning from foster care:

- Experience enhanced self-esteem and feelings of self-efficacy.
- Demonstrate cooperation in accomplishing tasks and objectives.
- Demonstrate the ability to develop and implement work schedules to accomplish specific tasks.

Table 10.2 lists example activities and objectives and explains how each activity addresses the goal.

Assessing and documenting outcomes

Assessing and documenting outcomes are as important in wellness-model-based horticultural therapy programs as they are in programs with therapeutic or vocational emphases. Outcomes assessment and documentation can provide information useful in gauging progress and development, identifying opportunities for change or improvement, modifying programs to improve outcomes, and keeping others on the treatment team up to date. Outcomes documentation can also aid in legitimizing horticultural therapy by improving the ability to quantify program-associated changes.

Table 10.2 Example goals, objectives, and activities applicable to a wellness model horticultural therapy program conducted for youth transitioning from foster care

Goal: Experience enhanced self-esteem and feelings of self-efficacy		
Activities	How activity addresses goal(s)	Example objective
Scouting the vegetable blocks for insects and disease	This activity involves enhancing self-esteem and feelings of self-efficacy (Goal 1) as the participant is given responsibility for contributing to solving a potentially big problem on the farm.	By the end of the week, use the internet to research the identity of the insect pest discovered in the eggplant and make recommendations for how the pest should be managed.
Participate in the annual regional farm tour	This activity involves enhancing self-esteem and feelings of self-efficacy (Goal 1) as the participant is being entrusted with the responsibility of representing the farm to the public.	For two, two-hour shifts over one weekend, serve as one of the farm's greeters during the annual regional farm tour.
Working at the farmer's market stand	This activity involves enhancing self-esteem and feelings of self-efficacy (Goal 1) as the participant is being entrusted with the responsibility of representing the farm at the farmer's market.	Work two, two-hour shifts at the farmer's market in a month.
Cooking and eating food harvested from the farm	This activity involves cooperation (Goal 2) in that the participant needs to interface with team members to choose the meal items and to prepare the meal. It involves planning and adhering to a time line (relevant to Goal 3) to get supper done on time. It involves enhancing self-esteem and self-efficacy (Goal 1) and a sense of accomplishment as the participant is eating food that he or she planted.	Work with your team to prepare a supper for tonight that contains at least three kinds of produce from the farm.

Outcomes vary by client and program. They should always be quantifiable and assessed on a regular basis. Documentation requirements and methods also vary by program. Common formats include checklists and summary forms, as described in Chapter 12.

In the hypothetical example described above of a horticultural therapy program involving children transitioning from foster care, outcomes are assessed and documented with the aid of a checklist, the Casey Life Skills Assessment (CLSA) tool, and a satisfaction questionnaire. First, the horticultural therapist documents participants' progress against goals and objectives after each session with a checklist developed by the horticultural therapist. The therapist and participant discuss progress at the monthly one-on-one meetings.

Treatment outcomes are also assessed and documented using the Casey Life Skills Assessment (Casey Life Skills 2011), a tool recommended for assessing outcomes in foster care. The CLSA is a computerized questionnaire that participants are asked to complete at the beginning of monthly one-on-one sessions with the horticultural therapist. Answers are instantly available for therapist and youth to review together to engage participants in the process of developing and refining goals.

Finally, at monthly one-on-one meetings, participants' satisfaction with the program is assessed with a five-item questionnaire developed by the horticultural therapist. The results of the satisfaction questionnaire may also become part of overall program evaluation and influence future program modifications, funding, and continuity.

Summary

The holistic concept of wellness as encompassing physical, emotional, mental, and social aspects of being and functioning underpins wellness models of care. Approaches to care that are guided by wellness models help participants achieve their potential, address the whole person (body, mind, spirit), and capitalize on participants' strengths. Wellness models also emphasize the importance of relationships with others and with the ecosystem. Horticultural therapy is uniquely suited to support wellness-based programming.

Horticultural therapy programs based in wellness models will likely play increasingly important roles around the world in health and wellness. Mainstream health care providers are beginning to see nature-based therapies as important contributors to wellness and health. The increasing popularity of so-called nature prescriptions for activities such as gardening or walking outside illustrates this shift in perspective (Sorgen 2013). In Scandinavian countries, workers are prescribed nature contact to help manage stress. Local, regional, and federal governments are increasingly supportive of nature-based approaches as important contributors to individual, community, and cultural health. "If we can have a fast food restaurant on almost every corner, then we can certainly have a garden," said New York City Mayor Bill de Blasio (AZ Quotes 2017). Nature-based programs, including gardening and nutrition education, are increasingly being offered in public and private schools and through churches (Schmutz et al. 2014).

Nature-based therapies including horticultural therapy have been identified as having a potentially important role in preventive health care in the context of the high prevalence in developed countries of habit-driven lifestyle diseases such as type II diabetes and obesity, and the increasing trends toward urbanization, disconnection from nature, and sedentary lifestyles (Haluza et al. 2014). The multifaceted, holistic nature of wellness-model-based horticultural therapy lends itself well to preventive health care, which requires engagement of both body and mind and necessitates changes in habits.

The increasing recognition of the potential role of nature-based approaches to health care provides a favorable context for establishing wellness-model-based horticultural therapy as a useful approach to enhancing health and wellness in the context of mainstream health care, where horticultural therapy is beginning to be perceived as a legitimate therapeutic modality. Horticultural therapy, unlike complementary modalities or allied therapies such as occupational therapy and physical therapy, is not yet widely recognized among health care providers and the public as an approach to wellness, and it is not directly reimbursed by insurance. More evidence from well-designed studies will help to quantify the benefits of horticultural therapy to general health and well-being. This effort will entail quantifying some of the salient but difficult-to-operationalize benefits of horticultural therapy—social, cultural, and spiritual improvements, for example—and increasing the numbers of participants in horticultural therapy studies in order to achieve adequate sample sizes for research. Advances in quantification such as new quality-of-life scales and innovative approaches to study design will help to address these challenges.

Key terms

Coping skill Method a person uses to deal with stressful situations.
Holistic Characterized by the treatment of the whole person, taking into account mental and social factors, rather than just the physical symptoms of a disease.
Self-efficacy One's belief in one's ability to succeed in specific situations or accomplish a task.
Traditional medical model A model that focuses on the physical and biological aspects of diseases and conditions and employs a problem-solving approach in which medical history, physical exam, and diagnostic tests are used to diagnose and guide treatment of a condition.

References

Casey Life Skills. 2011. "Providers Guide to Casey Life Skills." Accessed October 29, 2012. http://caseylifeskills.force.com/clsa_learn_provider.
Chalquist, C. 2009. "A Look at the Ecotherapy Research Evidence." *Ecopsychology* 1: 1–12.
Detweiler, M. B., T. Sharma, and J. G. Detweiler. 2012. "What Is the Evidence to Support the Use of Therapeutic Gardens for the Elderly?" *Psychiatry Investigation* 9: 100–110.
Haluza, D., R. Schonbauer, and R. Cervinka. 2014. "Green Perspectives for Public Health: A Narrative Review on the Psychological Effects of Experiencing Outdoor Nature." *International Journal of Environmental Research and Public Health* 11: 5445–5461.
Hettler, B. 1976. "Six Dimensions of Wellness Model." National Wellness Institute, Inc. Accessed September 3, 2017. http://www.nationalwellness.org/?page=six_dimensions.
Lewis, C. 1996. *Green Nature: Human Nature: The Meaning of Plants in Our Lives*. Champaign, IL: University of Illinois Press.
Louv, R. 2005. *Last Child in the Woods: Saving Our Children from Nature-Deficit Disorder*. Chapel Hill, NC: Algonquin Books of Chapel Hill, pp. 63–64.
Mandela, N. 1994. *Long Walk to Freedom: The Autobiography of Nelson Mandela*. Boston: Little, Brown and Company.
Niklassson, J. 2007. "Horticultural Therapy for Homeless People." *Examensarbete* 24: 1651–8160.
Ohly, H., S. Gentry, R. Wigglesworth, A. Bethel, R. Lovell, and R. Gardie. 2016. "A Systematic Review of the Health and Well-Being Impacts of School Gardening: Synthesis of Quantitative and Qualitative Evidence." *BMC Public Health* 16: 286.
Pollan, M. 2008. "Why Bother?" Accessed September 5, 2017. http://michaelpollan.com/articles-archive/why-bother/.

Quotes, A. Z. 2017. Accessed October 1, 2017. http://www.azquotes.com/quote/795887.

RHIhub. "Health Promotion and Disease Prevention Models." Accessed September 5, 2017. https://www.ruralhealthinfo.org/community-health/health-promotion/2/theories-and-models.

Savoie-Roskos, M. R., H. Wengreen, and C. Durward. 2017. "Increasing Fruit and Vegetable Intake among Children and Youth through Gardening-based Interventions: A Systematic Review." *Journal of the Academy of Nutrition and Dietetics* 117: 240–250.

Schmutz, U., M. Lennartsson, S. Williams, M. Devereaux, and G. Davies. 2014. "The Benefits of Gardening and Food Growing for Health and Well-Being." *Growing Health: Food Growing for Health and Well-Being*. Garden Organic and Sustain. Henry Doubleday Research Association.

Shiue, I. 2016. "Gardening Is Beneficial for Adult Mental Health: Scottish Health Survey, 2012–2013." *Scandinavian Journal of Occupational Therapy* 23: 320–325.

Soga, M., K. J. Gaston, and Y. Yamaura. 2017. "Gardening Is Beneficial for Health: A Meta-Analysis." *Preventive Medicine Reports* 5: 92–99.

Sorgen, C. 2013. "Do You Need a Nature Prescription?" Accessed September 3, 2017. http://www.webmd.com/balance/features/nature-therapy-ecotherapy#1.

Tweddle, A. 2007. "Youth Leaving Care: How Do They Fare?" *New Directions for Youth Development* 11: 9–10, 15–31.

Van Lier, L. E., J. Utter, S. Denny, M. Lucassen, B. Dyson, and T. Clark. 2017. "Home Gardening and the Health and Well-Being of Adolescents." *Health Promotion Practice* 18: 34–43.

Wilson, E. O., and S. Kellert. 1984. *Biophilia*. Cambridge, MA: Harvard University Press.

World Health Organization (WHO). 2005. "Constitution of the World Health Organization." *World Health Organization: Basic Documents*. 45th ed. Geneva, Switzerland: World Health Organization.

section four

Tools for the therapist

An assortment of tools is included in this section to guide readers in the application of professional practices to successful programs. Included are accommodations, adaptive techniques, and safety concerns to best serve the needs of all participants; strategies for assessment and documentation for horticultural therapy interventions; best practices for the development and management of programs; and the use of research-based practice. Note that the term *session* indicates the treatment time spent between the horticultural therapist and the client; the term *activity* indicates the task performed during the session.

chapter eleven

Considerations and adaptations to safely accommodate program participants

Susan Conlon Morgan

Contents

This chapter explores various program facilitation methods and techniques; tools and equipment; and activity, garden, and greenhouse accommodations used by horticultural therapists to engage varied participants in horticultural therapy programs. Safety standards that affect access to services are also addressed. The aim of applying these techniques is to foster independence, enable clients to access the benefits of gardening, and facilitate full engagement in the activities that are carefully selected to meet therapeutic goals.

The use of horticulture provides a stimulating, normalizing therapeutic activity, which can be delivered in indoor and outdoor settings and accommodated to suit a variety of client abilities, levels of functioning, and interests (Kennedy and Haller 2017). For the client to function as independently as possible, the horticultural therapist plans each session, chooses an activity, and carefully plots out each step within the activity. Thoughtful task analysis and planning of all aspects of activity delivery ensures that the planned intervention accommodates the individual participants' abilities and is compatible with client and group goals (Buettner and Martin 1995). In addition, this preparation allows for the horticultural therapist to identify strategies for accommodating participants during the session that are based on individual characteristics. This informs the structure and delivery of the activity, layout, and features of indoor and outdoor workspaces; number of support staff members needed; safety precautions; available tools and equipment; and use of supplemental materials, such as written instructions and visual aids, to name a few. See Exhibit 11.1 for examples of using accommodations and matching horticultural activities to the therapeutic goals and abilities of clients with spinal cord injury and traumatic brain injury in a rehabilitation setting.

In working with clients individually or in a group setting, the horticultural therapist sets the tone for the treatment session and organizes support staff in providing appropriate logistical support to clients throughout the session and watching for potential safety hazards. During client interactions in and outside session, the horticultural therapist must be flexible, constantly assessing individual engagement. Particularly within a group setting,

---

EXHIBIT 11.1

Technique: Matching Activities to Abilities and Needs for People with Spinal Cord Injuries or Traumatic Brain Injuries

Assessment

In the initial assessment, the horticultural therapist evaluates the client's cognitive and physical abilities to determine activities that are realistic for the client to complete successfully in therapy sessions. The assessment includes an interest survey to determine the types of tasks and projects that the client will enjoy.

Goals

Clients with traumatic brain injury and spinal cord injury in rehabilitation facilities have specific goals that are determined in conjunction with the medical and rehabilitation team. The horticultural therapy program supports these goals. The horticultural therapist develops objectives and provides horticultural and nature-based activities as a means to experience positive change and progress toward the designated goals.

Therapist

The role of the horticultural therapist is to consider client interest and abilities in providing activities. The appropriate match of activities maximizes the opportunity for the client to complete objectives and progress toward the goal.

(Continued)

EXHIBIT 11.1 (Continued)

**Technique: Matching Activities to Abilities and Needs for People
with Spinal Cord Injuries or Traumatic Brain Injuries**

Activities

From mixing soil, potting plants, and painting clay pots to outdoor gardening, horticultural therapy activities offer varied opportunities for clients to experience personal growth and achieve cognitive, physical, emotional, and behavioral goals. The therapist encourages clients to complete the activity to the maximum of their abilities, even if it is a slow process and the task is not completed in an exact manner. The focus is on the process, not the final product.

Activity 1: Mixing soil for clients with upper extremity weakness and limited fine motor skills.
  • Goals—move and use arms, prepare soil for planting project, follow verbal direction.
  • Materials—soil materials; water; lightweight mixing tool such as a plastic serving utensil, large spoon, or adaptive soil scoop.
  • Accommodations/setup—utilize a tabletop soil bin and adjustable table set to ideal height for client.
  • Options for tool modification—place spoon handle into palm band or utilize Velcro to secure tool to hand, wrapping material around handle to create wider grip area.
  • Steps—add water or perlite to a lightweight soil mix. Instruct client to blend and mix soil.
  • Note—premoisten the perlite to avoid dust.

Activity 2: Repotting plants
  • Goals—follow verbal and/or written direction in a multistep task, utilize judgment in sizing the new pot and amount of soil needed.
  • Materials—soil mix, soil scoop, potted plants, plastic pots.
  • Steps—remove plant from original pot, separate roots if rootbound, place soil mix in bottom of new larger pot, place plant in pot, fill to top with soil, water and mist plant.
  • Note—plants with colorful leaves and flowers or pleasant fragrance add sensory benefits.

Each activity has many benefits to provide the opportunity to achieve a specific goal in a physical or cognitive area; provide an experience to promote self-confidence, self-esteem, and positive change; and enjoy the benefits of a plant-rich environment and relaxation from nature experience. Assessment of the client's response to an activity may determine the need for change or modification during the session. The horticultural therapist should be aware of factors such as mood changes or recent physical or cognitive changes and be flexible in adjusting the materials, activity steps, or overall activity to maximize the benefits to the client.

Cathy Flinton, HTR
Horticultural Therapist, Hope Network

accommodating for a variety of abilities and levels of functioning in real time requires skill and advanced planning. In order to meet the individual needs of participants in a horticultural therapy program, the horticultural therapist must be able to prepare and tailor activities, modifying tasks to make activities accessible and enable clients' participation in the therapeutic activity (Kennedy and Haller 2017). Exhibit 11.2 offers an example of accommodations made by a horticultural therapist when working with clients who have dementia.

EXHIBIT 11.2

Technique: Accommodations for Working with People with Dementia

For many people with dementia or memory loss, gardening and working with plants was a part of early life. It was often done with family members, especially parents and grandparents. Due to this connection to long-term memory, the required skills for working with plants, arranging flowers, and gardening can remain a long time in life. Treatment goals for this population tend to be focused on maintaining abilities and life enrichment rather than on development and improvement of skills.

When serving individuals with Alzheimer's or other forms of memory loss, certain techniques help ensure a success. A top priority is to know as much of each person's life history as possible. This aids in developing programming that honors each individual's past interests, skills, experiences, strengths, and needs. Other techniques that are important when serving this population are to:

- Match horticulture activities with ability levels while remaining age appropriateness.
- Select plants based primarily on:
  - Reminiscence value—red geraniums, petunias, herbs, etc.
  - Nontoxic and without harmful thorns.
  - Interesting textures.
  - Bright colors and/or interesting leaf patterns.
  - Fragrance, for those who still retain olfactory senses.
  - Ease in growing.
  - Ease in propagating.
  - Potential for use in future sessions such as nature art and garden chefs.
- Adhere to safety precautions such as:
  - Using only nontoxic materials.
  - Avoiding the use of any items that are typically connected to dining, such as cups and spoons, as planting tools to avoid consumption of gardening materials.
  - Avoiding the use of the dining area for sessions to diminish confusion regarding the purpose of the activity.
  - Having adequate staff or volunteer assistance.
  - Counting sharps such as scissors at the beginning and end of sessions, and monitor their use.

(Continued)

> **EXHIBIT 11.2 (Continued)**
>
> **Technique: Accommodations for Working with People with Dementia**
>
> - Providing easily accessible garden containers and raised garden beds.
> - Providing smooth, wide walkways that are continuous loops leading back to a main patio to help the participant with navigation in the garden.
> - Use tools that are lightweight and can extend reach.
> - Provide individual horticultural therapy sessions for those who no longer function well in group programs.
> - Offer assistance as needed, such as verbal or visual cues, or hand-over-hand and/or other physical assistance.
>
> Horticultural therapy programming provides something for almost everyone dealing with dementia. Even those with advanced cognitive and physical impairment can take part in activities such as mixing soil, watering, and sensory visits. Challenging behaviors can drop away, the words to express thoughts can become more available, and many more benefits arise from horticultural therapy involvement.
>
> **Pamela A. Catlin, HTR**
> *Horticultural Therapy Consultant, Horticultural Therapy Institute Faculty*

Gathering client information

Information gleaned from client records and assessments, combined with listening and observing the client, inform the types of accommodations to use. Records provide important information about the person's medical or behavioral precautions and may reveal techniques that have been effective and enabled full participation in past situations. The therapist may also conduct formal and informal assessments, particularly to learn more about actual performance during horticultural activities. (See Chapter 12 for more about situational assessments.)

Observing and listening to the client during horticultural therapy sessions provides details and clues to any modifications to space, tools, or instruction that may encourage participation and/or enable independence. Through active listening, the horticultural therapist becomes aware of the challenges an individual is experiencing—such as a disruptive fellow participant, equipment that is difficult to use, or too much clutter in the workspace—and responds to curb or work through a frustrating situation before it escalates (Figure 11.1).

Cultural awareness

It is important to identify the culture of each client in an effort to better understand the client's perspective and bring awareness of the horticultural therapist's own cultural beliefs and potential biases, which may affect client interaction. Cultural competence is defined as the ability of a service provider, particularly in health care settings, to provide patients from diverse cultural and ethnic backgrounds with care and services that reflect their

Figure 11.1 Participant enjoys watering after he plants the garden, and benefits from assistance to overcome the challenges of limited strength and coordination at the Bucks County Audubon Society Healing Garden. (Courtesy of Nancy Minich.)

social and cultural needs (Betancourt et al. 2003; Georgetown University Health Policy Institute 2004; Vanderneut 2014). Recognizing each participant as an individual with a specific blend of beliefs, values, cultural experiences, personal background, and language helps the horticultural therapist to improve his or her ability to interact and engage with each client (Betancourt et al. 2002). See Exhibit 11.3 to gain further understanding of how cultural differences can influence the efficacy of the therapist's delivery of services among a diverse group of clients. Understanding the perspective of the client enables the therapist to be sensitive and respectful of the types of accommodations offered.

EXHIBIT 11.3

Perspective: Cultural Diversity

Cultural diversity has played an important role in the history of the United States. Easiest to identify are people from other countries, where language, religion, and education form attitudes and values unlike those in their new home. Variances are seen in the way people look; dress; and view family, work, nature, health, and well-being. Less obvious are cultural differences and ways of thinking that reflect age, gender, urban or rural upbringings, and physical and cognitive differences.

   Cultural variations can impede effective delivery of services, especially in health care, social work, educational, and governmental institutions. For horticultural

(Continued)

EXHIBIT 11.3 (Continued)
Perspective: Cultural Diversity

therapists, the need for recognizing and adapting to cultural differences depends on who you are working with and what you hope to achieve. In a way, horticultural therapy has an advantage in that plants are fundamental to all cultures and ethnic groups. They provide food, clothing, shelter, dyes, flavor, ornaments, and medicine. In every culture, plants inspire poetry, art, and literature, and are used symbolically in rituals of birth, death, and every life transition in between. Educational, health care, and social programs can use horticultural activities to bridge cultural differences and enrich program offerings for all segments of the population.

Over the past decade, many government and professional organizations have addressed the issues of diversity and what has come to be called cultural competency. Articles can be found in journals of law, nursing, social work, psychology, speech therapy, and other professions where clear understanding and communication determine outcomes. These articles offer suggestions on how one can become better at working with culturally diverse clients.

The keys are sensitivity and humility. Knowing your own cultural biases, values, and attitudes is an important first step. Learning about the people you are working with is obviously helpful and shows respect. We need to become good observers and curious about how different cultural perspectives might be reflected, including our own. And we have to appreciate this diversity and recognize that different points of view can enrich our experiences and add a positive dimension to the activity and program.

Being aware of cultural differences can make activities and programs more successful and help bridge and improve interactions between different groups living in our community.

Nancy Chambers, HTR
Director Glass Garden and Horticultural Therapist, Retired
Rusk Institute of Rehabilitation Medicine, NYU Medical Center

Communication techniques for accommodation

Listening and setting up receptive conditions

Active listening techniques are reasonably applied when communicating with and accommodating program participants in a horticultural therapy program. The setup of the activity and the room as well as the way the horticultural therapist presents the activity help to elicit participant response. Kennedy and Haller (2017) note that, before participants engage in the activity, they must feel at ease in the environment and be open to participation. This is naturally facilitated by the presence of plants in the setting because plants are commonly perceived as familiar and safe, as noted in previous chapters. To begin developing a rapport with participants, the therapist may ask open-ended questions for more detailed individual responses than are elicited by yes or no questions and invite opportunities for humor and storytelling to develop a more relaxed, upbeat atmosphere (Kennedy and Haller 2017). The use of these strategies with certain populations, such as individuals who are nonverbal or have difficulty processing social cues, should be considered and adapted appropriately for participants.

Prompts

Throughout interactions with clients, the horticultural therapist conveys information in the form of directions, feedback, supports, and recommendations. The methods used to guide clients are carefully considered to meet the needs and learning style of individuals, as well as the situation, environment, and surroundings. This section focuses on the types and use of prompts to foster and enable participation and task engagement. When engaging participants in activity tasks, the horticultural therapist regularly adapts his or her approach with participants of varying levels of physical, cognitive, and social functioning and employs a series of measured and progressive prompts (Brown et al. 2013). The Merriam-Webster Dictionary (2017) defines *prompt* as:

- to move to action: incite
- to assist (one in acting or reciting) by suggesting or saying the next words of something forgotten or imperfectly learned: cue
- to serve as the inciting cause of

Every effort should be made to use minimal interventions and empower individuals to participate in the activity in his or her own way. However, sometimes participants require additional prompts in order to engage in activities in a specific and directed way.

Some of these prompting techniques may be used in tandem with each other, or the use of only one technique may need to be utilized to enable participant response. Avoid unnecessary prompts, such as prematurely placing materials and tools within reach and giving verbal or physical cues that might signal participants to begin a task before it is appropriate or safe to do so.

Types of prompts, from least intrusive to most intrusive

The following descriptions of prompting interventions is derived from Alberto and Troutman (2016). The horticultural therapist should begin prompts in the least intrusive method possible and increase the level of intervention only as needed, to the extent of full, direct physical intervention (ErinoaksKids 2017). Care should be taken to respect an individual's personal preferences and space at all times. The following types of prompts begin with the least intrusive (verbal) described first and the most intrusive (physical) last (Figure 11.2).

Verbal information

The horticultural therapist may use verbal information to provide an explanation of the tasks to be completed and how they fit with the goals and objectives of the session. They may also verbally describe the task or activity by giving a sequence of step-by-step instructions or by giving one step at a time to be followed, depending on the skills of the participants. Verbal information may also be used to encourage or enable full participation. This may be as simple as a one- or two-word prompt to cue the client of a next step or may involve more detail.

Visual information

Visual cues include written step-by-step instructions or the use of pictures and symbols. If the participant does not respond to a verbal cue, offer a written list or pictures illustrating step-by-step instructions. Visual cues are highly effective in communicating the

Figure 11.2 Therapist uses prompts to encourage success. (Courtesy of Life Enrichment Center.)

sequencing of multistep activities and should be tailored to each participant, with care taken to understand how each best responds to instructions. Well-positioned gardening tools in a planting bed or floral arranging materials on a table also provide visual cues for participants to read the situation and intuitively take action. For individuals with cognitive or sensory processing disorders who have difficulty responding to verbal cues, visual cues can be used successfully to increase confidence and independence in task completion. They are also used effectively in tandem with verbal information to support or reinforce learning in noisy environments or when the client may be overwhelmed by too much verbal information.

Gestures and modeling

Gestural prompts involve careful body positioning and physical gestures, such as pointing or motioning with hands. Point or gesture to objects with one hand to signal the next step to participants. Indicate the placement of a tool or other object in a certain location through the horticultural therapist's body positioning within the workspace. The horticultural therapist may use verbal and visual cues in connection with gestures.

In modeling, the horticultural therapist shows how to do a step or series of steps, then waits for participants to take action and imitate the step(s) in response. Observation and active listening techniques help the horticultural therapist read participants' comfort level with step completion. Based on participant feedback, the horticultural therapist gives participants enough time to process and act as appropriate.

Task sharing

Task sharing is a form of partial physical intervention when the horticultural therapist works in tandem with the participant on a one-to-one basis. Working together, the

Figure 11.3 Task sharing at Shepherd Center. (Courtesy of Shepherd Center, Atlanta, Georgia.)

participant takes on one side of the task, while the horticultural therapist takes on the other side of the task. For example, the horticultural therapist may hold out the long stem of a heartleaf philodendron during a plant propagation activity, while the client uses scissors to cut the stem. The client's level of involvement may vary at first and, as confidence builds, the horticultural therapist may decrease the level of support. Task sharing is effective with participants who have weakened mobility or paralysis of limbs and may be used regularly in certain situations. If the participant is nonresponsive or needs additional support, the horticultural therapist may continue to the next level of intervention: direct physical prompt (Figure 11.3).

Physical prompt

Physical prompt involves direct physical intervention with the participant, such as hand-over-hand or hand-under-hand, or may simply be a touch to the arm or shoulder to prompt movement. Approaching the participant with care and respect, the horticultural therapist secures awareness and permission to take the individual's hand and together they work to complete one step of the task. Supplemental verbal cues serve to reassure participants and communicate action steps during physical prompts.

Fading the prompts decreases the horticultural therapist's intervention with increased participant independence. The horticultural therapist gradually decreases her or his level of direct intervention and uses previous, less intrusive prompts based on individual response. The horticultural therapist should offer positive reinforcement for all independent, unprompted participant responses, even more so than for prompted responses (ErinoaksKids 2017).

Design accommodations and accessible therapy space

In a horticultural therapy program, the plants, materials, tools, and techniques that a horticultural therapist employs are important in facilitating therapeutic interventions with participants. The environment where these interventions take place should be treated as an equally valuable therapeutic tool. Horticultural therapy activities take place in a variety of indoor or outdoor settings or a combination of both, including greenhouses, indoor workrooms, community gardens, dining areas, or outdoor therapeutic gardens. The design and

Figure 11.4 The Garden at The Centers for Exceptional Children proves a positive atmosphere for therapeutic interventions. (Courtesy of JoAnn Yates.)

organizational features of these settings establish the atmosphere where therapeutic interventions take place and can positively or negatively influence participants' experiences of these programs and places (Kavanaugh 1995; Rodiek et al. 2010; Rodiek and Schwarz 2012; Marshall and Gilliard 2014; Smith and Watkins 2016) (Figure 11.4).

The therapeutic environment, whether it be indoors or outdoors, must be physically and psychologically accessible and supportive for participants, as well as staff members, in order to cultivate productive outcomes in a horticultural therapy program (Hazen 2013; Ulrich 1992; Smith and Watkins 2016). Exhibit 11.4 addresses how the psychological accessibility of a therapeutic garden is as equally important as the physical accessibility of its design features for users. In fact, research studies have connected "poor design to anxiety, delirium, elevated blood pressure, increased need for pain medication, and longer hospital stays following surgery. Conversely, research has shown that good design can reduce stress and anxiety, lower blood pressure, improve postoperative courses, reduce the need for pain medication, and shorten hospital stays" (Ulrich 1992). Organizations, such as the US Green Building Council and Eden Alternative®, cite the need for incorporating plants and nature through sound design practices in residential, health care, and other public buildings and outdoor spaces in order to support the needs of people and enhance their quality of life. Well-designed green buildings and landscapes provide welcomed respite for visitors in otherwise sterile environments; reduce confusion, agitation, and aggression; improve spatial orientation; restore circadian rhythms; improve concentration and attentional capacity; increase opportunities for reminiscing; and enable independent use of the space, among numerous other benefits (Cooper Marcus and Barnes 1999; Rodiek and Schwarz 2012; Cooper Marcus and Sachs 2014; Waller and Masterson 2014; Winterbottom and Wagenfeld 2015). This section of the chapter explores some aspects of the purposeful design and manipulation of the clinical space used in horticultural therapy, with an emphasis on accommodating a variety of participants.

EXHIBIT 11.4

Perspective: Psychological Accessibility—A Perspective on Therapeutic Garden Design

Physical accessibility plays a critical role in the design of any healing or therapeutic garden. If the visitor cannot get to the garden or take part in its details, then it is difficult to reap the garden's benefits. Accessibility, however, is about more than just physical accommodation. Ensuring the psychological accessibility of a garden is also critical: Does the garden's design entice one to spend time there? Does it inspire the visitor to open him- or herself to its healing benefits? The most successful gardens incorporate both physical and psychological accessibility.

For example, consider the relationship between inside and outside as they relate to a garden. The garden needs to be visually accessible from indoors so people know it is there, and it needs to be physically accessible so they can get to it and move around comfortably once there. The *transition* between inside and outside is where psychological accessibility plays an important role. Many older adults or those with sensory issues struggle with a dramatic change in light or temperature as they move outdoors. A gradual, comfortable transition through the use of an extended threshold alleviates that discomfort and in turn encourages venturing out into the garden. This extended threshold might be an overhead structure or a stretch of shade trees against the building, and its purpose is to dilute the sudden environmental changes. An extended threshold can serve as a destination of its own, creating a more peaceful and calming introduction, both physically and cognitively, to the environmental changes as one moves from indoors to out, and vice versa.

The visual connection of spaces also plays an important role in the psychological accessibility of the garden. For example, can a visitor see a comfortable bench in an inviting location near the entrance that encourages him or her to step into the garden? This is especially important for older adults or those with physical impairments who might not feel they can walk very far without a place to rest. Once at that inviting rest point, are there new potential rest spots that encourage further, safe exploration of the garden?

Kaplan and Kaplan (1989) discuss the importance of offering opportunities for exploration and mystery in their work on landscape preferences. These components in psychological accessibility of gardens are important because, when used correctly, they can beckon one into the garden and promote a deeper connection to nature. Mystery in a garden is the promise that new information will be revealed if one walks further into the space. It promotes curiosity and anticipation, encourages exploration, and guarantees discovery. By revealing stopping points, vistas, interesting plants, and objects of folly through the use of visual links, interesting sounds, and other sensory-oriented cues, the visitor is drawn into the garden both physically and psychologically.

A primary purpose of healing and therapeutic gardens is to enable a rich connection to plants and nature. Once established, this connection creates a foundation for further healing work. When both physical and psychological accessibility are an intentional part of the garden's design, they boost the garden's benefits and facilitate those important nature connections.

Elizabeth R.M Diehl, RLA, HTM
Director of Therapeutic Horticulture, Wilmot Gardens at the University of Florida

Universal design

Universal design is a series of established principles for creating a physical space or product that is universally appealing and accessible to the widest possible range of people, regardless of ability, age, gender, cultural background, or preference. Horticultural therapists apply universal design principles in all aspects of program design and implementation, including growing and treatment spaces, programming, and tool selection (Connell et al. 1997; Marcus 2014; Winterbottom and Wagenfeld 2015).

The design and use of public facilities and spaces in the United States are regulated by the Americans with Disabilities Act (ADA). Since 1990, the ADA has prohibited discrimination toward Americans living with disabilities and set forth federal regulations for accessibility and accommodations within public spaces, facilities, and transportation. New structures and spaces must be built, and older buildings and spaces retrofitted, to be ADA-compliant. To ensure that ADA standards are met, it may be necessary to consult with an expert, such as a landscape architect, on the development or redesign of a horticultural therapy space (Winterbottom and Wagenfeld 2015).

Characteristics of a therapeutic environment

One of the tools used by horticultural therapy programs is a plant-rich garden that incorporates principles of therapeutic environments. The following characteristics, which apply to indoor and outdoor spaces, are essential in a therapeutic environment (Hazler and Barwick 2001; Rodiek et al. 2010; Hazen 2013).

> *Plants*—The inclusion of plant material is considered a necessary element in a therapeutic environment. Gardens described as therapeutic tend to be dominated by plants relative to hardscape features. Plants are essential in the clinical spaces used for horticultural therapy.
>
> *Universal Design and Accessibility*—The workspace should be designed to accommodate a range of individual users with varying levels of functioning and mobility, ages, abilities, and cultural backgrounds. Adaptable features, such as moveable tables and chairs, potted plants on rolling dollies, or raised garden beds and vertical gardening features that accommodate gardeners at seated or standing positions, help improve the universal functionality of the space. Pathways, ramps, and other similar features, which enable those who use wheelchairs and people with physical disabilities to move throughout the space, must be built to comply with ADA. The availability of shade structures, large trees, cooling and hydration stations, and plenty of seating provide respite opportunities for those who fatigue easily or have issues with regulating body temperature. See Exhibit 11.5 for an example of how a healing garden employed biophilic design features in its design and construction for use in horticultural therapy programming for adults with a range of intellectual disabilities and autism.
>
> *Safety and Security*—A secured environment is necessary in some horticultural therapy program sites to reduce the risk of elopement or the risk of harm to self or others. Elements such as locked doors and gates, walls, and fences can be softened or hidden through careful placement of plants and other landscape features; these help to provide security for users while not being obvious. Design features for intended user groups, such as continuous pathways that always lead back to the entrance for garden visitors with dementia, enhance their sense of security within the space and eliminate visitors getting stuck at the end of a path.

EXHIBIT 11.5

Program Example: Biophilic Design—A Healing Garden for Adults with a Range of Intellectual Disabilities and Autism

A healing garden was designed by the author at Bucks County Audubon Society (BCAS) at Honey Hollow specifically to support the needs of adults with a range of intellectual disabilities and autism. It was funded through a grant from a local, private foundation to fulfill a need in the community for individuals with similar disabilities using horticultural therapy. The healing garden was designed using biophilic design techniques, which incorporates nature and natural features and materials into the enclosed garden space. Individuals in the Garden Club built the garden over one summer. A ribbon-cutting ceremony was held in the fall to celebrate its completion and bring awareness to the community about biophilic design, horticultural therapy, and the abilities of individuals with autism and intellectual disabilities.

The mission of BCAS is to educate the people of greater Bucks County, Pennsylvania, about the natural world. Peaceful Living, in Chalfont, Pennsylvania, had been the recipient of a horticultural therapy program for individuals with a range of intellectual disabilities and autism, meeting weekly at BCAS for several years. Through volunteering with clients there, the author noticed how receptive the individuals were to gardening and nature-related activities. The Peaceful Living staff also reported less agitation and stress-related behaviors after returning from BCAS. With this positive response, the author developed a grant to fund the healing garden construction to address the specific needs of the type of individuals served by their program.

The grant goals were to build a healing garden in which the clients at Peaceful Living could work to connect with nature and gain the benefits that engaging with plants and nature can provide. Secondary goals were to develop an effective horticultural therapy program for adults with autism and intellectual disabilities in the region; develop a manual of horticultural therapy activities for use by others who wish to implement similar programs; and train others, specifically horticultural therapy interns. The garden offers sensory stimulation from the colors, textures, scents, textures, and sounds from the rustling of the surrounding trees, shrubs, birds, and insects visiting the garden. For the individuals who came weekly to the Garden Club, constructing the garden helped alleviate anxiety through the physical activity of using a wheelbarrow and spreading gradations of gravel for drainage under the engineered wood chips—an approved accessible surface—thus building self-esteem and leisure time interests through successfully completing tasks, improving socialization skills by working cooperatively, and improving gross and fine motor skills by repetitive planting and watering.

To keep the Garden Club participants interested on rainy days or in extreme heat, the group briefly harvested the vegetables and herbs in the two raised planters, worked on a floral craft project under the umbrella on the picnic table, or took a walk on the trails in the woods. They were always excited to see what grew from the previous week and proud to take things home with them. This twenty-foot-by-thirty-foot healing garden continues to be a restorative space for the individuals with autism and intellectual disabilities who partake in various horticultural therapy activities in that garden most of the year.

Nancy A. Minich, RLA, HTR, ASLA
Principal, NAM Planning and Design, LLC

Positive Distractions—Ulrich (1992) defines a positive distraction as "an element that produces positive feelings, effortlessly holds attention and interest, and therefore may block or reduce worrisome thoughts." A garden setting, plant-rich indoor environment, images of nature, or even a single houseplant naturally provide positive distractions for participants, boosting moods while reducing stress and anxiety. On the other hand, the therapeutic environment should minimize the number of negative distractions, including noises, objectionable odors, and interruptions by external sources, which can decrease concentration and focus and increase frustration (Ulrich 1992; Hazler and Barwick 2001).

Creature Comforts—Based on the unique site conditions and population served, considerations include temperature and climate control, availability of seating, good lighting, play areas, opportunities for sensory stimulation, restrooms, quiet spaces, and hydration stations. The availability of one or more of these creature comforts meet the needs of participants, enabling full participation in programming.

Social Support—Participants benefit from having access to other people who are helpful, empathetic, friendly, and supportive, such as family members, friends, volunteers, and staff members. A bench in an out-of-the-way section of the garden or plenty of open space in a common gathering area provide welcoming spots for visitors to engage comfortably with one another.

The horticultural therapist exercises creativity and flexibility in adapting the workspace to help clients feel safe and comfortable, enabling them to engage in gardening activities to work toward goals (Catlin 2017) (Figure 11.5).

Figure 11.5 Shade and seating accommodate a horticultural therapy group on a warm day. (Courtesy of Rebecca Haller.)

The necessity of organization

Another key aspect of treatment space in horticultural therapy is that it be well organized. Many client groups can perform more independently and with less stress in environments that are structured, clearly labeled, and free of clutter. For example, storage areas for tools and garden materials may be labeled and color coded, and may provide clear guidance for the location of objects used. Organized growing spaces are also enabling to most participants and may foster independence and access to the garden or greenhouse. It is important to consider the cognitive, physical, social, and emotional needs of the clients served. For someone with a mobility impairment, a cluttered space may literally impede access. For someone with a mental illness, disorganized spaces may cause confusion or anxiety. A person with autism may be overwhelmed by disarray, reducing her or his ability to focus on the garden or the activity (Figure 11.6).

Needs assessment for a horticultural therapy workspace

In order to understand how well a space accommodates the clients served in any particular program and to guide adjustments, an assessment is recommended. A needs assessment takes an inventory of what works and does not work in a space. It also identifies user needs and desires and current gaps in the characteristics of indoor and outdoor horticultural therapy spaces. The needs assessment may be conducted by the horticultural therapist as part of the initial or ongoing process for assessing current and desired uses of the environment by program constituents. Or it may be done by design team members, such as a landscape architect, during the early stages of the development or redesign process of the workspace. This systematic approach to documenting program and facility needs helps to make the case to decision makers for future budgetary planning and spending choices in workplace development. See Exhibit 11.6 for an example of how staff at Perkins School for

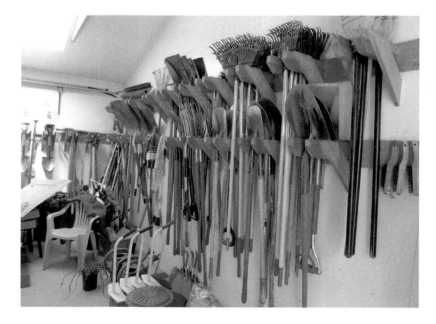

Figure 11.6 An organized storage space enhances accessibility. (Courtesy of Rebecca Haller.)

the Blind assessed the functionality of the workspace and made appropriate accommodations by analyzing the individualized needs and abilities of students, including those with low vision.

EXHIBIT 11.6

Program Example: Horticultural Therapy with Students Who Are Blind, Massachusetts

At Perkins School for the Blind, educational considerations require teaching varied skills to students who are blind, deafblind, or have multiple disabilities with a wide range of cognitive, intellectual, social, communicative, and physical challenges. An expanded core curriculum (ECC) is taught, which is intended to empower all students with disabilities to access their education and make their own choices throughout their entire lives. The environment and educational goals are adapted to meet the specific needs of each student. An individual education program (IEP) is developed to create an educational recipe, tailored for each individual student. Teams work together to promote growth and progress in academic and/or independent living skills, so each individual student will lead a fulfilling, productive life as an adult on graduation or completion of postsecondary educational endeavors.

Environmental factors are highly important within the horticultural therapy program at Perkins School for the Blind. During the past twenty-two years, I have come to appreciate the unique flexibility and adaptability within the horticultural program as teacher and horticultural therapist for students who are blind or have low vision. These students often have additional disabilities, which might even be considered their primary disability with vision or hearing loss as their secondary or tertiary disability. Environmental adaptations are essential to promote equal access for every student. The horticultural therapy program is unique in that it offers such a range of environmental access across each setting. During daily instruction from classroom to indoor sensory gardens or outdoors within raised bed or square foot gardens, we can divide groups, given appropriate staffing, to direct students with very individualized approaches within a range of settings. An environmental assessment is essential to manage classroom instruction, especially with such a range of special needs to address in each therapy session.

The space must be evaluated to accommodate individualized student needs that are communicatory, social, recreational, behavioral, physical, and auditory. Visual needs that involve lighting are also considered. In my own classroom, an extensive environmental assessment was done, and adaptations were initiated, to allow low vision students to function more effectively in the classroom. At the time of the assessment, there were four off-white walls, visual clutter, and ineffective lighting. Adaptations to lighting were made, with light fixtures that would bounce light emission upward toward the ceiling, to illuminate the space more efficiently. In addition, the electrician adjusted the light switch panel to allow the instructor to illuminate portions of the room or the entire room all at once. There was an additional fixture placed over the sink, with a separate light switch, to enable students to work on washing/cleaning tasks. Changes were made to paint colors and the front and back wall panels were painted with contrasting earth tones. It was essential to label

(Continued)

EXHIBIT 11.6 (Continued)

**Program Example: Horticultural Therapy with
Students Who Are Blind, Massachusetts**

all materials and equipment, with appropriate labels in Braille, large print, tactual/
visual markers, tactile symbols, Mayer Johnson symbols, and pictures, so any student could potentially identify her or his classroom location and materials. Spatial
needs are carefully considered and adapted for each student, so he or she can access
his or her workspace and materials on adaptable trays. Often, we utilize American
Printing House for the Blind trays, which are designed with a border.

As a teacher and horticultural therapist, I often instruct students in a wide range
of coursework, such as horticultural crafts production, floral bud vase design and
delivery, plant maintenance, greenhouse management, farmer's market, horticulture
delivery, recycling, plant production, horticulture sales, horticultural therapy,
science instruction, and horticulture vocational training both on-campus and in
community-based locations. Social and emotional, orientation and mobility, home
and personal management, communication, auditory needs, learning disabilities,
and recreation and leisure are other areas students must learn. We often work with
other teachers or therapists on our team to achieve common goals and objectives,
ensuring that skills are carried out throughout the entire team, which also involves
carryover into the clients' home environment.

Marion V. Myhre, M.Ed, TVI, HTR
Teacher and Horticultural Therapist, Perkins School for the Blind

Tools for horticultural therapy program use

Tools and various forms of assistive technology are often used in a horticultural therapy
program to help facilitate participants' engagement in tasks. Facilitation may be through
the way tasks are organized or through the positioning and use of tools and equipment
utilized by participants. Assistive technology includes products, devices, and services
that enhance individuals' abilities to accomplish tasks, including activities of daily living, which they may have previously been unable or had difficulty being able to do (US
Congress 2008). Adaptive tools are assistive technology devices that remove barriers and
enable participants' ability to complete tasks (Broach 2000).

Gardening tools and equipment are utilized as assistive devices for participants with
a range of abilities in completing horticultural tasks, such as planting, watering, weeding, raking, harvesting, and other activities. These tools may be adapted to make them
more accessible or functional for participants. Children's gardening tools and inexpensive
modifications to basic gardening tools can be quite useful and cost-effective. Techniques
that adapt the activity, such as the way in which an activity is organized by the horticultural therapist, the positioning of tools and equipment, and the design and setup of the
workspace, can eliminate the need for modified tools and enable the use of traditional
tools. See Exhibit 11.7 for an example of the workspace setup and other accommodations
made before and during a group horticultural therapy activity for adults with acquired
brain injury.

EXHIBIT 11.7

Program Example: Accommodations for Patients with Acquired Brain Injury

Horticultural Therapy Group Overview

Shepherd Center is a rehabilitation hospital for people with spinal cord injury and acquired brain injury in Atlanta, Georgia. The horticultural therapy group meets weekly for one hour and includes up to five patients, who have a Rancho Los Amigos Level of Cognitive Function scale of level five or higher (Hagen and Durham 1987). Patients with a score of five generally have disorientation to time, place, and/or person and may require maximum redirection and assistance. They may also exhibit inappropriate behavior at times and have difficulty with memory recall. Patients with a score of six generally have the ability to attend to highly familiar tasks in a nondistracting environment and may require moderate redirection. Typically, for these patients, long-term memory has more depth and detail than recent memory.

Some patients participate in several group sessions, while others may attend only once. A horticultural therapist facilitates the group, with additional support from a physical therapist, occupational therapist, or other support staff member for patients who require one-on-one assistance. Programs take place indoors or in the outdoor therapy garden with raised garden beds.

Therapeutic Goals

Cognitive goals include increasing attention span; following directions; addressing inattention, visual scanning, or neglect; focusing on word finding; improving short- and long-term memory; encouraging organizational and sequencing skills; and supporting decision-making skills.

Psychosocial goals include expression of feelings, opportunities for personal reflection and socialization through group interaction, anxiety and stress reduction, and opportunities for nonverbal patients to express themselves.

Accommodations

The neat and creative visual presentation of supplies is a key accommodation in capturing the patients' attention because, with most patients, attention span is limited. In addition, supplies are arranged to encourage visual scanning and reaching to address depth perception and upper extremity range of motion. Other accommodations include adjustable tables, which are necessary to accommodate different wheelchair heights and other physical challenges. Lap trays are handy when a patient does not have bilateral use of upper extremities due to a stroke. Towels on patients' laps give patients permission to make a mess, which in turn reduces stress and increases pleasure.

Patients are given extra time to process and respond to questions or choice-making opportunities during an activity. This may be addressed by limiting choices—"Is the plant red or pink?"—to help the brain find that particular pathway in order to answer the question. Patients may need verbal and/or visual cues, depending on individuals' attention challenges or aphasia. Patients may need to take short breaks

(Continued)

EXHIBIT 11.7 (Continued)

Program Example: Accommodations for Patients with Acquired Brain Injury

due to sensory overload or overstimulation. Lightweight and universally designed tools support physical challenges. Rotating turntables provide increased access when planting and decorating pots, especially for a patient who has experienced a stroke and who may not have bilateral range of motion.

Wendy Battaglia
Certificate in Horticultural Therapy
Horticultural Specialist, Shepherd Center, Atlanta

Recent trends in the design and marketing of gardening tools for the general public have focused on the development of products with universally designed characteristics that help to make them more appealing to and adaptable for a wider range of people (Fleming 2013). Traditionally heavy and awkward gardening tools are being replaced by tools made of more colorful, sturdy, and lightweight aluminum, plastic, or composite materials, which reduces user fatigue and increases physical independence. Design features are focused on their ease and flexibility of use and other characteristics, such as color and novelty, which make them more appealing.

Features of tools for horticultural therapy program use

With the characteristics and abilities of participants in mind, the horticultural therapist should select tools based on several criteria. Ergonomic, safety, and other design features may enhance or detract from individuals' use of tools and should be used with careful consideration.

Grips and Handles—Nonslip, cushioned, or angled tool handles of various sizes and lengths enable ease of use for participants, notably individuals with weakened grip and upper body strength or limited range of motion. Cushioned or ergonomically designed grips reduce fatigue for users and increase confidence levels and time spent working on gardening activities. Inexpensive foam pipe insulation, which is cut to fit and slipped onto tool handles, offers a larger and more cushioned grip.

Reach—Tools with long handles allow for easier access to garden tasks on the ground, overhead, or just out of reach. Window boxes, hanging baskets, containers, and raised gardening beds bring plants to an elevated level, reducing the risk of falls and allowing use of lightweight handheld tools. Telescoping, extendable reach handles can be adjusted to fit the situation and user.

Weight—Tools vary in their quality and type of construction materials, including wood, plastic, aluminum, and composite materials. Construction materials determine whether they are light, moderate, or heavy in weight, each offering an advantage in different situations. For example, lightweight tools are ideal for use by participants with limited range of motion and upper body strength or individuals working with tools for a prolonged amount of time, such as in vocational settings. Heavy tools may be necessary for individuals who have gait or balance issues where they can be used for leverage during certain tasks.

Color—The use of brightly colored tools makes them more fun and appealing to use, while also being practical as they are easy to spot in the garden. Tools with colors that contrast from their surroundings are more visible to some individuals with partial vision loss. Differently colored tools can also be used by a therapist in prompting participants to use certain tools for certain tasks.

Size and Portability—The size of tools used in horticultural therapy programming is as varied as there are ways to engage clients with them. Smaller handheld tools for seed sowing or container-planting activities can be used for practicing fine motor skills with individuals recovering from a stroke, traumatic brain injury, or spinal cord injury. Larger tools for soil preparation or digging tasks, such as planting trees and shrubs, can be used for working out aggression for veterans dealing with anger management issues. For equipment and objects that are heavy or bulky to move, lighten the load by transporting them with wheelbarrows or other garden carts with wheels.

Blades—Tools with sharp blades and other objects that can be used as or fashioned into a weapon are generally handled and stored in a space with restricted access. They should be kept sharpened and in good working condition. For some programs, the therapist might provide safer alternatives, such as safety scissors, or organize tasks so that participants do not have to cut or prune things.

Labeling and Storage—Tools should have an assigned storage location in order to properly maintain them and keep workspaces tidy and organized. Labeling for tools and other program materials may include color-coding systems and stenciled, handwritten, or printed labels for easy storage. A traced outline of tools on the walls of storage areas helps participants know where to place tools at the end of a session and enables a quick visual inventory. Tools can be organized and categorized based on their uses or availability of suitable storage space. Durable, labeled bins and baskets, with or without lids, aid in easy, organized storage of program tools and materials. Items can be locked or kept out of reach and sight as needed.

Potential stigmas for use of specialized tools

Specially designed assistive devices and modifications to traditional gardening tools enable participants to engage more independently in horticultural activities. However, a culturally competent therapist should recognize that these tools may carry a stigma for participants and use adaptive tools after careful consideration and other less intrusive interventions have been made available. For some individuals, the stigma associated with using bulky or unusual tools is that they appear different from other tools, which in turn can cultivate the participant's feeling of being different or separate from others or bring attention to differences in ability. The horticultural therapist's careful attention to the setup and delivery of the activity as well as the positioning of tools and objects within the workspace can mitigate the use of certain assistive devices and redirect focus on the strengths and abilities of clients for task completion (Fleming 2013). Using readily available tools that are ergonomically and universally designed may be a better option than applying obvious modifications.

Examples of tools used in horticultural therapy programming

Tools should be used for their intended functions, and modifications should enable the participants to use tools properly. Improper use of tools can result in injury or damage to equipment. Keep in mind that a tool that works for a person who is standing may be inappropriate and awkward for someone who is in a seated position due to its intended use, weight, construction materials, and other design features (Adil 1994). The following are some examples of tools and adaptations for horticultural therapy programming:

> *Tools for Digging, Planting, and Cultivating*—Individuals who have difficulty bending or stooping in order to work at ground level may use tools with long handles from a standing position. Telescoping handles can be adjusted to the user's standing or sitting height. Raised planters and gardening beds can also bring gardening activities to a more comfortable, elevated level for participants
>
> *Watering Devices*—Creative methods can be employed to accomplish watering and accommodate people with diminished strength, range of motion, and energy and other impairments. Utilize rolling carts to help transport heavy filled watering cans if the water source is at a distance from the workspace. In-ground landscape irrigation systems may be costly to install and maintain, but soaker hoses or do-it-yourself drip watering kits for potted plants are inexpensive alternatives where appropriate. Colored hoses can be used as visual cues for program tasks and safety awareness. Watering wands, water shut-off valves, and other hose attachments help to keep watering tasks tidy and are ideal for extending reach and watering concentrated areas, such as containers, hanging baskets, and hard-to-reach spots in the garden.
>
> *Plant Propagation Tools*—Seed tapes can be bought or handmade to make seeding more manageable and precise for individuals who have limited range of motion and fine motor skills. Long-handled, walk-behind precision seed planters enable individuals with difficulty bending or stooping to plant seeds more precisely in vegetable garden rows. Jumbo easy-grip tweezers or handmade or store-bought dibble sticks of various sizes and lengths are used to build hand strength and coordination skills during propagation tasks. Soil blocks and peat pots are useful alternatives to plastic pots, which can be prohibited for use in certain settings such as mental health facilities. Everyday objects may be modified to suit programming needs, such as notched pieces of cardboard for seed sowing precision or a carefully positioned polyvinyl chloride (PVC) pipe as an extended reach device to drop seeds in place on the ground.
>
> *Pruning Tools*—Pruning tools with ratcheting action and rolling, ergonomic handles are useful for individuals with limited hand strength and muscle control. The handle sizes of pruners or secateurs are best matched to the hand sizes of participants.
>
> *Tools for Garden Access*—The therapist may use tools and equipment that offer easier kneeling, bending, stooping, sitting, and other physical movements. Cushioned kneeling pads, garden kneelers, and seats that are convertible or that roll on wheels; kneepads; and adjustable seating enable participants to work close to the ground. Hanging baskets on adjustable pulley systems provide easier access for watering and maintenance activities with minimal strain. Raised outdoor planters, tabletop and vertical gardening systems, and grow carts with adjustable lighting timers elevate gardening activities to individuals in seated and standing positions (Figure 11.7).

General safety precautions and recommendations for specific settings

Gardening and related activities within a horticultural therapy program have associated inherent risks, such as working with gardening tools or in an unpredictable outdoor

Figure 11.7 Raising a planter allows gardening from a seated, comfortable position. (Courtesy of Pam Catlin.)

setting; these activities require staff members who work in these settings to recognize and minimize the potential risks and safely and successfully accommodate participants. Clients, particularly vulnerable populations who are not able to speak up for themselves, have the right to work in an environment that is welcoming and safe. This is directly affected by the way the workspace is organized, the activity is presented, and the selection of participants in the program. (See Chapter 13 for safety considerations in managing a horticultural therapy program.)

An established safety protocol communicates to program constituents that they are valued and cultivates an environment where people know that their safety is a top priority. The elements to include are often informed by the risk management guidelines of the agency in order to maintain certain accreditation and licensing standards as well as additional agency-specific standards. These standards are designed to enhance quality of care and services for clients as well as to meet governmental mandates, insurance requirements, and professional standards. In addition, for registered horticultural therapists, these are established as part of the standards of practice through the American Horticultural Therapy Association (AHTA; 2015). When establishing horticultural therapy safety procedures, it is important to meet with various agency staff members, including treatment team members and risk management staff members. These staff members can provide invaluable information to help minimize risk and ensure that rules and regulations are met.

Safety precautions for horticultural therapy programs

The following list offers several safety considerations that a horticultural therapist plans for when establishing a program. Unique conditions appear within each setting, and the horticultural therapist must use common sense and consult with others on how to handle certain situations. Common potential hazards and associated safety considerations include:

Disruptive Participants—On occasion, individuals can become overstimulated, angry, or disruptive and pose a threat to others in a program. A clearly communicated plan of how to deal with a disruption and the use of conflict management techniques should be in place prior to program participation. Remove disruptive individuals or move the group to a secondary location away from the person who is acting out if other interventions have not worked or there is immediate risk of harm to self or others.

Plants—Plants are the primary tools of a horticultural therapy program, yet it is important that the horticultural therapist knows the various characteristics and uses for each plant prior to program use. Some plants have qualities that make them unsafe or unpleasant to use with certain populations. Some have properties that can counteract or compromise current medical treatments if ingested. There are a variety of electronic and print resources provided by universities, research institutions, poison control centers, and other reputable sources that can be consulted for researching the potential hazards of plants.

- *Poisonous Plants*—Plants that are poisonous or have varying levels of toxicity may be unsafe to use with individuals who have impulse control issues, have cognitive impairments, or are prone to self-injury or putting nonfood items in their mouths. Limit or avoid use of these plants within the program.
- *Caustic Plants*—Some plants have parts or emit sap caustic to the skin, which can cause rashes or other symptoms when participants touch them. Caution should be exercised in using these plants with individuals who have skin sensitivities. Provide latex-free medical gloves and long sleeves for handling.
- *Sharp Plants*—Plants and natural objects, such as cacti, certain ornamental grasses, rose stems, and even pine cones, have sharp edges or parts that can cut or seriously injure participants. If, after careful consideration, sharp plant materials are to be used, pointed leaf tips, thorns, or offending parts can be removed, or participants trained in their safe handling.
- *Plants That Cause Allergies*—The pollen from many wind-pollinated herbaceous plants and trees can aggravate allergies and worsen symptoms of hay fever. With input from treatment team members, a threshold for pollen count may be determined for taking participants outdoors. Some clients and staff members may have the perception that all plants cause allergies, which may be a psychological deterrent to participation in activities.
- *Plants with Strong Odors and Fragrances*—Plants that emit odors, such as roses (*Rosa*), lilacs (*Syringa*), scented geraniums (*Pelargonium*), and *Citrus*, can provide an enjoyable sensory experience for many participants and be utilized in a variety of activities. Based on cultural backgrounds, personal preferences, and experiences, individuals may have positive, negative, or neutral responses to certain fragrances. These odors can be too strong, offensive, and even nauseating for some individuals with olfactory sensitivities, including some individuals with autism spectrum disorder and those who are taking medications that heighten sensitivity to strong odors or who are undergoing chemotherapy. Plants with strong odors may cause participants to experience headaches, migraines, breathing difficulties, and vomiting, among other symptoms. Generally, however, plant fragrance is a positive asset to a horticultural therapy program.

- *Edible Plants*—The use of edible plants, such as vegetables, herbs, and those used in making teas and other consumables, is popular in many horticultural therapy activities. However, precautions should be taken when using edibles with clients with dietary and medical restrictions. Participants may have allergies to certain food items, or edibles may negatively interact with certain medicines. In addition, the growing and preparation of edible plants used in food items may be regulated per agency standards or public health codes and regulations, particularly if they are to be distributed or sold to participants, staff members, and the general public.

Potting Media—Potting media range from garden soil to soil-less blends and may be inorganic to organically based, or a mixture of both. They may also contain a variety of added fertilizers, amendments, and fertilizers. Attention must be paid to the use of various forms of potting media with certain individuals, including those who are immunocompromised or are prone to putting nonfood objects in their mouths. The use of sterilized soil-less media and gloves for handling materials is often recommended in these situations. Dust masks and premoistening the materials can prevent the inhalation of materials that produce dust, such as perlite and vermiculite. Proper storage and management of potting media can also help with infection control.

Trip Hazards—Uneven surfaces, wet areas on greenhouse or garden pathways, grade changes, garden hoses, and cords stretched across entryways are examples of potential trip hazards to consider when working with different populations. Even shadows and multicolored paving surfaces on pathways can cause unsafe situations, particularly for individuals with vision loss who could interpret the change of color as a change in grade or a hole in the ground.

Stairs and Ramps—Clients with mobility and gait issues may find it difficult to move throughout a workspace where there are stairs, ramps, and elevated surfaces. The design of workspaces should enable participants' access and freedom of movement, not limit it, and be appropriately addressed with administration and facilities staff members. Lifts, appropriately graded ramps and pathways, sturdy railings, and other features can be added to increase mobility throughout the space.

Tools—The various tools used in a horticultural therapy programs vary in their design features, uses, and sizes. They can also be used as weapons against another person or for self-harm. In selecting tools for program use, it is important to analyze the qualities that make them easily weaponized or hazardous and difficult or successful to use, such as their weight and construction materials, and characteristics of the population who will be using them.

Sharp Objects and Breakables—For individuals who have cognitive impairments, difficulty grasping objects, or potential for injury to self or others, it may be necessary to limit the availability of objects with sharp blades, such as pruners or scissors. Similarly, the use of materials that are made of substances that can be made into sharp weapons, such as glass vases, clay pots, or certain plastics, may be prohibited. Alternative tools such as safety scissors or materials that have been cut prior to the session minimize the need for tools and reduce frustration for participants. Instead of using objects made of glass, ceramic, terra cotta, and other breakable materials, the horticultural therapist employs materials that are shatter resistant or difficult to break, such as some forms of plastic, acrylic, polyresin, composite materials, concrete, foam, metal, or wood. Exhibit 11.8 offers an example of the safety considerations and restrictions of certain tools and materials used with horticultural therapy participants in a mental health setting.

EXHIBIT 11.8

Technique: Safety Considerations in a Mental Health Setting

Silver Hill Hospital is a nonprofit psychiatric hospital in New Canaan, Connecticut, which treats adolescents and adults living with a range of psychiatric and addictive disorders. They offer inpatient treatment as well as outpatient and transitional living services. The horticultural therapy program takes a therapeutic and wellness approach. Since this was a fairly new program to the hospital and each unit has different rules, there is constant communication with the direct care and treatment teams. These teams include the senior residential counselors, residential counselors, nurses, psychiatric technicians, the director of nursing, nursing supervisors, the director of social work, and a recreational therapist.

In an acute care setting, it is essential to carefully evaluate how horticultural therapy activities are conducted and the plants to be used. No strings or poisonous plants are permitted, and plastic materials must be approved by the nursing director. In the example of an herbal sachet activity, strings from the organza bags are removed. Various approaches such as sewing or double-sided tape were initially taken to secure bags closed but proved to be time consuming or were not long-lasting. Since patients are allowed to use hair bands for personal grooming, the sachets are tied with hair bands.

Since patients could be suicidal, be experiencing a psychotic episode, or have cognitive impairments, it is important to use plants that are nontoxic or, if consumed, would only cause mild stomach upset or diarrhea. There are various online and print resources to consult on plant toxicity, including from the California Poison Control (2018) and University of Wisconsin Health (2012), which offer extensive lists of plants and their varying levels of toxicity.

When determining what plastic is approved for use, check out materials already in use within the unit may be considered. It is possible to repurpose clean plastic food utensils to cultivate soil, food serving trays as plant saucers, and thick plastic pots for the plants. Strong plastic that does not break easily can be used along with peat pots. For plant propagation activities, clear plastic salad bowls, vases, or storage containers can be turned upside down to make mini greenhouses.

When working with groups, it may be necessary to regularly redirect patients who are verbalizing war stories, drug-related conversations, or thoughts of harming someone or oneself. Some individuals are triggered by pointy objects and scissors, so use child scissors with blunt ends. When there is a need for multiple scissors, such as during an individual floral arrangement group, make sure to have numerous staff members in the room to monitor use. It is helpful to have one staff member specifically assigned to watch the scissors or listening for verbalization of scissor misuse and redirect as necessary. If helpers, such as volunteers and other treatment team staff members, are not available for a session, notify the unit staff that staff members should be present in the group room in order for the group to begin the session.

Erin Backus, MS, CRC, HTR
Horticultural Therapist, Silver Hill Hospital

Moveable Objects—Objects, such as furniture, potted plants, gardening tools, and garden statuary, that are not fixed or permanent features in the workspace may serve as potential hazards for tipping over, shattering or breaking, or enabling an individual's elopement. Careful placement and use of sturdy, heavyweight objects made of high quality, shatter-resistant materials are advised. They should be secure in place or a locked area while not in use, with access limited to times when adequate supervision can be provided. A brake feature on objects with wheels, including wheelchairs, walkers, utility carts, or rolling greenhouse benches, when parked prevents unexpected movement and falls.

Storage—It may be necessary to keep program tools, equipment, plant material, and other related supplies in locked storage to limit access. Per safety protocol, agency regulations, or recordkeeping policies, the program facilitator or support staff members may be required to keep a written inventory log that is inspected and updated regularly—particularly after each session.

Hideouts and Blocked Viewing Areas—The height and density of plant materials, walls, garden sheds, furniture, and other structures should be considered when designing indoor and outdoor spaces for certain populations. It may be necessary to avoid use of or modify certain workspace features to ensure visibility from all points of view and monitor for falls, harmful or delinquent activity, and other potential risks.

Electrical Hazards—Limit access to unprotected electrical sockets and equipment, exposed wiring, and other potential electrical hazards and maintain these to proper standards. Lighting must also be properly maintained, with attached electrical cords appropriately stored and light bulbs changed out in a timely manner and kept secure. Avoid overloading electrical systems with too many plugged-in items or insufficient wiring. Watch for water near electrical features, such as grow-light systems.

Water and Watering Systems—Monitor hoses stretched out across pathways and place caution signs or brightly colored safety cones nearby. Monitor irrigation watering times and run only as needed to avoid unnecessary puddling and water runoff. Remove standing water and provide proper drainage to avoid cultivating mosquito breeding areas and slippery surfaces.

Water Features—From birdbaths and a bubbling fountain to a pond or water fountain, water features can be a central therapeutic element within a space and provide opportunities for sensory stimulation. It is important to consider the characteristics of water features, such as water depth, risk of drowning, stability, likelihood of spray drift, drainage systems, electrical wiring, and pumping systems, in relation to participants and their ability to access these features. A routine maintenance plan should be established to maintain water features to meet infection control and other safety standards, with documentation properly updated and maintained.

Cooling and Hydration Stations—Young children, older adults, and individuals with certain medical conditions or who are taking certain medications may have difficulty with regulating body temperature and become overheated or dehydrated, particularly in outdoor settings. It is advised to limit heat exposure to short intervals, use fans or misting systems, position individuals under overhead shade structures or trees, and provide hats and other wearable sun protection for participants as appropriate. Drinking water and liquids should be readily available. Communicate with fellow treatment team members, such as nursing staff, to make sure appropriate drinking liquids, such as water with thickening agents, are available for participants.

Outside Time Parameters—Special consideration to the amount of time individuals spend outdoors should be addressed with fellow treatment team members. Outdoor time may need to be monitored for individuals who have breathing issues or are immuno-compromised, prone to infection, photosensitive, or temperature sensitive, including some older adults, and individuals who have autoimmune disorders, are medically fragile, or are undergoing chemotherapy. Alternative indoor activities that complement outdoor experiences should be planned during inclement weather.

Weather—In order to prepare for a variety of considerations, the therapist should consult the weather forecast, identify a threshold for temperature and weather conditions to determine where the program takes place, and have a backup activity or alternative meeting location planned in the event of inclement weather.

Access to First Aid and Medical Staff—In health care and residential communities, it is important to have medical and other assistive staff nearby or on hand in the event of medical emergency. In other circumstances, the horticultural therapist may need to have support staff on call, as well as a first aid kit and a communication device on hand to call for help. Per agency protocol, it may also be required that all staff members and volunteers be certified in first aid techniques and properly trained on how to handle medical emergencies and use certain lifesaving equipment like a portable defibrillator device or an epinephrine auto-injector.

Chemicals and Related Substances—Household cleaners, fertilizers, and pesticides used on plants in program activities or as part of overall facility maintenance are among the manufactured products and chemicals that may be used or stored in a program workspace. Chemicals and related materials and equipment should be handled and stored according to governmental and organizational regulations and guidelines. An integrated pest management (IPM) program is often recommended as a strategy to use for insect and plant disease control, particularly in greenhouse production and garden management. In IPM, chemical use is the last resort after other nonchemical strategies are exhausted. In programs that utilize greenhouse and nursery production facilities, the program facilitator and other staff members may be required to become licensed pesticide applicators. Material safety data sheets (MSDSs) and other related records will likely be required to be maintained in a central location. It is also important to be aware of the chemical composition of materials often utilized on horticultural therapy programs, such as floral preservatives, floral foam, and art supplies like glue and paint. For example, floral foam is often used in cut flower arranging activities, but it is nonbiodegradable and often contains toxins like formaldehyde. When possible, use alternative products or techniques to complete the activities, such as floral frogs, pebbles or marbles, or tape grids in vases and other containers. Note that organic products are not necessarily less toxic than inorganic ones.

Summary

This chapter outlined a range of accommodations and adaptations for activity setup, tool use, and design features of workspaces and therapeutic environments, which can be used to promote individual engagement and independent task performance. Active listening, prompts, and other communication techniques enable the horticultural therapist to accommodate participants of various ages, abilities, and cultural backgrounds in horticultural therapy programming. An established safety protocol cultivates a safe atmosphere and provides safeguards to protect program constituents, including participants, staff members, and volunteers.

Key terms

Active listening Also called empathic listening; a technique in which one listens with full attention to another person and responds in an appropriate manner in order to create mutual understanding in a given interaction.

Biophilic design techniques Design techniques that reference the concept of biophilia, which is defined as the innate human desire to connect with nature. Biophilic design techniques involve incorporating design elements that integrate greenery, views of nature, natural light, and various natural materials and cultivate experiences of nature into the built environment.

Cultural competency The ability of a service provider, such as the horticultural therapist, to provide clients from diverse cultural and ethnic backgrounds with care and services that reflect their social and cultural needs.

Task analysis A carefully analyzed, written series of the necessary steps that must be taken in order to accomplish a specific task. The horticultural therapist uses a task analysis to determine the series of steps a client should take in order to complete a task. Task analysis is used by the therapist in activity planning and client engagement.

Universal design A series of applied principles for creating a physical space or product that is appealing and accessible to the widest possible range of people, regardless of ability, age, gender, cultural background, or preference.

References

Adil, J. R. 1994. *Accessible Gardening for People with Physical Disabilities: A Guide to Methods, Tools, and Plants*. Bethesda, MD: Woodbine House.

Alberto, P. A., and A. C. Troutman. 2016. *Applied Behavior Analysis for Teachers*, 9th ed. London, UK: Pearson Education.

American Horticultural Therapy Association (AHTA). 2015. "2015 Standards of Practice for Horticultural Therapy." Accessed September 1, 2017. http://www.ahta.org/assets/docs/ahta%20standards%20of%20practice.pdf.

Betancourt, J. R., A. R. Green, and J. E. Carrillo. 2002. *Cultural Competence in Health Care: Emerging Frameworks and Practical Approaches*. New York: The Commonwealth Fund.

Betancourt, J. R., A. R. Green, J. E. Carrillo, and O. Ananeh-Firempong. 2003. "Defining Cultural Competence: A Practical Framework for Addressing Racial/Ethnic Disparities in Health and Health Care." *Public Health Reports* 118(4): 293–302.

Broach, E., J. Dattilo, and M. Deavours. 2000. "Assistive Technology." In *Facilitation Techniques in Therapeutic Recreation*, 3rd ed., ed. J. Dattilo, and A. McKenney. State College, PA: Venture Publishing, pp. 99–132.

Brown, L., E. Callaway, W. DuBroc. et al. 2013. "Dallas Arboretum Volunteer Training Resource Manual." Document presented to the (former) Dallas Arboretum Therapeutic Horticulture program.

Buettner, L., and S. L. Martin. 1995. *Therapeutic Recreation in the Nursing Home*. State College, PA: Venture Publishing.

California Poison Control System. "Plants." Accessed March 27, 2018. https://calpoison.org/topics/plant.

Catlin, P. A. 2017. "Activity Planning: Developing Horticultural Therapy Activities and Tasks." In *Horticultural Therapy Methods: Connecting People and Plants in Health Care, Human Services, and Therapeutic Programs*, ed. R. L. Haller, and C. L. Capra. Boca Raton, FL: CRC Press/Taylor & Francis Group, pp. 37–62.

Connell, B. R., M. Jones, R. Mace, J. Mueller, A. Mullick, A. Ostroff, E. Sanford, J. Steinfeld, M. Story, and G. Vanderheiden. 1997. *The Principles of Universal Design.* Accessed September 1, 2017. https://projects.ncsu.edu/ncsu/design/cud/about_ud/udprinciplestext.htm.

Cooper Marcus, C., and M. Barnes. 1999. *Healing Gardens: Therapeutic Benefits and Design Recommendations.* Hoboken, NJ: John Wiley & Sons.

Cooper Marcus, C., and N. Sachs. 2014. *Therapeutic Landscapes: An Evidence-based Approach to Designing Healing Gardens and Restorative Outdoor Spaces.* Hoboken, NJ: John Wiley & Sons.

ErinoaksKids, Centre of Treatment and Development. 2012. "Prompting and fading." Accessed September 1, 2017. https://www.erinoakkids.ca/getattachment/Resources/Growing-Up/Autism/Applied-Behaviour-Analysis/ABA-for-Families-Prompting-and-Fading.pdf.aspx.

Fleming, L. 2013. "Recent Trends in Adaptive Gardening Tool Use in HT Settings." *AHTA News Magazine* 41(1): 12–13.

Georgetown University Health Policy Institute. 2004. "Cultural Competence in Health Care: Is It Important for People with Chronic Conditions?" Issue Brief Number February 5, 2004. Accessed August 18, 2017. https://hpi.georgetown.edu/agingsociety/pubhtml/cultural/cultural.html.

Hagen, C., and P. Durham. 1987. "Levels of Cognitive Functioning." In *Rehabilitation of the Head Injured Adult: Comprehensive Physical Management,* ed. Professional Staff Association of Rancho Los Amigos Hospital. Downey, CA: Rancho Los Amigos Hospital.

Hazen, T. 2013. "Therapeutic Garden Characteristics." *AHTA News Magazine* 41(2): 3.

Hazler, R. J., and N. Barwick. 2001. *The Therapeutic Environment: Core Conditions for Facilitating Therapy.* Buckingham, UK: Open University Press.

Kaplan, R., and S. Kaplan. 1989. *The Experience of Nature.* Cambridge, UK: Cambridge University Press.

Kavanaugh, J. S. 1995. "Therapeutic Landscapes: Gardens for Horticultural Therapy Coming of Age." *HortTechnology* 5(2): 104–107.

Kennedy, K. L., and R. L. Haller. 2017. "Working with Program Participants: Techniques for Therapists, Trainers, and Program Facilitators." In *Horticultural Therapy Methods: Connecting People and Plants in Health Care, Human Services, and Therapeutic Programs,* ed. R. L. Haller and C. L. Capra. Boca Raton, FL: CRC Press/Taylor & Francis Group, pp. 63–93.

Marcus, C. C., and N. A. Sachs. 2014. *Therapeutic Landscapes: An Evidence-Based Approach to Designing Healing Gardens and Restorative Outdoor Spaces.* Hoboken, NJ: Wiley.

Marshall, M., and J. Gilliard. 2014. *Creating Culturally Appropriate Outside Spaces and Experiences for People with Dementia: Using Nature and the Outdoors in Person-Centred Care.* London, UK: Jessica Kingsley.

Merriam-Webster Dictionary. 2017. Prompt. Accessed on August 13, 2017. https://www.merriam-webster.com/dictionary/prompt.

Rodiek, S., et al. 2010. "Access to Nature for Older Adults: Promoting Health through Landscape Design." Accessed September 1, 2017. https://www.asla.org/2010awards/564.html.

Rodiek, S., and B. Schwarz. 2012. *Outdoor Environments for People with Dementia.* New York: Routledge/Taylor & Francis Group.

Smith, R., and N. Watkins. 2016. "Therapeutic Environments." From the Therapeutic Environments Forum, AIA Academy of Architecture of Health. Accessed on September 1, 2017. https://www.wbdg.org/resources/therapeutic-environments.

Ulrich, R. S. 1992. "How Design Impacts Wellness." *Healthcare Forum Journal* 35(5): 20–25. Accessed September 1, 2017. https://www.researchgate.net/publication/13177406_How_Design_Impacts_Wellness.

United States Congress. 1998. Assistive Technology Act of 1998. Accessed August 30, 2017. https://section508.gov/assistive-technology-act-1998.

University of Wisconsin Health American Family Children's Hospital. 2012. "Common Plants: What's Poisonous and What's Not?" Accessed May 8, 2018. https://www.uwhealth.org/files/uwhealth/docs/pdf/poisonous_plants.pdf.

Vanderneut, E. 2014. "Cultural Competency vs. Cultural Humility." *AHTA News Magazine* 42(4): 17–18.

Waller, S., and A. Masterson. 2014. "Contact with the Natural World within Hospital Care." In *Creating Culturally Appropriate Outside Spaces and Experiences for People with Dementia: Using Nature and the Outdoors in Person-centred Care*, ed. M. Marshall, and J. Gilliard. London, UK: Jessica Kingsley, pp. 68–78.

Winterbottom, D., and A. Wagenfeld. 2015. *Therapeutic Gardens: Design for Healing Spaces.* Portland, OR: Timber Press.

chapter twelve

Assessment and documentation strategies for horticultural therapy intervention

Barbara Kreski

Contents

The importance of documentation

The process of using horticultural therapy on behalf of a client has a number of steps. These steps should be documented, formally or informally, in order to guide the process and to communicate to any reader the changes and progress the client makes over time. Documentation has several purposes, and the purpose may dictate the format of the documentation. Some purposes are:

- Maintain a consistently professional approach to one's work.
- Create and preserve a written, legal document of services provided.
- Illustrate and monitor client progress.
- Communicate with treatment team members, including the client.
- Fulfill regulations of third-party payers and/or federal, state, or local laws.
- Provide data for research or program evaluation and improvement.

The format of documentation may be determined by a specific organization, but these general guidelines apply universally. The record should be neat, well-organized, in black pen (if handwritten), succinct, and completed on the same day as service was delivered. The focus of the writing should be on the client's behavioral response that illustrates the conclusion regarding the client's status and functioning at that time. Avoid terms that reflect bias such as *well-behaved*, *fair*, or *lazy*. See Table 12.1 for examples of objective versus judgmental phrases.

Table 12.1 Technique: Nonjudgmental documentation (Use objective descriptions rather than judgmental terms.)

Objective	Judgmental
Listened to many suggestions before agreeing to participate	Stubborn; resistant, uncooperative
Needed three brief rests during a twenty-minute activity	Lazy; slow; manipulative
Sat apart from group	Anxious; aloof; troubled
Finished work quickly but skipped some steps	Impulsive; manic; sloppy

Electronic records are being used with increasing frequency and require training specific to the hardware and software used. Electronic records use numerical codes to represent diagnoses and phrases. They are clear and precise, but with some systems, it is difficult to express nuance.

Initial assessment and documentation

The first time the therapist meets with the client, she or he begins learning about the person, assessing the client's strengths and weaknesses, and understanding the dreams and goals of the individual. Prior to this initial meeting, the therapist should read any written records and attend any team meetings where the client is discussed. This is particularly important for understanding any medical or safety concerns prior to the first session with the client. Assessment can be done by observation of the client while engaged in a task (called a situational assessment) by noting the outcome of an assigned task (called a work sample), by a checklist that the therapist has developed to simplify and organize those observations, or by using a standardized assessment tool. One standardized assessment tool that is very straightforward and simple to administer is the General Self-Efficacy Scale, which is a ten-question test designed to assess whether one feels optimistic that he or she can cope with a variety of demands in life (Schwarzer and Jerusalem 2004). This can be useful in a number of settings where feeling positive about engaging with the world at large is a goal of a client in a treatment setting.

Developing one's own initial assessment tool to supplement a review of records and standardized tests can be very helpful and specific to the program, its participants, and its resources. Start by brainstorming what you want to know about a new client, what kind of activities are generally appropriate for beginners, and what types of issues or goals the setting lends itself to. After the brainstorming session, begin to cluster similar items together, refining and rephrasing and ultimately formatting so that the result is either a checklist or a series of questions with space to answer each. It is helpful to keep the document as concise as possible. Be sure to include a space where any safety precautions can be noted prominently. Be sure to also leave a specific spot for the date, client's name, and the name of the person completing the assessment. After becoming familiar with the document, the therapist may find that a numeric scale or a Likert scale can be developed so that each answer can be represented by a number and then graphed. This makes progress easy to assess, document, and communicate. Information to be included in the initial assessment may be gathered from a review of written records, through an interview with the client, or through observation.

Goals and objectives

The assessment leads logically to the formation of at least one goal with one or more objectives. In this text, goals are long-term and articulate something to be achieved through several steps or series of efforts. Objectives are the measurable, achievable, and realistic steps toward the long-term goal. They are to be achieved in a relatively short, specific time frame (Haller 2017).

Sometimes, there is an overarching goal of a program and all participants are working toward very similar objectives to reach it. For instance, a program may exist primarily to increase social contacts and interaction as a significant part of human wellness. A horticultural therapy program where new immigrants are welcome to have tea made from leaves harvested in the garden each week would support the goal of increasing social contacts through participation in the garden tea parties. Objectives of the participants can be measured by recording participation and length of time that participants engage if the session is open-ended. Documentation that shows increases in these numbers demonstrates that these objectives are contributing toward goal achievement.

In most instances, however, the horticultural therapist works closely with the client to formulate an objective that is compatible with the needs and interests of the client. Writing the objective down in a structured format helps ensure that it is clear, concise, and meaningful. One format that is often used is the SMART goal. Each letter reminds the writer of a component of a well-written statement, and these components also apply when writing objectives. (Haller and Capra 2017).

- S stands for *specific*. Focus the objective on just one thing. Remember that it is not possible or helpful to try to cover every possibility in one statement.
- M stands for *measurable*. Make sure the objective includes a unit of measure. For example, number of repetitions, number of steps in the instructions, time on task, or incidence per session. This will make it easy to assess baseline and progress.
- A is for *attainable*. Think about the conditions for the client, including his or her physical and mental health, culture, and stage of life. Be sure the objective can be achieved given these conditions as well as the interest and motivation of the client. For example, if the client has significant dementia, the objective should be in terms of retention of skills rather than developing new ones that depend on memory.
- R stands for *relevant*. The client should care whether she or he reaches the objective. The objective should relate to what the client needs or wants to do.
- T stands for *time*. Define the time that achieving the objective is estimated to take. This should be in terms of how much horticultural therapy intervention will take place. Achieving the goal in one month may mean just four sessions of one hour every week. Or it could mean sixteen one-hour sessions every weekday in a month. It is important to clarify.

Clarity and brevity are of utmost importance when documenting the status of a client. It is important to consider who will be reading the documentation. Readers will be interested in whether what has been done with and for the client has had an impact on the client's health or functioning. They are not interested in the specifics of the therapy choices (e.g., type of plant) but will focus on measurable markers of progress. Remembering this can help in forming goals and objectives. Making them clear, measurable, and relevant to the client will lead logically to documentation that delineates progress. Examples of objectives that have been rewritten to follow the SMART format can be found in Table 12.2.

Table 12.2 Technique: Using the SMART system to write objectives

Not SMART	SMART
Joe will water the plants daily.	By [date], Joe will use a half-gallon pitcher to water each of sixteen hanging baskets without stopping for a break for three consecutive days.
Dan will improve his fine motor skills.	After participating in four horticultural therapy sessions, Dan will use his thumb and first two fingers to pinch spent flower blossoms from pansies in the container garden without verbal or physical reminders of how to position his hand.
Marie will learn to focus on her work.	By [date], Marie will place one sunflower seed per cell in a sixty-four-cell tray with no errors in twenty minutes.

Collecting data

A well-written objective can also help determine how to record progress. These steps can help:

- Distill one key function or action that is essential for the client to achieve in order to lead to a stated goal that involves some aspect of an improved quality of life.
- Decide on a few key markers that show the client's progress.
- Estimate how many sessions it will take for the client to achieve the first marker of progress.
- Determine a measurable action that reflects the status of the ability of the client to function in a way that is closer to the objective. Examples: number of social interactions per session, number of steps in written directions followed independently, number of seedlings transplanted, length of time able to work without a rest.
- Develop a system to measure progress at regular intervals. This can be a simple way to count what is being measured with tally marks. Be sure to note the beginning and ending time of the observation.

Table 12.3 demonstrates one format that therapists can use for their own records to organize information about each client. The form includes the goal, the objective, and the

Table 12.3 Technique: Measuring progress

Client name: John Doe Age: 58 Work: Bookkeeper	Goal: John Doe will be able to return to work for four hours/day within two months. Precautions: Diabetic; watch for adverse reactions to changing blood sugar levels

Objective: John Doe will show better concentration by working at a multistep task for fifteen minutes without a break by his fourth HT session.

Date of session	Start time	Number of breaks (pause in work of more than 20 secs.)	End time	Objective achieved?
	10:32	IIII II	10:44	No
	10:44	IIII I	11:02	No
	10:34	III	10:49	No

measurement that will be taken at regular intervals to document progress. Using tally marks to record observed instances of the target behavior is an efficient way to record measurements.

Exhibit 12.1 shows a form that was created by a therapist to record information about a particular session as well as some of the interactions of the group members. This would

EXHIBIT 12.1

Technique: Group Documentation

An adaptation of a therapist-created form (Backus 2015):

Location: Green Chimneys—Boni-Bel	Dorm/Class: Starzyk	Date/Time: March 24, 2015
Staff: Ms. Erin, Ms. Deirdra, Charley, Ms. V, Ms. Starzyk		Program: Maple syrup products

	K	S	Ki	N	W	R	Comments:
Participated in program	x	x	x	x	x	x	Everyone participated
Helped others	x	x	x	x	x	x	Worked together to dry molds
Followed directions	x	x	x-	x	x-	x	
Carried out activity without complaining	x-	x	x	x	x	x	
Asked appropriate program-related questions	x	x	x	x	x	x	
Was safe with tools	x	x	x	x	x	x	Wooden spoon, hot syrup, molds, stove pots
Demonstrated proper horticulture tasks							
Learned something new	x	x	x	x	x	x	Sugar on the snow
Showed appropriate behaviors	x	x	x	x	x	x	

Notes: This was the first session back at the Boni-Bel location. **W and Ki:** went with Charley and Ms. V to help move firewood and fix a pipe on the Sugar Shack, then they joined us in the back room for the candy program. They were both energetic and helped with the extra tasks. **Ki** correctly answered questions about each individual step of candy making process. **S:** did five of seven tasks that were asked of him; he moved about the room rather than sitting during the short video. **N:** worked under supervision of Ms. Deidra and accomplished the three tasks assigned to him. **R:** interacted with peers by cooperating on two tasks, using both verbal and nonverbal communication. **K:** waited until given individual encouragement to engage in tasks. He stated that he was tired but successfully executed a responsible task on his own. He also advocated for himself, stating his preference for work location. Consider fostering more initiative for initiating tasks among participants in future sessions.

Erin Backus, MS, CRC, HTR
Horticultural Therapist, Silver Hill Hospital

not be entered into a client's record but rather serves to track information. Individual performance and notes could be transferred into each person's file. It also may help the therapist recall details of the session itself.

In order to use documented data for a retrospective research project, it must be in a format commonly used by researchers. One such format that holds promise for horticultural therapists is the World Health Organization Disability Assessment Schedule (WHODAS). This instrument is relatively brief (twelve to thirty-six questions), can be read to a client or self-administered, and has validity and reliability across cultures. The schedule rates function across six domains:

- *Cognition*: understanding and communicating.
- *Mobility*: moving and getting around.
- *Self-Care*: hygiene, dressing, eating and staying alone.
- *Getting Along*: interacting with other people.
- *Life Activities*: domestic responsibilities, leisure, work, and school.
- *Participation*: joining in community activities.

It is sensitive to changes in function that may occur after intervention, so it is appropriate for use as an initial assessment as well as an intermediate and final assessment of progress (Üstün et al. 2010).

Writing an initial note

When therapists have completed an initial assessment, they document that assessment in an initial note. This note should include the time period and date of the initial assessment; the assessment tool used, if any; notation of any precautions or coexisting conditions that may affect treatment; and the goal statement. In some organizations, the entire assessment document becomes part of the client's permanent record. More often, the therapist documents the conclusions and the basis for those conclusions rather than entering every answer on an assessment tool into the record; for example, "The client indicates that he has no difficulty using his fingers in fine motor tasks per WHODAS 2.0; however he was unable to turn on a faucet independently due to weakness of his hand." The note must be signed (Waldon 2016).

Progress notes

A progress note is written at regular intervals during the time a client participates in horticultural therapy. This type of note documents progress or lack of progress that the client is making toward a stated objective. In many health care settings, a progress note must be written into the client record each time a client is seen for horticultural therapy. Other settings, such as vocational or prevocational programs, require less frequent documentation; the therapist should comply with the standards of the facility. Frequency also depends on the type of objective, as some may require a periodic record of behavior at structured times throughout a horticultural therapy session or day of vocational training. A progress note should be written in reference to the goal statement and the objective. Exhibit 12.2 is an example of a progress note. Additional examples may be found in *Horticultural Therapy Methods: Connecting People and Plants in Health Care, Human Services, and Therapeutic Programs*, Second Edition, by Rebecca Haller and Christine Capra (2017).

EXHIBIT 12.2

Technique: Progress Note

Date___: John Doe has participated in three horticultural therapy sessions in one week, working toward his goal of increasing his concentration. John Doe has improved from requiring seven breaks while engaged in a task for twelve minutes to requiring three breaks while engaged in a task for fifteen minutes. He is on track to achieve his goal of zero breaks during his fourth, twenty-minute session.

Mary Smith, HTR

An important detail of formatting is to leave no blank spaces. Instead, draw a single line through that space so that nothing can be added at a later date. To correct an error, draw one line through and initial, such as: ~~mistake~~ MS (AOTA Official Document 2013).

Instead of, or in addition to, actual narrative notes, the record of progress toward goals and objectives may be recorded in chart form, using yes or no, scales, or other methods to indicate actions of the client. Yet another way to ascertain progress is to use the initial assessment tool once again for *reassessment* in order to compare results. Note whether improvement has occurred. If it has not, the therapist should work with the treatment team and the client to decide whether there will be a change in the approach or a change in the goal, and then record that modification to the treatment plan.

Outcome or discharge notes

Outcome notes are written at the time the goal is accomplished or when the client stops participating in horticultural therapy. Observations or reassessment indicating progress toward goal achievement should be recorded. If the client is leaving the facility or horticultural therapy for any reason, this becomes the discharge note, whether the goal has been achieved or not. If the therapist has discussed recommendations with the client regarding the use of any horticultural therapy techniques in the future, this should also be recorded (American Speech-Language-Hearing Association 2017).

Some settings do not maintain written records on individual clients. This may be the case in an informal, community-based wellness group focused on maintaining social contacts and/or mobility. In other cases, another staff member is responsible for individual documentation, and the horticultural therapist is responsible for conducting the session. In these cases, the therapist should create her or his own documents in order to plan the most effective sessions that incorporate knowledge gained from past experiences.

Another reason for keeping records is to create a collection of data that may prove useful in grant requests, research, defending the existence of a program, or articulating the value of a program. Also, these notes allow for clear and accurate communication regarding the events of a session if questioned later by a client, family member, or staff member. The practice of writing regular notes assists the therapist to be a reflective practitioner. Taking time to note successes and disappointments in working with clients is critical to growing therapeutic skills. Reviewing documentation over time can reveal trends that may not be clear otherwise (Figure 12.1).

Figure 12.1 Recording client actions. (Courtesy of Rebecca Haller.)

Documentation and client privacy

Various laws determine how written records are kept, how long they are stored, and who can read them. Since these laws are modified over time and different laws exist in different countries, it is important for each therapist to know and comply with the laws that pertain to his or her work. If that work is based within any type of facility or broader program, the therapist can ask for training and adhere to the protocols of that program. In the United States, therapists should inform themselves and adhere to the Health Insurance Portability and Accountability Act (HIPAA) (Health and Human Services 2017).

Ethical practice everywhere recognizes that it is important to respect the client's privacy and never to share anecdotes, issues, or even the fact that a client is participating in horticultural therapy with anyone outside the team working together on the client's behalf. It is the choice of the client or her or his guardian whether and with whom to discuss any aspect of her or his life.

In practice, safeguarding client privacy means:

- Never discuss a client in a public place or with anyone who is not part of the treatment team.
- Lock written records in a drawer or storage area when they are not in use.
- Do not photograph a client without written permission.
- Be careful that no identifying information is included in any social media or electronic post, including email.

Maintaining privacy is part of treating each person with dignity and respect, which is the basis of an effective therapeutic relationship.

Summary

Writing down an initial assessment, a goal and objective, steps showing progress, and the final status of a client is an essential discipline. This documentation helps to focus and facilitate the work of both the client and the therapist. It is a way to communicate with other team members. It provides a different perspective that enables therapist to think critically about their work and strive to improve. Finally, written records are valuable sources of data to support grant requests or reports, and to attest to the effectiveness of a program or to inform a research study.

Key terms

Goal A long-term change in behavior or skill that may be accomplished by achieving component steps.

Likert scale A questionnaire designed so that the response is selected along a scale. The format of a typical five-level Likert item is, for example: (1) Strongly disagree, (2) Disagree, (3) Neither agree nor disagree, (4) Agree, (5) Strongly agree.

Objective A statement of a measurable behavior or skill to be achieved within a specific time period.

Situational assessment A systematic observation and recording of work behavior and performance done as the subject completes a task in an actual or simulated environment.

Standardized assessment tool A method of determining comparable skill or knowledge where all subjects answer the same questions and the results are rated by the same system.

Work sample An example of the final product after an attempt to follow prescribed procedures.

References

American Speech-Language-Hearing Association. 2017. "Documentation in Health Care." *Practice Portal, Professional Issues*. Accessed July 6, 2017. https://www.asha.org/PRPSpecificTopic.aspx?folderid=8589935365&section=Overview.

AOTA Official Document. 2013. "Guidelines for Documentation of Occupational Therapy." *American Journal of Occupational Therapy* 67(Supplement 6). Accessed May 8, 2017. https://ajot.aota.org/article.aspx?articleid=1853060.

Backus, E. 2015. "Group Program Evaluation. Green Chimneys Program." Document shared upon request, September 25, 2017.

Haller, R. L. 2017. "Goals and Treatment Planning: The Process." In *Horticultural Therapy Methods: Connecting People and Plants in Health Care, Human Services, and Therapeutic Programs*, 2nd ed., ed. R. L. Haller, and C. L. Capra. Boca Raton, FL: CRC Press/Taylor & Francis Group, pp. 27–36.

Haller, R. L., and C. L. Capra. 2017. *Horticultural Therapy Methods: Connecting People and Plants in Health Care, Human Services, and Therapeutic Programs*, 2nd ed. Boca Raton, FL: CRC Press/Taylor & Francis Group.

Health and Human Services. 2017. Health Information Privacy. accessed January 25, 2018. http://www.hhs.gov/hipaa/index.html.

Schwarzer, R., and M. Jerusalem. 2004. "General Self-efficacy Scale." In *Compendium of Quality of Life Instruments*, ed. S. Salek, (Vol. 6, Section 2A:1). Cardiff, Wales: Centre for Socioeconomic Research, Cardiff University and Haslemere, UK: Euromed Communications.

Üstün, T. B., N. Kostanjsek, S. Chatterji, and J. Rehm. 2010. *Measuring Health and Disability: Manual for WHO Disability Assessment Schedule (WHODAS 2.0)*. Geneva, Switzerland: WHO Press, World Health Organization.

Waldon, E. 2016. "Clinical Documentation in Music Therapy: Standards, Guidelines, and Laws." *Music Therapy Perspectives* 34: 1.

chapter thirteen

Tools for program management

Emilee Vanderneut

Contents

Horticultural therapy is a dynamic, client-centered field that can be seamlessly integrated into nonprofit and private organizations, compliment the efficacy of traditional treatment interventions, and offer clients empowering accommodation and access to the natural world. This chapter discusses horticultural therapy program management and introduces the tools that will facilitate the effective coordination of programs in a large variety of settings. Ultimately, the people served will benefit from the skillful application of these methods if the program is supported, well-integrated into the organization, and sustained through changes in administrations and circumstances.

This chapter will discuss:

- The unique management role of a horticultural therapist
- How to develop program goals
- How to manage staff, volunteers, and interns

- Management and safety of programming sites
- Effective marketing considerations
- Strategic program evaluation
- Budgeting considerations

Program management: Roles and functions

Managing a successful horticultural therapy program is a complex and dynamic process. Regardless of job title, virtually every horticultural therapist needs management skills for success. The manager of a horticultural therapy program is typically tasked with coordinating the interplay between the clinical and organizational demands of the program while at the same time navigating the aesthetic, logistical elements of the physical programming site. Horticultural therapists play a crucial role and can bridge a horticultural therapy program with its desired outcomes and performance goals. As in all organizations, the primary function of program management is to facilitate the process of developing and implementing the overarching mission and goals of a program. Note that the word *facilitate* was used instead of the word *create*. The word *facilitate* suggests that someone is guiding and providing a venue for a collaborative conversation rather than creating something independent of outside influences (Grange 2008). Horticultural therapists are typically the central hubs of communication for their programs and are responsible for ensuring that staff members, volunteers, interns, clients, and the surrounding community are working toward program goals as a unified team. It is important to communicate the function and value of the program to the outside community while keeping lines of communication open internally within the program. In addition, horticultural therapists are responsible for organizing and evaluating the program so that it ensures an effective delivery of services to clients. Program managers monitor the program budget and make financial decisions based on program goals and mission (Lewis et al. 2012).

One of the unique roles that a horticultural therapy program manager assumes is the role of educator, both inside and outside the organization in which the program exists. Providing information about the purpose and implementation of the emerging field of horticultural therapy bolsters the success of a program while making services accessible to various populations. Educating internal and external parties involved with the program is a crucial technique for gaining investment and comprehension of how a horticultural therapy program can benefit specific target populations. Education is also often one of the best tools that a program manager can employ when attempting to secure program funding. For example, potential funders with knowledge of horticultural therapy–specific research statistics, case studies, and narratives make informed decisions on the impact that programs have on target populations. Financial investment is related to awareness that organizations with horticultural therapy programs have benefited from the intervention on therapeutic, aesthetic, and public relations levels. Horticultural therapists can also inform potential funders and organizations on the merit of maintaining a flexible approach to program management. Adapting to changing client demographics and the priorities of the organization and funders is a crucial element in maintaining a program's longevity, relevance, and effectiveness through time (Johnson et al. 2003). Exhibit 13.1 describes how a long-standing horticultural therapy program has successfully adapted to changes over the years (Figure 13.1).

EXHIBIT 13.1

Perspective: Program Continuity and Sustainability

For more than fifty years, the Horticultural Therapy program at Rusk Institute of Rehabilitation Medicine at New York University's Langone Health Medical Center pioneered programs and techniques to improve the lives of a wide range of individuals, both in a clinical setting and in the larger community. In 2012, Hurricane Sandy laid waste to the gardens, greenhouse, treatment facilities, and equipment. Many observers assumed that was the end of the horticultural therapy program at Rusk.

In fact, the program continues, with many of the same staff and most of the programs ongoing. Obviously, the ability to adapt to the loss of the gardens and physical infrastructure was critical. But equally important in ensuring continuity of programming was the unwavering support of hospital administrators, clinicians, patients, and community groups, which recognized the value that horticultural therapy contributed to helping them achieve their own goals.

It's hard to plan for a hurricane, but many other factors can cause a disruption or termination of even the best programs: Program staff and organization managers can change; new priorities and objects can be adopted; budgets can be cut; staff can be downsized; program hours can be cut; or garden space can be repurposed for new, potentially more profitable, uses. During my tenure as director of horticultural therapy at Rusk Institute, every effort was made to build bridges with important constituencies inside and outside the medical center and to keep these constituencies up to date on what we were doing and achieving. We knew our work was important, but we had to demonstrate that value to those whose support we needed.

Engagement was one way we built support. We invited families to participate in our children's activities, worked with preschool and high-school teachers to use our facilities and integrate our expertise for their classes, and sought out community leaders to identify needs our services could help meet. We encouraged nurses, doctors, and other therapists to observe or take part in patient horticultural therapy activities. We became a horticultural resource for the institution and community.

Outreach was another way to build support. We offered internships to horticultural therapy students and created hands-on vocational training curricula for community organizations working to place people with disabilities in horticulture jobs. Our staff taught in certificate programs at New York and Chicago Botanical Gardens. We opened our Children's Garden to local families, whose kids played alongside the young patients of Rusk. And we held a special community festival each year that was hugely popular. Health care organizations are often isolated from their communities; the garden can become a bridge to bring the two together.

In addition, we made sure our expertise was known and appreciated. Our staff members undertook research projects in partnership with other clinicians to quantify or otherwise demonstrate the benefits patients received from taking part in horticultural therapy. These findings and other articles penned by our staff were published in professional journals and often presented at conferences.

Yes, this is a lot of work. And yes, we probably had more staff and resources than many horticultural therapy programs. But a small team or even a single practitioner can do at least some of the above to build support and help ensure her or his program

(Continued)

EXHIBIT 13.1 (Continued)

Perspective: Program Continuity and Sustainability

continues beyond any change in personnel or other disruption. The key is clearly defining the value the program brings to the wider organization or community and communicating that value to key decision makers.

Nancy K. Chambers, HTR
Director Glass Garden and Horticultural Therapist, Retired,
Rusk Institute of Rehabilitation Medicine, NYU Medical Center

Figure 13.1 John Murphy, Director of Bullington Gardens, explains the horticultural therapy program to visitors. (Courtesy of Christine Capra.)

Many horticultural therapists have strong horticultural backgrounds, recognizing the importance of collaboration and curiosity as well as of flexibility and resilience in the face of unknown factors such as weather and seasonal variables. These lessons and experiences hold deep value and are the cornerstone of horticultural therapy theory and practice. It is only fitting that the values that guide horticultural therapists' work with clients also guide their style of leadership and management in horticultural therapy programs. Program management is much like tending a garden: Consistency and persistence are key.

Communication tools for program management

Open communication is a program management tool that provides an opportunity to gain insight into the needs and strengths of a program's stakeholders. Strong communication encourages personal investment in the program and helps to shape the manner in which everyone will move toward their common goal as a team (Johnson et al. 2003).

The principle of maintaining open communication can also be applied to the program development stage of a horticultural therapy program. Program development paired with an atmosphere that allows for client input and engagement results in a positive outcome for all parties involved. Communication with the client population eliminates unfounded assumptions about client needs while therapeutically incorporating the client into the program development process. When paired with current research and evidence-based practices, this combination can be used to address the needs of a wide variety of populations.

Consider the difference between the potential engagement and attendance rates in the following examples. A horticultural therapist has been hired to start a horticultural therapy program that will provide relevant services to individuals who have experienced a spinal cord injury causing paralysis from the neck or waist down. The horticultural therapist can visually see that the group of clients that require services have physical challenges associated with their diagnoses because they all use wheelchairs. The clients provide preliminary input regarding the new accessibility challenges that they face in their homes and in the community. Which of the following examples demonstrates the program management skill of open communication?

Example 13.1

The horticultural therapist designs a stellar horticultural therapy program that rigorously targets the specific physical health and accessibility needs of the population. A therapeutic garden is created amid a terrain park where the participants of the program can practice real-life scenarios that they will face in the community with their wheelchair—curbs, gravel, slopes, etc. The garden activities that are employed are designed to bolster the participants' confidence levels and encourage the patients to find new ways to accommodate and adapt to their new circumstances.

Example 13.2

The horticultural therapist assembles a focus group of individuals who have spinal cord injuries and who live in the community that the program will serve. They co-create a list of needs and desires that they would like the program to address. Through the process, the horticultural therapist discovers that the top two identified needs are (1) the clients are concerned about their accessibility challenges as it pertains to reentering the workplace, and (2) the intense grief they experience as a result of being seen as disabled in the eyes of their families and the community as a whole.

Example 13.2 best exemplifies true open communication as it pertains to program development and client need. In this second scenario, the horticultural therapist worked harder to gain the insight that the clients had to offer than the therapist did in Example 13.1, but the hard work ensured the relevance of the services that would be developed.

Client engagement and attendance are two major factors involved in securing and maintaining funding for a program. Funders want to know that clients are invested and benefitting from a program before they provide funds. The program in Example 13.2 captures the investment of the target population and in turn provides a better opportunity for the program manager to provide relevant interventions; it also demonstrates the programmatic value to funders. (See Chapter 14 for more discussion of using evidence-based practice for program development.)

The skill of active listening and unmitigated curiosity often go hand-in-hand and is a program management tool that most successful program managers use daily. Active listening is more than just listening to the words being spoken, it requires truly being curious and committed to the message that a person is trying to convey. One can ensure that he or she is actively listening by summarizing, asking questions, and clarifying what was said (Rogers and Farson 2015).

When a horticultural therapist is asked to provide services for an organization, it can be very helpful to first determine why the organization is pursuing horticultural therapy and how it envisions the intervention being implemented into its programming. Organizations may seek horticultural therapy services because they need or want to increase billable hours, get a warranted public relations boost, or satisfy an unmet need that their current programming does not provide. These motivators will be foundational information points throughout the program development process, but they are mere starting points for how the program will be realized through time. The role of the horticultural therapist is to broaden the agency's understanding of what horticultural therapy is and how it can be applied into a specific setting. Therapists expand on the organization's vision and describe the specific, evidence-based benefits that will be provided to the clients served and the organization's site as a whole. This is a time in the program development stage when the definition, efficacy, and legitimacy of horticultural therapy is communicated. From this point, the horticultural therapist asks a series of detailed questions that pinpoint the type of services required, and the professional methods in which those services will be delivered. Active listening facilitates the act of two entities mutually working toward a goal as a team.

Some of the basic questions that can be helpful throughout this mutual exchange during program development are as follows:

- What does the organization hope to gain from introducing horticultural therapy into its current programming?
- What are the general demographics (age range, gender, race/ethnicity, education level, income, etc.) of the population the organization serves?
- What is the most common diagnosis, co-occurring diagnoses, or general challenges of the patients served?
- What strengths, benefits, and support systems do the clients possess?
- What services does the organization currently provide for clients?
- How does the organization assess and track the progress that clients make through time?
- Does the organization have a gap in services that could be bridged by horticultural therapy? If so, ask for details about the gap.
- How does the organization view the importance of family and community engagement through the treatment process? What are the current policy and practices on family/community engagement?

Consider the following example. A horticultural therapist has been hired by an organization whose website and marketing materials advertise that it serves clients who have the primary diagnosis of a traumatic brain injury (TBI). The therapist has done as much research about the organization as possible from an outside perspective and is expecting (assuming) to work with clients who have been involved in severe accidents, combat, etc., that caused extreme injury to the brain. However, when the horticultural therapist meets with administrators, he or she learns that 20% of the clients who are receiving services have had at least one limb amputated as a result of an accident or military combat. Based on limited information on the agency's website, the inclusion of patients who have limb amputations would not have been initially assumed by the horticultural therapist. The amputation of one or more limb, paired with the effects of a traumatic brain injury, causes the horticultural therapist to reconceptualize what services will be required. The organization explains that originally it hadn't expected such a large percentage of their clients to be amputees but found that an increasingly large gap in available inpatient services existed for patients with these co-occurring needs. In an attempt to bridge this gap, and to extend the financial longevity of its program, it adapted to the needs of the market and planned to update their marketing materials to reflect these changes in services rendered.

This information would greatly affect the design and management of a horticultural therapy program because these amputation-related challenges add an expanded range of physical and psychological factors, which will require accommodation by way of revised activities; techniques; and special adaptive tools, artificial limbs, and/or wheelchairs. As with many other programs, horticultural therapy programs need to continually adapt and shift based on the varied and diverse needs of clients. Similarly, they must adapt to changing demographics of the populations served by an organization. Open communication strategies include talking with other treatment and caregiving staff members; observing current programming; and, if appropriate for the setting and length of stay, requesting detailed information about the specific participants. Good communication strategies are critical in ensuring that the organization's and ultimately the clients' needs are met.

How employment models affect program management

In the United States, horticultural therapy employment tends to fall within one of three models, with the therapist acting as an employee, contractor, or consultant. The employment model in which a horticultural therapist works has an impact on the manner in which program budget, marketing, staff, volunteers, interns, and overall therapeutic site maintenance is managed.

Horticultural therapists who work as *employees* tend to have overhead program and labor costs covered by the site's general operating budget. In this scenario, the horticultural therapist might be responsible for managing and reporting the financial resources that exist above and beyond general operating costs. Regardless of the source or type of funding that is secured, horticultural therapists almost invariably play a role in collecting, assessing, and distributing the data required to maintain continual financial resources or gain new funding for the program. Horticultural therapists who are employed by an organization may share the responsibility of training and managing volunteers and interns with the human resources department. In this employment model, the horticultural therapist works as a member of the larger team of professionals and is required to adhere to policies and support the mission as a whole. This model requires a great deal of interdepartmental communication and coordination to ensure continuity of high-quality services.

Horticultural therapists who work on a *contractual* basis require many of the same management tools that those serving as an employee need. In the contract employment model, the terms of the contract dictate the management style that the horticultural therapist employs. If the horticultural therapist has a long-term, ongoing contract, she or he might have quite a bit of influence over how the budget is created and managed through time, as well as how volunteers and interns are managed. If the contract is short term and/ or occurs only on a seasonal basis, the horticultural therapist has the opportunity to influence the organization on how the budget and marketing of the program might best be carried out, but the organization ultimately assumes responsibility and control over those elements. The short-term nature of this type of contract also affects the type of program that is created and potentially what kind of program and treatment goals are targeted.

Horticultural therapists who provide *consultation* services typically visit an organization for one or more consultations, meet with a variety of staff members, and provide advice about starting or managing a program based on their expertise in the field. They also may be asked to develop horticultural therapy treatment space. Of course, any recommendations that are made by the consultant about program management are subject to the organization's approval. A consultant might provide guidelines and templates for the staff to work within and then help monitor the staff's progress through time. In this case, the consultant may be helping to establish a therapeutic horticulture program if there is not a horticultural therapist on staff.

Collaboration with an interdisciplinary teams

In a horticultural therapy setting, an interdisciplinary team is a diverse group of professionals who work together to effectively assess, treat, empower, and collaborate with their clients to reach measurable treatment goals. Horticultural therapists that are employees may collaborate with interdisciplinary teams that consist of professionals such as social workers, doctors, rehabilitation therapists, psychologists, psychiatrists, vocational rehabilitation counselors, occupational therapists, and administrators. In many settings, key participants on the team are the client being served as well as family members. Sometimes, a horticultural therapy contractor, on the other hand, is not an official member of the treatment team and may not be allowed to access confidential client/patient records. However, treatment outcomes and direct care observations that are made by contractors in a horticultural therapy session can be shared and used by a treatment team as needed. For the most effective programming, it is advisable to seek contractual agreements that allow access to records and encourage open interdisciplinary communication and collaboration.

Communication with treatment teams conveys client progress and challenges, provides an opportunity for the team to see unique responses often found in horticultural therapy treatment, and facilitates continuity and collaborative opportunities. In addition, working on a diverse team presents an incredible opportunity to develop the breadth and depth of one's clinical skills and provides a well-rounded perspective and approach to evidence-based, client-centered treatment. Each team member has a different priority in regard to the treatment of clients, which presents a benefit to the management of a program (Grange 2008). A psychologist may have excellent assessment tools on hand as well as the skills and credentials to interpret them accurately. A social worker may have a large network of relevant service providers within the community to which he or she can connect clients seamlessly. Taking advantage of the expertise of an interdisciplinary team is a productive use of time and an excellent way to efficiently meet the needs of clients that extend beyond the range that horticultural therapy is meant to reach.

One major element of being effective on an interdisciplinary team is being willing to actively listen to others while also contributing one's own voice in the interest of the client's success. It is important to build meaningful relationships with the key players on the team and cultivate those relationships just as one would a garden—consistently and strategically. Working with a team of professionals is a complex balancing act of person-alities, treatment philosophies, and personal and professional motivators (Gamble and Gamble 2013). It is customary for all team members to keep the entire team abreast of the client's progress by documenting and evaluating gains and obstacles to achieving treatment goals throughout the treatment process.

Developing program goals

It is important to note that program goals and treatment goals serve two very different functions but often get mistaken for each other. Program goals should be consistent with the overarching mission of the program and specifically address the desired impact on the participants served, the financial goals, and the program's long-term vision. Program goals are developed to guide the outcomes, systems, and finances of the program in which the treatment team functions. In contrast, treatment goals are developed to address the specific target behaviors that clients identify as areas that require growth. The following are a few examples.

Program goals and objectives that address the desired impact on clients

Vocational horticultural therapy program goal: To empower clients with the skills required to gain employment in the landscaping industry.
  Objectives:

- Within a calendar year, two classes of fifteen clients will complete a fifty-hour voca-tional landscaping curriculum and physically demonstrate mastery of skills by the ninetieth day of each three-month course.
- At least ten students per year will secure a paid landscaping position in the commu-nity within ninety days of graduating from the program.

Wellness horticultural therapy program goal: To facilitate social opportunities within a garden community
  Objectives:

- Three "Garden Buddy" one-hour sessions will be held for adults over the age of sixty-five years old each week for the duration of twelve weeks.
- Market "Garden Buddy" volunteer opportunities through social media outlets once a month for four months to develop the participant and volunteer pool and to increase community engagement.

Therapeutic horticultural therapy program goal: To support the clients' right to engage in meaningful, ability-appropriate physical activity.
  Objectives:

- Administer an initial ability assessment by a registered horticultural therapist within three days after clients attend program orientation.
- Provide a minimum of one opportunity per day for clients to participate in horticul-tural therapy programming in the enabling garden or greenhouse to improve ambu-lation and balance.

Program goal that addresses finances

Program goal: To be financially self-sustaining and independent of grant funding in five years.

    Objectives:

- Perform an audit of current program expenditures and overhead costs by January 30 and June 30 annually.
- Increase program marketing efforts within the next six months by:
  - Contacting ten referring agencies to increase the number of clients in the program.
  - Making at least three social media updates a month.
  - Holding an informational open house to showcase the program.

Program goal that addresses long-range vision

Program Goal: To diversify and increase the number of programs offered to clients.

    Objectives:

- Expand program offerings to include a group that addresses children.
- Add an outpatient program to the mix of current inpatient services in horticultural therapy.
- Hire an additional horticultural therapist to work with outpatient clients.

A logic model is a helpful tool to explore and articulate program goals and facilitate planning of resources, inputs, and intended outcomes. It assists in determining the allocation of financial and human resources to support and achieve most efficiently the mission of the program. Logic models provide a map by which to monitor and track how financial resources are affecting the clinical and operational elements of a program and are also useful in budgeting and program evaluation. Exhibit 13.2 describes the logic model and its use. Exhibit 13.3 provides an example of its organizational and logistical use in a therapeutic horticulture program in the United States, outlining a program that is managed by a trained horticultural therapist and executed by a team of Master Gardeners. Table 13.1 is an example of using a logic model to explain the therapeutic horticulture program at a veterans hospital in the United States.

Internal and external communication

The success of any type of horticultural therapy program depends on professional, mission-driven internal communication and external public relations. Internal communication involves any interaction that takes place between representatives of the program and the individuals who work at the site where the program is being conducted, ranging from the clinical staff to the administrators of a site. An example of a relationship that typically requires a lot of internal communication is the interaction between the grounds department and the horticultural therapy program. The therapist may need to coordinate and follow up on the delivery of gardening materials, manage watering schedules, or report facility safety issues that have arisen that the grounds department is responsible for addressing.

EXHIBIT 13.2

Techniques: Using a Logic Model

The better we design and manage a horticultural therapy program, the better we can serve our clients and enjoy long-lived treatment outcomes and program success. Just as a landscape design shows the physical layout of the plants, hardscapes, and water sources in a garden, a logic model is the road map for the inputs, activities, and measurable impacts that can occur in that space.

A Logic Model

- Is a simplified picture of a program, initiative, or intervention that is a response to a given situation.
- Shows the logic relationships among the resources that are invested, the activities that take place, and the benefits or changes that result.
- Is the core of program planning, evaluation, program management and communications.

Planning

A logic model serves as a framework and process for planning to bridge the gap between where you are and where you want to be. It provides a structure for clearly understanding the situation that drives the need for an initiative, the desired end state, and how investments are linked to activities for targeted people in order to achieve the desired results. The model can connect the garden activities to the targeted outcomes for the client or population and ensure alignment with the goals and mission of an organization.

Program Management

A logic model displays the connections between resources, activities, and outcomes. As such it is the basis for developing a more detailed management plan. During the course of implementation, a logic model is used to explain, track, and monitor operations, processes and functions. It serves as a management tool as well as a framework to monitor fidelity to the plan.

These models can help identify the shovels, seeds, watering cans, and other supplies needed for a targeted audience to successfully engage in the planned gardening endeavors. Also, it can help to map garden maintenance activities into programmed activities as part of the sustainability plan for the garden.

Communications

Communication is key to the success and sustainability of a program. Often communicated as a simple and clear graphic representation, as seen in Figure 13.2, a logic model helps communicate some details about programs or initiatives, whether it be to program staff members, donors, or key stakeholders.

(Continued)

EXHIBIT 13.2 (Continued)

Techniques: Using a Logic Model

Program: Veterans Hospital Garden Program Logic Model
Background Information: Green Cities: Good Health- Stress, Wellness & Physiology; Green Cities: Good Health- Healing & Therapy; Barnicle, HortTechnology 2013

| Inputs | Outputs | | Outcomes -- Impact | | |
	Activities	Participation	Short	Medium	Long
Master Garden Volunteers					To fulfill President Lincoln's promise "To care for him who shall have borne the battle, and for his widow, and his orphan" by serving and honoring the men and women who are America's Veterans.
VA hospital staff			Voluntary attendance during scheduled garden time		
Time	Design and deliver programming weekly, to include garden maintenance activities	Master Garden Volunteers	Contribution to conversation during scheduled garden time	CLC patients increase socialization	Food Safety, Security & Health
Money					*Food Safety, Security and*
Donations		Horticulture educators		CLC patients increase participation in positive diversionary activities	*Health* focuses on the availability of and access to nutritious, affordable and safe food, and decision-making regarding healthy behavior and access to medical care.
Tools	Obtain plants and gardening supplies	CLC patients (hospital residents)	Active participation during scheduled garden time	RT has weekly program	
Expendable gardening supplies				Staff able to evaluate patients for progress on indvidual needs	
Garden space	Support participants to engage in accessible gardening spaces	Hospital staff - Recreational Therapist (RT)	Use of garden outside of scheduled garden time by residents, staff, and visitors		Thriving Youth, Families & Communities
Educational resources			Master Garden volunteers answer garden questions		*Thriving Youth, Families and Communities* focuses on the conditions that support and enhance community members' growth and support, civic engagement and community cohesion.
Gardening resources					
Research					

Assumptions
CLC patients and staff will engage in program. Volunteers have know-how to engage participants; participants have little knowledge on how to garden, staff and visitors support program, chemical free garden; VA staff will address medical needs of participants during program.

External Factors
Weather, funding, donations, pest problems

Here are the links shown in Background Information above:
http://depts.washington.edu/hhwb/Thm_StressPhysiology.html
http://depts.washington.edu/hhwb/Thm_Healing.html
http://hortiech.ashspublications.org/content/13/1/81.full.pdf

Figure 13.2 Graphic representation of a logic model for a veteran's hospital garden program. (Courtesy of Mike Maddox.)

Successful garden-based programs should be viewed by all staff members as an integral part of the mission and not as an additional chore or add-on to their duties. Engaging key personnel in the creation of a logic model is critical. The engagement process aligns the purpose and activities in the garden with the organizational mission, resources, and the needs for individuals utilizing the garden space, and it is an important way of defining how the space can be utilized by staff.

It is also important to include volunteers, grounds crew members, and custodial staff members in a conversation on how the garden is to be utilized, as they may have a role, too. A logic model can also serve as a historical document and can be shared with new staff members and community stakeholders for the purpose of communicating how a garden has changed through time in relation to the evolving needs of the clients served (Taylor-Powell 2003).

Mike Maddox
Master Gardener, Program Director, University of Wisconsin–Extension

EXHIBIT 13.3

Technique: Use of Logic Model for Veteran's Hospital Garden Program

Editors' note: this exhibit outlines the items in the model depicted in Figure 13.2.

Inputs

- Master Gardener volunteers
- VA hospital staff
- Time
- Money
- Donations
- Tools
- Expendable gardening supplies
- Garden space
- Educational resources
- Gardening resources
- Research

Outputs

- *Activities*
  - Design and deliver programming weekly, to include garden maintenance activities
  - Obtain plants and gardening supplies
  - Support participants to engage in accessible gardening spaces
- *Participation*
  - Master Gardener volunteers
  - Horticulture educators
  - CLC patients (hospital residents)
  - Hospital staff—recreational therapist

Outcomes

- *Short term*
  - Voluntary attendance during scheduled garden time
  - Conversation during scheduled garden time
  - Active participation during scheduled garden time
  - Use of garden outside scheduled garden time by residents, staff members, and visitors
  - Master Gardener volunteers answer garden questions
- *Medium term*
  - CLC patients increase socialization
  - CLC patients increase participation in positive diversionary activities
  - Recreational therapist has weekly program
  - Staff able to evaluate patients for progress on individual needs

(Continued)

EXHIBIT 13.3 (Continued)

Technique: Use of Logic Model for Veteran's Hospital Garden Program

- *Long term*
  - To fulfill President Lincoln's promise, "To care for him who shall have borne the battle, and for his widow, and his orphan" by serving and honoring the men and women who are America's veterans
  - Food safety, security and health: focuses on the availability of and access to nutritious, affordable, and safe food, and decision-making regarding healthy behavior and access to medical care
  - Thriving youth, families, and communities: focuses on conditions that support and enhance community members' growth, civic engagement, and community cohesion

Assumptions

- CLC patients and staff members will engage in program
- Volunteers will have knowledge and skill to engage participants
- Participants have minimal knowledge on how to garden
- Staff members and visitors support program
- Chemical-free garden
- VA staff members will address medical needs of participants during the program

External factors

- Weather
- Funding
- Donations
- Pest problems

Background information

- Green Cities: Good Health-Stress, Wellness and Physiology; Green Cities: Good Health-Healing and Therapy; Barnicle, HortTechnology 2013
- http://depts.washington.edu/hhwb/Thm_StressPhysiology.html
- http://depts.washington.edu/hhwb/Thm_Healing.html
- http://horttech.ashspublications.org/content/13/1/81.full.pdf

Mike Maddox
Master Gardener, Program Director, University of Wisconsin–Extension

Table 13.1 Technique: Horticultural therapy operating budget

Income	Amount	Notes
Programming fees	_____	Number of sessions per fee per session
Other funding	_____	Grants, government, private contributions, fundraising events, corporate contributions, and sales
In-kind donations	_____	Noncash donations of goods
Total income	_____	

Expenditures	Amount	Notes
Labor costs		
Wages	_____	Full-time and seasonal employees
Employee benefits	_____	Medical insurance, retirement benefits
Payroll taxes	_____	Taxes paid by employer for employees
Contractor fees	_____	Contractor fees
Overhead costs		
Garden maintenance	_____	Tree and turf care, other maintenance outside program activities
Facility expense	_____	Rent or mortgage
Utilities	_____	Electricity, gas, water, sewage, trash, snow removal
Communication	_____	Internet, phones
Office	_____	Space, equipment, and supplies
Insurance	_____	Liability, etc.
Marketing	_____	Printing, exhibits, etc.
Vehicle use	_____	Cost of fuel, miles, or other expense
Direct program costs		
Plant materials	_____	Woody plants, perennials, annuals, vegetables, and herbs
Growing media	_____	Garden soil, compost, fertilizers, potting mix, etc.
Equipment and tools	_____	Garden hand and power tools, gloves, hoses, etc.
Containers	_____	Pots, packs, flats, hanging baskets, etc.
Activity supplies	_____	Including markers, journals, paper, etc.
Miscellaneous	_____	Labels, stakes, row covers, plant supports, etc.
Total expenditures	_____	

The horticultural therapist may also play a role in cultivating the external image of the program. The major goal behind the majority of external communications is to continually convey the presence, worth, relevance, and efficacy of the horticultural therapy program. Many hospitals and human services organizations employ marketing professionals, who are important collaborators in public relations, and usually carefully manage and control all marketing materials or efforts. In these situations, the horticultural therapist works closely with the marketing team.

Internal communication strategies

Every programming site is unique and comprised of unpredictable, varying components. Horticultural therapy program managers must find creative solutions to address the specific requirements of each site. As such, the following internal communication strategies may or may not be relevant to *every* horticultural therapy program, but they should offer a starting point from which to discover strategies that might work best for a particular program.

> *Create a calendar of horticultural therapy sessions and events.* Post it in a central physical location or share it digitally with co-workers and others who need this information. This calendar should indicate the date, time, and location of each horticultural therapy session and any special events that will be held in conjunction with the program. Special events may include fundraising events, farm-to-table dinners, small or large volunteer days, appreciation or awards ceremonies, or any event that is connected with the horticultural therapy program. This calendar will inform those who share space with the horticultural therapy program about what the program is doing and will help keep the presence of the program visible in an organization or programming site in a more tangible way.
>
> *Have a strategy for unplanned absences.* There will inevitably be a day when the horticultural therapist will miss work due to illness or to attend a meeting that cannot be held apart from session hours. In these cases, another staff member or intern may need to step in and manage clients or perform maintenance tasks on the garden, such as watering, in the absence of a horticultural therapist. For a more seamless transition, it is helpful to make a step-by-step list of tasks that need to be performed on a daily basis. The list should be explained to key personnel, with demonstrations of the tasks to someone at the programming site so that she or he can comfortably perform them in the horticultural therapist's absence. Leave the task list in a digital file and/or physical location that is easily accessible.
>
> *Send email updates on program achievements.* The successes of a horticultural therapy program belong to everyone contributing their time and energy to make it possible. It is good practice to share exciting news by sending periodic communications to staff members at the programming site, volunteers, and clients and their families. Examples of these types of communications are:
> * An announcement of how many clients completed the program
> * The awarding of a grant to the program
> * The installment of a new garden
> * Hiring of a new staff member or intern
> * The success of a fundraising event
> * Thank you to staff members, volunteers and/or clients

Provide treatment updates. Where relevant, send a brief, yet descriptive email or other communication to the client's treatment team when clients have notable breakthroughs or setbacks in their treatment. This information is important in order to build on the gains that the client has made. It is also an opportunity to inform the treatment team on the variety of ways that horticultural therapy affects clients. It is crucial to note that confidentiality communication protocols of a particular organization should be observed at all times. For example, in a hospital setting, writing the name of a patient in an unsecured email to a co-worker would be considered a violation of the patient's privacy. Conversely, if the program is located on a family farm, the protocols for internal communication could be less stringent, yet client confidentiality should be maintained and respected.

External communication strategies

External communication can encompass contact with referral sources, potential clients, individuals who want to interview program staff, and the methods in which the program will be marketed to the community. How this communication is carried out depends on the horticultural therapy program's referral mechanisms, the type of program, and the organization in which it operates. It makes a difference whether a program functions as one of many interventions within a large institution and is represented individually on the institution's website, or if it is completely independent and governed uniquely by the horticultural therapy program staff. There is no one-size-fits-all when it comes to marketing and engaging with the external stakeholders of a horticultural therapy program.

The easiest place to violate client confidentiality inadvertently is through external interactions. To protect confidentiality, any picture, image, or written correspondence between the horticultural therapy program and the outside world must be reviewed for personal identifying client information. Some clients are pleased to be named or pictured in marketing materials or on the program website; others are not. No identifying client information can be released without the client's written consent.

Develop a program narrative. A program narrative is a crucial element for community engagement, funding resources, and attracting the clients that the program strives to serve. It serves a purpose in developing and maintaining a favorable public image for a horticultural therapy program. In essence, the program narrative should answer the questions: Why does this horticultural therapy program exist? Who does it serve? and Why should it continue to exist? The horticultural therapist is often the one who develops the narrative that brings the public image of the program to life. This may include heart-lifting stories about how clients have been empowered, how the program is benefiting the community, or tangible examples of how land has been improved and utilized. The narrative should be modified as the horticultural therapy program grows and develops in order to continually reach the audience that is being targeted.

Proactively educate the public. A horticultural therapist should not assume that most people in a community know what horticultural therapy is or how it can be applied in various settings. Following are a few ideas of ways and reasons to reach out to the community. It is useful to build on this list as it pertains to the program's specific client demographics.

- Contact local information hubs, programs, and agencies. Inform them about horticultural therapy and the mission of the program. Include a detailed description of the types of clients that the program serves so that if they come upon people who may require such services, they have the confidence to refer them to the program.
- Email, mail, or hand-deliver informational brochures to targeted individuals or groups and provide them with flyers that can be distributed to potential clients.
- Reach out to local news sources, which may want to cover an interesting new story; offer to write a press release.
- Develop a positive relationship with local public gardens, Master Gardeners, and others in horticulture businesses to gain enrichment, volunteer, and vocational opportunities for those served.

Develop attractive, professional marketing materials and a website. Marketing materials are the "face" of a horticultural therapy program and should convey important program information accurately and succinctly. Producing high-quality marketing materials ensures the professional image of the program. One must take care to market the program in line with the motivators and needs of the population in which they hope to reach. It is important to work closely with the organization's marketing department when one exists. This will ensure that the program is accurately represented in marketing materials.

Develop a short video that summarizes the program. A picture is worth a thousand words. Creating a short professional quality video that explains the services or products that a business or organization has to offer condenses a large, nuanced concept into small, bite-sized pieces that an audience can easily digest. An attractive video has the ability to capture the spirit of a program in ways that still-life images and text cannot. Including an interview with a successful client in which she or he explains how the program has helped her or him is one way to bring the mission and outcomes of a program to a level with which most people can connect. Videos can be posted online and linked to web pages that already attract the type of client or consumer that is being targeted. They can also be sent to potential funders who may benefit by gaining a clearer picture of what the program visually looks like and specifically how their money will have an impact on the program. As always, obtain written consent before using any client images or names.

Use social media. Establishing an online presence for any business or service-providing entity is effectively non-negotiable in the technology-driven culture, and horticultural therapy programs are no exception. Take time to research the most up-to-date, relevant social media interfaces available. Establishing an informative website and social media presence is a crucial element in selling the program's narrative, recruiting new clients, and informing potential funders and community members about the program's impact on the community. Keeping blog posts and social media feeds updated on a regular basis is another way to engage the target audience. In essence, the act of sharing program information or links through social media creates an exponential advertising venue at very little cost to the program. It is essential to maintain professionalism, maintain client confidentiality, and adhere to organizational policies in any program or personal use of social media.

Managing program staff

Effectively managing program and related staff is essential for the smooth operation of a horticultural therapy program. Primary to other considerations is to develop a well-defined culture and vocabulary with members of the program so that everyone feels confident when articulating and implementing the mission and methods of the program both on and off site. Mobilizing a team of professionals requires that the horticultural therapist has clear program goals that are regularly revisited and measured through time, as previously described. This outcome-oriented technique also helps to maximize productivity by catching weaknesses early in the process and providing an opportunity to problem-solve as a team proactively (Johnson et al. 2003).

It takes more than the horticultural therapist to keep a horticultural therapy program running smoothly. Simultaneously managing a program and a programming site takes time and energy; it is important that everyone involved knows what role they are responsible for playing. Horticultural therapy program managers collaborate with other clinicians, grounds workers, teachers, custodians, administrative assistants, fellow horticultural therapists, and program aids. They also regularly build relationships with individuals from referring agencies or individuals who transport clients to the programming site. At some point, garden materials need to be purchased and delivered, messes that were caused by indoor programming need to be cleaned, and transportation staff need to drop off clients. Coordinating these moving parts needs to be addressed daily. It is the program manager's job to communicate both small and large program needs to staff members and ensure that tasks are accomplished.

Managing interns

An intern is a student who is interested in gaining hands-on work experience in a particular industry. Every intern has unique educational goals that need to be met. It is the intern's responsibility to communicate her or his educational goals to the internship supervisor, and it is the supervisor's job to ensure that the opportunities to meet those goals are provided. Although interns can lighten the workloads of staff, they also add some extra responsibilities that must be considered. Interns require direct supervision and guidance as they navigate their internship site, just as paid employees do. Interns often struggle with balancing the time and energy demands of their personal lives, school, internship, and possibly a job all at once. It is important that these stressors are validated and accommodated as much as possible.

When working with horticultural therapy interns, it is important to remember that every student is coming from a unique educational, horticultural, and personal background, all of which should affect the responsibilities that are given to them. A horticultural therapy intern who has a strong horticultural background may need extra training and support to develop proficient clinical skills. An intern who comes from a social services background may not have professional experience as a horticulturist and require more instruction in order to master gardening skills. Choosing an intern with the educational background and level of horticultural experience that fits each particular horticultural therapy program is crucial to the success of the program and to the intern's education.

Horticultural therapists may interact with a wide range of interns from disciplines other than horticultural therapy. In a physical rehabilitation center, a horticultural therapist may share patients with a physical or occupational therapy intern and consult with that intern on a regular basis. In a hospital, a horticultural therapist may interact with social work, psychology, or psychiatry interns in order to address client treatment goals as a team. Every interaction that a horticultural therapist has with an intern is another opportunity to illustrate the efficacy of horticultural therapy to these future professionals.

Interns bring a fresh perspective to the program and can be enjoyable colleagues with whom to collaborate. People are more productive when they are doing something they love and in which they feel invested. Supervisors should empower interns to identify their professional interests by giving them the freedom to explore ideas and techniques that they think might be effective within the program parameters. Interns need the opportunity to design sessions and activities so that they have the chance to understand the entire process of creating a successful intervention. Each and every horticultural therapist has something unique to offer the field of horticultural therapy. It is a program manager's job to ensure that every member of the team, including the interns, feel valued, informed, and supported as they grow professionally.

Managing volunteers

Many horticultural therapy programs engage volunteers to assist with various aspects of site maintenance, programming, and other support. Therefore, maintaining an effective volunteer base is often high on a program manager's list of priorities. Volunteers support the program, but do not take the place of a paid therapist who has the skills to run sessions and develop programming. Volunteers add a rich diversity of personalities, skill sets, and perspectives to clients and program staff members alike. They can play a large variety of helpful roles, and every volunteer assignment should be matched to individual desires and ability levels. For example, some may prefer to work on garden maintenance; others may desire to interact directly as a program aid alongside the horticultural therapist. Some volunteers might have a passion for performing administrative duties or preparing a banquet hall for a program fundraising event. Some volunteers might be paired with a client who has particularly acute special needs that require one-on-one attention, and others may be assigned to several clients in a group and provide general assistance with an activity. Programs that serve individuals who have several co-occurring diagnoses benefit from having extra eyes and ears on each individual client. This not only provides a higher quality service but also ensures a higher degree of safety and security for the clients. No matter the type of the patient served in a program, volunteers can fortify a program's ability to reach the varied individual needs of each client through meaningful interpersonal connection. Strategically placing volunteers where the program needs match the volunteers' skills adds value both to the program and to the volunteer's experience.

The cornerstone of a healthy volunteer base is a solid training or orientation program, and clear communication throughout the volunteering experience. Volunteers, especially those who have direct client contact, should be oriented to the program goals, desired treatment outcomes, existing program staff members, safety precautions, policies and procedures of the program, and overall organization. In orientation, volunteers should be informed of a client's right to self-determination and confidentiality (Connors 2012). The concept of written consent must be explained regarding the use of any photos taken in the garden, and volunteers should understand that they should not refer to a client by name outside the programming site. They should be informed of what type of clothing is appropriate and what they should bring, such as water, sunscreen, a hat, and appropriate footwear. When a volunteer arrives for a scheduled shift, she or he should receive clear instructions about what she or he will be doing that day and how to perform the tasks according to program standards.

Often, volunteers have never worked in a horticultural therapy setting and may be unaware of the best ways to help a participant during programming. Two of the most common missteps volunteers make are patronizing elders or cognitively impaired clients with an infantilizing tone and jumping in too quickly to relieve a client who appears to be struggling before the client has an opportunity to adapt to a potentially manageable struggle. Both examples derive from a well-intentioned desire to connect and care for the client, but sometimes it can limit the client's improvement in the long run. Demonstrating preferable ways of connecting respectfully with clients is highly advised.

Program managers should find meaningful ways to express gratitude and appreciation for all the work that volunteers contribute to a program. This could be accomplished through volunteer appreciation events or service awards. Other programs may send handwritten cards expressing gratitude for volunteer contributions. Regardless how the program shows appreciation, it is helpful to always keep in mind that gaining, and maintaining, volunteers is an effective way to provide a variety of services while enriching the program by the presence of people of mixed backgrounds and skill sets (Gamble and Gamble 2013).

Management of space used for horticulture

A horticultural therapist is required to have diverse skills and knowledge. In addition to varied clinical skills involved with providing effective interventions, proficiency in horticulture is also critical for success. The professional image and functionality of any horticultural therapy program typically rely on the program manager's ability to control the logistics, standard of quality, and safety of the indoor or outdoor spaces used for horticulture. How involved the horticultural therapist is in garden operations depends on the employment model and arrangements. For example, an employee may serve as a manager and have duties that include responsibility for garden oversight. A contractor might only provide programming but may also contract to provide garden maintenance and oversight for the organization. The gardens are the clinical space used for horticultural therapy, so generally it is advised that some oversight of those spaces be included in the role of the horticultural therapist (Horowitz 2012) (Figure 13.3).

Logistical and quality control

Common logistical and quality control duties that may be performed by a horticultural therapist are as follows (Figure 13.4):

- Establish and protect the standard of excellence of the horticulture space.
- Collaborate with the programming site's grounds manager to develop a common goal and to discuss who is responsible for various ongoing tasks.
- Design gardens, greenhouses, sunrooms, and other indoor programing spaces as needed, and/or work closely with professional landscape designers.
- Coordinate the construction of structures.
- Coordinate the planting/installation of gardens, including ordering materials and coordinating vendor deliveries.
- Develop a maintenance plan for the gardens and/or programming spaces and ensure that those plans are executed.

Figure 13.3 Garden and greenhouse as clinical space for horticultural therapy at Alnarp, SLU in Sweden. (Courtesy of Anna Maria Palsdottir.)

Figure 13.4 The horticultural therapy program manager at Recovery Ventures Inc. ensures quality horticulture outputs in order to enhance therapeutic outcomes. (Courtesy of Eugene Jones.)

- Manage clients and any volunteers to plant gardens and maintain horticulture space.
- Communicate with any staff members at the programming site who may share the horticulture space with the horticultural therapy program.
- Develop a schedule where all usages for the space are documented on a calendar.
- Rectify issues that diminish the aesthetics or functionality of a garden in a timely manner.

Site safety

The horticultural therapist may be responsible for personally directing others to maintain the safety and security of horticultural therapy programming sites. Again, this is advisable because it allows the practitioner to have some control over the clinical space used for treatment. This responsibility may be as simple as keeping sidewalks swept and free of major stumbling blocks or as complex as developing a solid action plan in case of medical emergencies, accidents, or severe weather threats. Depending on the program and setting, specific procedures may need to be put into place. For example, forensic, memory care, or psychiatric settings may require elopement prevention and containment procedures. Virtually every site should plan ahead for how clients or staff members would get medical attention if an injury were to occur. Staff members and volunteers should have a method in which to call for support staff or emergency services at all times. In many cases, the use of a cell phone, landline phone, or two-way radio are advisable and should have emergency contact numbers programmed prior to use.

Securing the safety of a site might also include teaching clients how to use tools properly to prevent injuries, keeping hoses and other equipment off walkways, and setting boundaries and rules with clients regarding tool safety. No matter the setting, it is always advantageous to develop a dangerous tool checkout sheet so that sharp/dangerous tools can be accounted for before and after every horticultural therapy session. This measure will not only keep clients and staff safe but also helps program staff prevent the loss of valuable tools. Being aware of a client's mobility challenges and ensuring that he or she has the accommodations required to garden safely is also a safety issue that needs to be addressed. If a client struggles to stand up from a crouching position, aides need to be available to assist that person or a mobility device needs to be employed. A client's safety must be ensured for programming to take place (Figure 13.5).

Finally, the program manager should be aware of and approve any natural or chemical products before they are applied to indoor plants or outdoor programming areas. Fertilizers, pesticides, and herbicides can be extremely dangerous and cause health hazards to those working in the programming site. The use of pressure-treated lumber in raised beds should be avoided, as the chemicals used in the treatment process are known carcinogens and will leach into the soil. (Additional safety issues, including plant choices, are discussed in Chapter 11.) Maintaining a general awareness of all these elements ensures the safety of stakeholders and the surrounding community.

Program evaluation

Program evaluation entails systematically collecting, analyzing, and utilizing data to examine the effectiveness and efficiency of a program. Evaluation and continued improvement are the backbone of a successful horticultural therapy program. As discussed in previous chapters, horticultural therapy is a process of establishing treatment goals with

Figure 13.5 Instruction on proper tool use. (Courtesy of Christine Capra.)

clients and then methodically working toward those goals, recording what progress has been made. This process ensures that the interventions used are targeting the desired behaviors. Evaluating a horticultural therapy program allows decision makers to determine whether the program is functioning as intended (Vedung 2017). Some of the program elements that evaluation can explore are:

- Determine to what extent the program goals and objectives are being met.
- Determine at what level of quality the program is functioning.
- Identify strengths and weaknesses of the program.
- Find correlations between practice and results.
- Manage resources effectively.
- Justify current and/or future program funding.
- Provide statistical data to administrators and funders.
- Document program accomplishments.
- Determine participant satisfaction.

Some of the most common program evaluation techniques are:

- Review of client documentation—summaries of the aggregate data, which track very specific behaviors through the treatment process to determine the amount and rate of change in program participants.
- Cost-benefit analysis—a process used to determine options that provide the best approach to achieve benefits while preserving savings and energy expenditures.
- Surveys—methods of gathering information from any stakeholders by asking specific written or verbal questions with the intent to identify patterns or previously unknown information with the responses.

- Observations and reviewing program records—ways of personally observing patterns of client responses, successful sessions, and a variety of other programming through observation or by reviewing the record.
- Exit interviews—interviews of clients leaving a program to determine how they experienced various elements of the program, what they would change, what they found valuable, etc.

Many of these evaluation tools are used in conjunction with one another and can provide a large amount of information that would not have otherwise been discovered. All of the data that is collected can be used to consistently move a horticultural therapy program forward so that it aligns with the program mission and needs of the clients.

For example, the data that was collected from the veterans in the scenario mentioned in Exhibit 13.4 can be used in many different ways to benefit the program. The horticultural therapist can analyze and use the data from a survey to improve the effectiveness and relevance of the services rendered. The data could also be used in the event that financial resources for the program needed to be obtained. Providing clear, succinct data that

EXHIBIT 13.4

Program Example: Military to Civilian Careers

A vocational and therapeutic horticultural therapy program that taught post-9/11 military veterans how to translate their military experience into civilian careers and to learn small-scale sustainable farming techniques used the following strategy as one component of program evaluation. Each participant was given a survey prior to starting the six-month program, again halfway through the program, and again at the exit interview when he or she completed the program. The survey asked the veterans questions about how confident they felt performing a series of job skills and farming tasks on a scale from 1 to 10, such as communicating with customers and co-workers, staying focused while at work, amending soil, planting, harvesting produce, etc.

Evaluation of the series of surveys that were collected after the first six-month program indicated that every veteran had experienced a significant increase in confidence when employing the skills that they practiced in the program. The only exception to the increase in confidence was the question that asked whether they felt more or less confident interacting with the general public who visited the farm. This data confirmed successful targeting of the mission of the program in many ways but that it would be desirable to find a way to strengthen the program as it pertains to interaction with the community. Several community engagement and communication activities were built into the second six-month program, and the confidence levels for the community interaction question increased dramatically.

The data that was collected in this survey was also used throughout the treatment process to indicate how each individual client was progressing in the program. Some clients indicated a lower level of confidence in some tasks at the three-month assessment period, so the horticultural therapist was able to work closely with those people on the skills they felt needed continued support.

Emily Vanderneut, BSW

indicate that a population is responding favorably to treatment is a compelling reason to fund a program. Putting evaluation measures into place from the start to the finish of the program will inevitably benefit the program in the long run and contribute to the longevity of the program.

Program budget development and management

Creating and managing an operating budget for a horticultural therapy program involves setting financial goals based on the needs and mission of the program, determining ongoing expenses, anticipating program income, as well as monitoring and reporting on the budget. When it comes to developing a program budget, function always needs to inform design. Horticultural therapy programs come in many types and sizes and, as such, each budget will be tailored to the individual needs of each program. The type of industry in which the program operates will greatly affect the way that a budget is developed and managed. The fundamental differences between the administrative styles and budgetary approaches of institutions such as a correctional facility versus a community-based organization are vast. Organizations also serve clients of various demographics, geographic locations, and needs. All factors affect budget management. Furthermore, budgets are affected by the size and type of the horticultural therapy program, and the employment model in which the horticultural therapist works.

Before creating an operating budget, it is important to develop a clear idea of what an ideal program looks like in the given setting. Developing an ideal for the program provides a goal that can be scaled back or phased in as needed, depending on the financial resources provided by a programming site or secured through grants or other sources.

Some budgeting terminology is essential to understand, including *labor, overhead,* and *direct program costs,* because every program will manage these costs in a variety of different ways.

- *Labor costs* involve any cost associated with employing human labor. This includes employee wages, contractors, medical benefits, payroll taxes, and any other benefits that an employer chooses to offer an employee, including a pension, paid sick leave, or vacation time.
- *Overhead costs* are ongoing expenses that are required to keep a business functioning. These include expenses such as electricity, internet, phone, water/sewage/garbage, insurance, rent/mortgage, office supplies and machines, etc. Horticultural therapy programs that utilize a climate-controlled greenhouse also need to anticipate the costs of heating and cooling that space. Often, overhead costs for horticultural therapy programs are absorbed into the budget of the overall organization and are not necessarily items that will be included on the program-specific budget sheet.
- *Direct program costs* are ongoing expenses that are typically required to keep a program functioning with high-quality services. These costs include expenses related to ongoing horticulture such as planting a garden (soil, plants, fertilizer, etc.), specific program materials (flower pots, potting soil, fairy garden plants), garden tools, topsoil, mulch, indoor plants, supplies, or any other tangible good needed for horticultural therapy sessions.

Part of determining the ongoing costs of a horticultural therapy program is to ascertain if the organization or site in which the programming will occur will absorb any overhead and/

or direct program costs. In many cases, established programming sites such as a hospital or a nursing home will absorb the overhead costs of a horticultural therapy program because those costs are already being accounted for in the organization's budget. It is assumed that the organization will provide funds to pay the horticultural therapist for labor unless other arrangements have been made. Once the financial responsibility of the labor, overhead, and program costs have been assigned to the overall organizational budget or the horticultural therapy program, the horticultural therapist will have the information necessary to list the fixed and projected expenses of the program.

Anticipating the amount of income that a program has the potential to generate needs to be factored into the budget. Some programs may be financially sustained through grant funding alone, others may be entirely supported by the programming site/organization, and others by billing outside payers for services. Some programs may be supported by several of these financial means at once. In addition, income may be obtained through the sale of plants or produce when appropriate for programming, such as in vocational training. Any revenue that a program generates needs to be forecasted and reflected in the operating budget.

Regardless of where program funding originates, monitoring spending and income is paramount to maintaining a balanced program budget. Harnessing spending and income data in financial reports provide information related to financial goals as well as a succinct financial picture for current and future funders.

When physically creating an operating budget on a spreadsheet or in a computer program, consider the layout and content pictured in Table 13.1. Note that actual program budgets differ considerably and include various items in each category, as well as additional line items besides those included in the example.

The budget shown in Table 13.1 is separated into two main categories: income and expenditures. The income section includes all sources of income that are earned or are granted by outside funders for the fiscal year. Sources of income include but are not limited to program fees, grants, private funders, government funding, corporate contributions, and in-kind contributions. Projected expenditures include the program labor, overhead, and direct program material costs that are anticipated for the fiscal year. Additional columns might be added to show the actual income and expenses. Yet another column for "Difference" would keep a running total of whether the horticultural therapy program has over- or underestimated the costs and income potential for that fiscal year. Having access to current Difference totals allows the horticultural therapist to determine whether current and future program projects or goals are achievable, sustainable, and reliable through time.

An effective management technique is to monitor and strategize how to best use funds to support the program's mission statement. Being proactive and forecasting potential financial needs is a key factor in long-term program sustainability. As always, keeping the needs of the organization and the clients of the program in mind while making budgeting and other program decisions is a sure way to provide effective services that foster client engagement and program sustainability.

Summary

A horticultural therapist with effective management skills creates sustainable programs in the best interest of the clients served. The program management tools discussed in this chapter provide guidelines for established and emerging practitioners. The responsibility of the clinical, administrative, public relations, and operational components of a program were addressed in detail, with the goal that horticultural therapists gain an in-depth understanding of effective program management.

The horticultural therapist's unique role broadly influences the development of program goals while orchestrating goal development, evaluation, and implementation. This is accomplished through creating an atmosphere of open communication, using active listening, and working within an interdisciplinary team to design a program that responds to the needs of the client population. It is the responsibility of the therapist to initiate collaboration on program goals and foster an environment that encourages contributions and ideas from investors, clinical team members, clients, and families.

In addition to program goals, the horticultural therapist typically takes on administrative responsibilities that include a range of tasks focused on the interaction between the program and the entities that affect it. The success of a horticultural program is closely tied to financial considerations, including attracting investors through effective marketing and information and developing relationships with organizations within the community. Often the therapist acts as the liaison between the program and those that may benefit from its services, so savvy program management will foster connections that increase interest in horticulture-based programs. Oversight of the internal components of a program includes the design and use of programming sites, supervision of staff, and budgeting. The ability to manage components of a horticultural therapy program is a developed skill and is central to a successful program. Through education, collaboration, and experience, effective program management can positively affect the lives of client populations in a way unique only to the practice of horticultural therapy.

Key terms

Cost-benefit analysis A process by which the benefits of an item, decision, or action are determined to be worth its cost. This type of analysis assists in comparing the various courses of action that an organization might take.

In-kind contributions Noncash items, such as goods or services, that are donated to an organization at no cost.

Intern A student who gains hands-on work experience in a particular industry by completing a predetermined number of unpaid or paid work hours with an organization.

Program budget A financial document that depicts the annual costs and income of a program. A program budgets reflects the priorities and needs of the program and serves as a record of whether a program is meeting its financial goals and obligations.

Program evaluation A process by which the effectiveness and efficiency of a program are measured and analyzed with various assessment tools.

Program goals Statements derived from the mission of a program that form the framework of what a program intends to accomplish.

Program objectives Specific, measurable steps that are strategically developed to achieve program goals actively and efficiently.

Program narrative A tool by which a program projects its desired image to the public. A compelling, informational snapshot of facts that teach the public how to interpret a program. Often used when marketing or securing funding for a program.

References

Connors, T. 2012. *The Volunteer Management Handbook: Leadership Strategies for Success.* 9–20. Hoboken, NJ: John Wiley & Sons.

Gamble, T., and M. Gamble. 2013. *Leading with Communication: A Practical Approach to Leadership Communication.* Thousand Oaks, CA: SAGE Publications, pp. 63–70, 183–189.

Grange, G. 2008. *Effectiveness of Interdisciplinary Team Dynamics on Treatments in a Behavioral Health Environment.* New York: Universal-Publishers, pp. 1–2.

Horowitz, S. 2012. "Therapeutic Gardens and Horticultural Therapy: Growing Roles in Health Care." *Alternative and Complementary Therapies* 18(2): 78–83. doi:10.1089/act.2012.18205.

Johnson, L., D. Zorn, B. Kai, Y. Tam, M. Lamontagne, and S. Johnson. 2003. "Stakeholders' Views of Factors That Impact Successful Interagency Collaboration." *Exceptional Children* 69: 195–209. doi:10.1177/001440290306900205.

Lewis, J., T. Packard, and M. Lewis. 2012. *Management of Human Service Programs.* Belmont, CA: Brooks/Cole.

Rogers, C., and R. Farson. 2015. *Active Listening.* Mansfield Centre, CT: Martino Publishing, pp. 3–12.

Taylor-Powell, E., L. Jones, and E. Henert. 2003. "Enhancing Program Performance with Logic Models." Accessed September 28, 2017. https://fyi.uwex.edu/programdevelopment/logic-models/.

Vedung, E. 2017. *Public Policy and Program Evaluation.* 103–119. New York: Routledge, pp. 103–119.

chapter fourteen

Research applied to practice

Barbara Kreski

Contents

Program planning

When horticultural therapists consider what format, focus, or activities they will include in a program for an individual or for a group of people sharing similar goals, they want to be sure of selecting those with an expectation of being effective. How do they make their decisions? In this chapter, the concept of evidence-based practice and ways to ground practice in research will be discussed.

What does it mean to be evidence-based?

Every field that attempts to affect the quality of life for individuals is challenged to base its efforts on sound evidence. A field or an intervention is considered to be evidence-based if it has been studied according to reliable and valid methods that have established the intervention to be helpful. Evidence-based practice is a systematic approach to developing a plan of action for health care providers (Karkada 2015). The ideal model for practice is to combine a thorough knowledge of research findings with all of the experiences of the horticultural therapist as well as the preferences and circumstances of the client (Ackley et al., 2008). The route to more widespread acceptance and use of horticulture as therapy is to establish a body of sound research studies and to use those findings to support and shape practice.

Clinicians in every discipline struggle to align their day-to-day work with the conclusions of research. Some disciplines are developing clinical practice guidelines that summarize best practices in their field. However, even with this rigorous approach, every field falls short of consistent adherence to evidence-based practice. This does not excuse horticultural therapists from reading and applying current literature to their daily work, but it does acknowledge that even fields with a greater quantity and/or quality of research must make a concerted effort to implement findings. It is the hallmark of a professional to make that effort.

Steps to evidence-based practice

The route to developing evidence-based practice has been broken down into steps by many authors since the introduction of the term by Sackett (1996). The following is a clear and concise compilation of steps adapted from the *International Journal of Nursing Research and Practice*.

1. Develop a habit of questioning. Become accustomed to asking why and how. For seeking information about a clinical or practice issue, the PICOT system helps to organize and frame the question (Riva et al. 2012):

    P *Population* refers to the sample of subjects you wish to recruit for your study. There may be a fine balance between defining a sample that is most likely to respond to your intervention (e.g. no co-morbidity) and one that can be generalized to patients that are likely to be seen in actual practice.

    I *Intervention* refers to the treatment that will be provided to subjects enrolled in your study.

    C *Comparison* identifies what you plan to use as a reference group to compare with your treatment intervention. Many study designs refer to this as the control group. If an existing treatment is considered the gold standard, then this should be the comparison group.

    O *Outcome* represents what result you plan on measuring to examine the effectiveness of your intervention. There are typically a multitude of outcome tools available for different clinical populations, each having strengths and weaknesses.

    T *Time* describes the duration of your data collection.

2. Search for and collect evidence. In other words, when using the internet to seek answers to a question, the words generated by PICOT can be entered as search terms.

3. Critically appraise the evidence. The research that is collected must be analyzed to determine if it is valid, reliable, and applicable to the question or to practice. The next section of this chapter elaborates on how to conduct a critical appraisal of a research study.

4. Integrate the evidence. Combine what has been learned through research with knowledge from experience, education, and what is known of the client's preferences and circumstances. Form and implement the care plan or approach.

5. Evaluate the outcome. Evaluating and documenting the outcome helps to develop a personal foundation of evidence regarding what works and the circumstances that are pertinent.

6. Disseminate results. Share what has been learned by writing an essay or an article, by forming a small study group for the purpose of sharing ideas, or by reporting outcomes to colleagues. Most complex research studies begin with anecdotal observations and experiences. Capturing and sharing findings from research and practice advances the field.

All research is not equal

It may seem that all studies that have been published in any way are worthy bases for decisions. However, the quality of published research ranges from very poor to very good. Citing poor-quality research to defend one's practice does more harm than good because it implies that there is no better basis for the therapy. Using a list of questions to conduct what is called a *critical appraisal* of an article is a good way to begin to sort out the strongest evidence on which to base practice.

Questions to ask when critically appraising a research article (Young and Solomon 2009)

- Is the study question relevant? Before going any further, the practitioner should determine whether the article has meaning for the issue being considered. There is no need to appraise articles that are irrelevant.
- Does the study test a stated hypothesis? If there is no hypothesis, the article may be another type of writing rather than a research study. Other types of writing found in journals include editorials, responses to previous articles, essays, and letters. This checklist of questions will not apply if that is the case.
- Does the study add anything new? The study may reinforce a previous study by repeating the methods or increasing the number of participants. While some research is truly ground-breaking, most studies further our understanding in small steps.
- What type of study is it and what level of evidence does the type of study have? Different types or designs of studies have different levels of evidence associated with them (Misso et al., 2008). The more rigorous the research method, the higher the level of evidence. High levels of evidence mean that more professionals have confidence in the results. There are variations in descriptions of the levels, but one basic hierarchy is listed below. Note that the higher the level of evidence, the lower the number associated with it, with systematic review as the highest level shown in Exhibit 14.1. However, research on horticultural therapy with single subjects or small sample sizes may be useful to provide examples of applied practice.
- Was the study design appropriate for the research question? Some designs test the subjects before and then again after an intervention. Another design is to have one group of subjects experience an intervention while a second, control group does not. There are several other designs as well. One thing to check is that there is random assignment of variables.

EXHIBIT 14.1

Technique: Levels of Evidence

Levels of evidence are assigned to studies based on the quality of their design and methods, validity, and applicability to patient care. In the hierarchy below, Level 1 is the highest quality of evidence and Level 5 is the lowest. Note that all levels have a legitimate place in guiding treatment planning and practice.

Level 1:   Systematic review or meta-analysis of all relevant random control studies.
Level 2:   Large, multi-site, random control study.
Level 3:   Well-designed, controlled studies without randomization.
Level 4:   Single descriptive trial, such as a case study.
Level 5:   Expert opinion based on experience.

Adapted from Sackett 1989

EXHIBIT 14.2

Technique: Key Sections of Research Reports

Title: Subject and what aspect of the subject was studied.
Abstract: Summary of paper: The main reason for the study, the primary results, the main conclusions.
Introduction: Background regarding why the study was undertaken.
Methods and Materials: Specifics of how the study was conducted.
Results: The outcomes of the methods undertaken.
Discussion: Implication of the results. What is the significance of the results? In this section or in a separate section, the limitations of the study are addressed.

Hooganboom and Manske (2012)

- Did the study methods address the most important potential sources of bias? Bias exists when the results are skewed by how study participants were selected, by how data was collected, or through the researcher's analysis or interpretation. If the methods are not sound, bias can influence the results.
- Are there any conflicts of interest? Note whether the researcher might have a reason to prefer one conclusion over another. For example, does the funding for the study come from a source that would profit from a particular result?
- Were there any confounders? In simplest terms, a confounder is a variable that was not accounted for and that may influence the results. For instance, age can be a confounder if the subjects in the control group are much older or younger than the ones in the test group. Most studies include a section on limitations or future directions. The issues that the author is aware of can be found in that section. Exhibit 14.2 briefly describes the common sections of most peer-reviewed journal articles.
- Was the study performed according to the original protocol? Sometimes, researchers find that they must make a change in how they had planned to conduct their study after they have already begun. This is worth noting because such changes often influence the result.
- Do the data justify the conclusions? Is the conclusion believable based on the evidence presented or is the conclusion too sweeping and general?

Examples of applying research in practice

Horticultural therapists are responsible for planning individual or group experiences to improve or support some aspect of life that benefits the participants. As covered in other chapters, the therapist considers the current status, needs, and interests of the clients, and the input of other team members when deciding what to do. The decision-making process is influenced by past experience and is grounded in familiarity with the pertinent research. Exhibits 14.3 and 14.4 illustrate horticultural therapy program development, and the use of research in that process.

Pertinent research may be found in the *Journal of Therapeutic Horticulture*, but many other resources can be useful as well. The therapist can use the participant's diagnostic label

EXHIBIT 14.3

Program Example: Using Research to Inform Program Development

The Chicago Botanic Garden (CBG) Nature Preschool collaborated with the Green Bay Early Childhood Center in Highland Park to create Nature Buddies: an inclusionary, nature-based educational partnership. Hosted at the Regenstein Learning Campus and in the Buehler Enabling Garden at CBG, the two schools combined classrooms once a month to provide students of differing abilities the opportunity to come together to learn through their curiosity and engagement with nature in a sensory-rich and accessible environment.

Outdoor educational experiences are often unintentionally developed with only the typical child in mind. Alternatively, Nature Buddies uses local resources to train its instructors in inclusive, mindfully designed lessons plans for varying educational levels and uses accessible gardens to accommodate diverse populations.

Research shows that nature provides an equalizing learning space. Children with disabilities offer multiple perspectives and styles of learning for typical students, and inclusionary programs allow students to share their interests and teach one another. Garden- and play-based activities provide opportunities for students to break out of the traditional classroom setting to study multiple subjects simultaneously through engagement and discovery.

Nature Buddies seeks to be an example of a type of inclusive outdoor education; a model for other programs to expand and mold. The program was led by experienced nature educators, horticultural therapists, and licensed special education instructors. After one school year, the program will be evaluated, and best practices will be shared globally through presentations, videos, and documentation. Nature programs looking to become universal in their audiences will be able to learn from this pilot program.

Program Goals

- Harness nature's ability to benefit all young children's cognitive, physical, language, and social development; offer responsibilities and opportunities for each child, not found in typical classroom settings.
- Expand students' awareness and respect of all people and abilities, increase empathy, and build bonds of trust and friends.
- Create an example of inclusive nature education.
- The Nature Buddies program is grounded in the following:
  - Young children develop science-understanding best when given multiple opportunities to engage in science exploration and experiences through inquiry (Bose et al. 2009; Gelman et al. 2010).
  - Prosocial behaviors and quality of social interactions increase when children spend time in nature (Acar and Torquati 2015).
  - Gardening promotes responsibility, patience, and cooperation (Kahn and Kellert 2002).
  - Nature is a classroom equalizer, promoting group work, and allowing for visual and immersive learning (Coyle 2010).

**Adapted from a program information sheet
provided by Katherine Knight, HTR**

EXHIBIT 14.4

Program Example: Piloting a Horticultural Therapy Program for Veterans

As the Director of Gardens Education at the University of Tennessee, it is my responsibility to manage project initiatives that have a direct impact on my department. The Dean of the College of Agriculture encouraged me to explore the establishment of a horticultural therapy program for veterans. When developing new horticultural therapy programs, it is vital to use evidence-based practices to ensure that horticultural therapists design client-centered programming. Evidence-based practice is the use of research, clinical expertise, and client values to make health care-related decisions (Thyer 2004).

Research

Research gives therapists the knowledge of programs that have been implemented before. It also provides a broader understanding of each specific client population.

Research found that horticulture-focused programs for veterans provided a sense of purpose, reduced stress, improved self-esteem, increased social engagement, increased physical activity, and improved career outcomes (Atkinson 2009; Anderson 2011; Fleming 2015). Detweiler (2015) reported that horticultural therapy programs help veterans modulate stress and improve overall quality of life.

Clinical Expertise

Knowledge and experience as a horticultural therapist create a unique understanding of how clients respond to interventions. Using information gathered from research and professional experience allows the horticultural therapist to tailor the program to meet the specific needs of clients (Wise 2015).

Client Values

One key component to client-centered programming is understanding client values. Without research, therapists may have preconceived ideas of a population. This reduces dignity and a client's right to self-determination.

These three components, research, client expertise, and client values, were utilized when developing a pilot horticultural therapy program for veterans called Veterans Experiencing Growth through Garden Interactive Experiences (VEGGIE). The pilot development was made possible through a grant from a local foundation. Through Phase I funding, two horticultural therapy programs serving veterans were visited. Cate Murphy, HTR and founder of TALMAR Inc. in Baltimore, Maryland, runs a year-long program for veterans called Breaking New Ground. The second site we visited was Chicago Botanic Garden. Alicia Green, HTR from Chicago Botanic Garden, runs its veteran programs. Those include a fourteen-week vocational training program for veterans called Veteran Internship Program (VIP), and a monthly program for veterans who have mental health issues.

(Continued)

EXHIBIT 14.4 (Continued)

Program Example: Piloting a Horticultural Therapy Program for Veterans

The pilot program's impact was measured using an assessment found in research called the Quality of Life Enjoyment and Satisfaction Questionnaire—Short Form (Q-LES-Q-SF) (Detweiler 2015). Dr. Gina Owens, associate professor of psychology at the University of Tennessee, also suggested the use of the Depression Anxiety and Stress Scale (DASS) to measure additional impacts. The pilot consisted of a five-week program that met for ninety minutes twice a week. During this time, the focus was on horticulture topics and therapeutic goals. The information gathered from the pilot and research was used to create a Phase II grant request from the local foundation to fully implement a year-round program for veterans. The use of evidence-based practice is a critical component to horticultural therapy program development. Use of research, clinical expertise, and client values allows horticultural therapists to create programs that meet the needs of each specific client population.

Derrick R. Stowell, MS, HTR, CTRS
Education Director, University of Tennessee Gardens

Gina Owens, PhD
Associate Professor and Director of Training,
Department of Psychology, University of Tennessee

plus *best practices* as a search term to find interventions that have been vetted by research. Often, a horticultural therapy session can be planned that compliments other treatments. For instance, if the therapist will be working with a classroom of students who have been diagnosed with autism spectrum disorder, using the term *autism best practices* to search databases will yield several articles that survey the research and make recommendations for working effectively with the students. Many of these articles may not specifically address the field of horticultural therapy; however, understanding the characteristics of other successful interventions is helpful when planning a course of treatment (Westlund 2014).

The most current research may provide results that point to *preliminary conclusions*. This is the case when a research question is in its early stages of being investigated. The first few studies may have some weakness in methodology, such as a small sample size or a control group that is not randomized. They are often qualitative in nature, meaning they are more subjective and/or exploratory, and the field of responses may be more open-ended (Aickin 2007). These early studies are valuable because they set the stage for further, more rigorous investigation. They establish that the results merit further study. The following paragraph from a preliminary study describes both potential application of the research to efforts to promote physical activity in older women as well as noting that future studies will be strengthened by having more participants and a longer duration (Park et al. 2017):

In conclusion, the 15-session gardening intervention as a low- to moderate-intensity physical activity may have improved HDL cholesterol, blood pressure, and markers of immune activation, such as TNF-α and RAGE expression. These study results show the potential of gardening as a physical activity for improving health in elderly women. Future studies need to have a longer duration and a larger sample size to analyze the effect of a long-term gardening intervention on maintaining or improving health conditions of the elderly.

In addition to familiarizing oneself with diagnoses, a therapist can research to learn about the culture of the group participants. While each individual is unique, a thorough background in traits that a cultural group shares will be of great help when planning sessions and setting goals. Ethnic and national groups are often recognized as taking pride in their culture. It is just as important to learn about other cultural categories such as veteran status, religious affiliation, or gender identity. Symbols, preferred colors, foods, and customs are very often related to the cultivation of certain plants.

A final example of using research to inform practice is referring to published measures rather than the therapist's estimates. This type of research is quantitative, meaning it uses measurable data to establish facts and patterns (Aickin 2007). If the participants are elders, for example, and the therapist wants to provide opportunities for moderate exercise by engaging them in gardening, it would be safer to research what constitutes moderate activity for this age group rather than relying on intuition or common sense. Sin-Ae Park of Konkuk University in South Korea researched exercise intensity of specific gardening activities using measures of metabolic energy consumed (Park 2011). Her results are summarized in Table 14.1.

Table 14.1 **Research example: Exercise intensity of various gardening activities in children, adults, and the elderly (Park et al. 2011)**

	Children	Adults	Elderly
Activity		Exercise intensity	
Digging	High	High	Moderate
Raking	High	Moderate	Moderate
Weeding	Moderate	Moderate	Moderate
Mulching	Moderate	Moderate	Moderate
Sowing	Moderate	Moderate	Low
Harvesting	Moderate	Moderate	Low
Watering	Moderate	Moderate	Low
Mixing soil	Moderate	Moderate	Low
Planting transplants	Moderate	Moderate	Low

Summary

Not every horticultural therapist will conduct publishable research in her or his career. However, each is charged with learning as much as possible about the people she or he serves and the ways to address their needs that have proved most effective. Although many professional journals require a fee in order to access full articles, a good research librarian can provide them at no cost. Botanical gardens, universities, and professional organizations can also be helpful resources for finding and accessing research and information. Staying current on the research enriches practice and earns respect from other professionals.

Key terms

Clinical practice guidelines Statements made by a professional association that recommend certain types of approaches to patient care based on sound evidence of effectiveness.

Confounders Variables that were not controlled or accounted for that influence the result of research, negatively affecting its usefulness.

Critical appraisal Systematically examining research to judge its quality, and its value and relevance.

Evidence-based practice Integration of a therapist's experience, patient values, and the best current research to plan and execute therapy.

Meta-analysis A statistical method of combining the results of a number of different studies to create one, larger, more significant and precise report.

Qualitative research Research that collects, classifies, analyzes, and interprets data generally without using mathematical models and without a hypothesis. This is most useful in discerning patterns and relationships among observations and/or interviews. It is often employed in the early phases of exploring a topic and may be a prelude to designing a quantitative study based on the patterns and data discerned.

Quantitative research Research that starts with a hypothesis and works deductively through mathematical models to either confirm or disprove the hypothesis. Often, the hypothesis is formulated based on information collected via qualitative research. Quantitative research can establish the probability that a cause-and-effect relationship exists between two variables.

References

Acar, I. H., & Torquati, J. 2015. "The Power of Nature: Developing Prosocial Behavior Toward Nature and Peers Through Nature-Based Activities." *Young children* 70(5): 62–71.

Ackley, B., G. Ladwig, B. A. Swan, and S. Tucker. 2008. *Evidence-Based Nursing Care Guidelines: Medical-Surgical Interventions*. 7. St. Louis, MO: Mosby Elsevier.

Aickin, M. 2007. "The Importance of Early Phase Research." *The Journal of Alternative and Complementary Medicine* 13(4): 447–450.

Anderson, B. 2011. "An Exploration of the Potential Benefits of Healing Gardens on Veterans with PTSD." Master's Thesis. Accessed on January 15, 2017. http://digitalcommons.usu.edu/cgi/viewcontent.cgi?article=1057&context=gradreports.

Atkinson, J. 2009. "An Evaluation of the Gardening Leave Project with PTSD and Other Combat Related Mental Health Problems." The Pears Foundation. Accessed on January 15, 2017. https://www.researchgate.net/profile/Jacqueline_Atkinson/publication/265575473_AN_EVALUATION_OF_THE_GARDENING_LEAVE_PROJECT_FOR_EX-MILITARY_PERSONNEL_WITH_PTSD_AND_OTHER_COMBAT_RELATED_MENTAL_HEALTH_PROBLEMS/links/55094b960cf26ff55f852b50.pdf

Bose, S., G. Jacobs, and T. L. Anderson, 2009. "Science in the Air." *Young Children* 64(6): 10–15. http://www.naeyc.org/files/yc/file/200911/BosseWeb1109.pdf.

Coyle, K. J. 2010. "Back to School: Back Outside: Create High Performing Students." Report, National Wildlife Federation. https://www.nwf.org/en/Educational-Resources/Reports/2010/09-01-2010-Back-to-School-Back-Outside.

Detweiler, M., J. Self, S. Lane, L. Spencer, B. Lutgens, D. Y. Kim, M. H. Halling, T. C. Rudder, and L. P. Lehmann. 2015. "Horticultural Therapy: A Pilot Study on Modulation Cortisol Levels and Indices of Substance Craving, Posttraumatic Stress Disorder, Depression, and Quality of Life in Veterans." *Alternative Therapies in Health & Medicine* 21(4): 36.

Fleming, L. 2015. "Veteran to Farmer Programs: An Emerging Nature-Based Programming Trend." *Journal of Therapeutic Horticulture* 25(1): 27–48.

Gelman, R., K. Brenneman, G. Macdonald, and M. Roman. 2010. "Preschool Pathways to Science: Ways of Doing, Thinking, Communicating and Knowing About Science." Baltimore: Brookes Publishing.

Hoogenboom, B., and R. Manske. 2012. "How to Write a Scientific Article." *International Journal of Sports Physical Therapy* 7(5): 512–517.

Karkada, S. 2015. "Evidence-Based Practice (EBP)." *International Journal of Nursing Research and Practice* 2(2): 1–2.

Kahn, P. H., and S. R. Kellert. 2002. Children and Nature: Psychological, Sociocultural, and Evolutionary Investigations. Cambridge, MA: MIT Press.

Misso, M., C. Harris, and S. Green. 2008. "Development of Evidence-Based Clinical Practice Guidelines (CPGs): Comparing Approaches." *Implementation Science* 3(45). doi:10.1186/1748-5908-3-45.

Park, S.-A., A.-Y. Lee, H.-G. Park, K.-C. Son, D.-S. Kim, and W.-L. Lee. 2017. "Gardening Intervention as a Low- to Moderate-Intensity Physical Activity for Improving Blood Lipid Profiles, Blood Pressure, Inflammation, and Oxidative Stress in Women over the Age of 70: A Pilot Study." *HortScience* 52(1): 200–205.

Park, S.-A., K.-S. Lee, K.-C. Son. 2011. "Determining Exercise Intensities of Gardening Tasks as a Physical Activity Using Metabolic Equivalents in Older Adults." *HortScience* 46(12): 1706–1710.

Riva, J. J., K. M. P. Malik, S. J. Burnie, A. R. Endicott, and J. W. Busse. 2012. "What Is Your Research Question? An Introduction to the PICOT Format for Clinicians." *Journal of the Canadian Chiropractic Association* 56(3): 167–171.

Sackett, D. L. 1989. "Rules of Evidence and Clinical Recommendations on the Use of Antithrombotic Agents." *Chest* 95: 2S–4S.

Sackett, D. L., W. M. C. Rosenberg, J. A. Muir-Gray, R. Brian Haynes, and W. Scott Richarson. 1996. "Evidence-Based Medicine: What It Is and What It Isn't." *British Medical Journal* 312: 71–72.

Thyer, B. 2004. "What Is Evidence-Based Practice?" *Brief Treatment & Crisis Intervention* 4(2): 167–176.

Westlund, S. 2014. *Field Exercises: How Veterans Are Healing Themselves Through Farming and Outdoor Activities.* Gabriola Island, BC: New Society Publishers.

Wise, J. 2015. *Digging for Victory: Horticultural Therapy with Veterans for Post-Traumatic Growth.* London, UK: Karnac Books.

Young, J., and M. Solomon. 2009. "How to Critically Appraise an Article." *Nature Clinical Practice, Gastroenterology & Hepatology* 6: 82–91.

Index

Note: Page numbers in italic and bold refer to figures and tables respectively.

9781138308695